T0280726

The climatic record in polar ice sheets

The climatic record in polar ice sheets

A study of isotopic and temperature
profiles in polar ice sheets based on
a workshop held in the
Scott Polar Research Institute, Cambridge

Edited by
G. de Q. ROBIN

CAMBRIDGE UNIVERSITY PRESS

Cambridge

London   New York   New Rochelle

Melbourne   Sydney

CAMBRIDGE UNIVERSITY PRESS
Cambridge, New York, Melbourne, Madrid, Cape Town, Singapore,
São Paulo, Delhi, Dubai, Tokyo, Mexico City

Cambridge University Press
The Edinburgh Building, Cambridge CB2 8RU, UK

Published in the United States of America by Cambridge University Press, New York

www.cambridge.org
Information on this title: www.cambridge.org/9780521153645

© Cambridge University Press 1983

This publication is in copyright. Subject to statutory exception
and to the provisions of relevant collective licensing agreements,
no reproduction of any part may take place without the written
permission of Cambridge University Press.

First published 1983
First paperback printing 2010

A catalogue record for this publication is available from the British Library

Library of Congress Catalogue Card Number: 82-9661

ISBN 978-0-521-25087-0 Hardback
ISBN 978-0-521-15364-5 Paperback

Cambridge University Press has no responsibility for the persistence or
accuracy of URLs for external or third-party Internet Web sites referred to in
this publication, and does not guarantee that any content on such Web sites is,
or will remain, accurate or appropriate.

CONTENTS

CONTRIBUTORS

General Editor
Dr G. de Q. Robin, Scott Polar Research Institute,
   Cambridge CB2 1ER

Dr G. S. Boulton, School of Environmental Sciences,
   University of East Anglia, Norwich NR4 7TJ
Professor W. F. Budd, Department of Meteorology,
   University of Melbourne, Parkville Vic. 3052,
   Australia
Mr J. A. Campbell, Department of Meteorology,
   University of Melbourne, Parkville Vic. 3052,
   Australia
Dr D. J. Drewry, Scott Polar Research Institute,
   Cambridge CB2 1ER
Dr D. Jenssen, Department of Meteorology, University
   of Melbourne, Parkville Vic. 3052, Australia
Mr S. J. Johnsen, University of Iceland, 3 Dunhagi,
   Reykjavik, Iceland
Dr C. Lorius, Laboratoire de Glaciologie, 2 rue Tres-
   Cloitres, 38031 Grenoble Cedex, France
Dr W. S. B. Paterson, P.O. Box 303, Heriot Bay,
   B.C. Canada V0P 1HO
Dr D. Raynaud, Laboratoire de Glaciologie, 2 rue Tres-
   Cloitres, 38031 Grenoble Cedex, France
Dr N. J. Shackleton, Godwin Laboratory for Quaternary
   Research, University of Cambridge, Free School
   Lane, Cambridge CB2 3RS
Professor J. Weertman, Department of Materials Science
   and Engineering and Department of Geological
   Sciences, The Technological Institute, Evanston,
   Illinois 60201, USA
Dr I. M. Whillans, Institute of Polar Studies, The Ohio
   State University, 125 South Oval Drive, Columbus,
   Ohio 43210, USA
Mr N. W. Young, Department of Meteorology,
   University of Melbourne, Parkville Vic. 3052,
   Australia

PREFACE

Prior to 1973, the study of temperature profiles in polar ice sheets developed independently of the corresponding study of the isotopic profiles, although both were obtained from the same boreholes and both recorded aspects of past climate. A grant from the Royal Society enabled the editor of this volume to bring together an international group of experts to the Scott Polar Research Institute, Cambridge, in January 1973, and again in March 1973, for a workshop to integrate studies of isotopic and temperature profiles in polar ice sheets.

The following took part in the workshop: from 1–14 January 1973, the main participants were D. Jenssen (SPRI and Melbourne), S. J. Johnsen (Kφbenhavn), C. Lorius (Grenoble), W. S. B. Paterson (Ottawa), G. de Q. Robin (SPRI; convenor). Others included N. J. Shackleton (Cambridge), C. Neal (SPRI), J. A. Campbell (SPRI and Melbourne; computing assistant). From 12–24 March, the main participants were W. F. Budd (Melbourne), D. Jenssen, S. J. Johnsen, C. Lorius, N. Reeh (Kφbenhavn), G. de Q. Robin, J. Weertman (Northwestern University). Others included P. Gudmandsen (Technical University of Denmark), C. W. M. Swithinbank (British Antarctic Survey), C. Neal, J. A. Campbell, and G. S. Boulton (Norwich).

The purpose of the workshop was to study the relationship between isotopic δ values and temperature suggested by various authors such as Epstein and Mayeda (1953), Scholander et al. (1962) and Dansgaard (1964). If the isotopic record from an ice core provided a true record of past climate, this record of past surface temperature could be incorporated in a model of the flow of an ice sheet to calculate temperature–depth profiles. The results could then be compared with observed temperature–depth profiles, and with earlier calculations of Robin (1955), Zotikov (1961) and others that were based on the assumption that dimensions and flow of an ice sheet, as well as the climate, were in a steady state. An improved match could indicate that the isotopic–temperature relationship based on present-day observations had also applied in the past. Although on physical grounds this would seem likely, confirmation that the isotopic–temperature relationship also applied in the past was necessary because so many processes affect the isotopic δ values of falling snow, and because of the obvious importance of the isotopic profiles as climatic records.

Interpretation of both isotopic and temperature profiles depends on the flow and past behaviour of ice sheets as well as other parameters. At the second workshop meeting, it was decided that the outcome of the study should be published as a monograph, and that this should cover not only the model calculations initiated at the workshop but also discuss the related data used in unravelling the past history of the ice sheet. A general plan for the monograph drawn up at that time has been followed with only limited modification. A list of contributors, including some glaciologists who had not taken part in the workshop, was tentatively agreed and the present editor accepted responsibility for the monograph.

Although progress with the monograph was slower than originally planned, due to delays in editing following earlier delays with model computations and some written contributions, there have been compensations. Our understanding of a number of topics covered in this monograph is now greater than in 1973. Most sections have been revised or reviewed to incorporate important advances and, if there are omissions in this respect, the responsibility lies with the editor. Other sections that were planned but not completed have been written by the editor to complete the coverage of the monograph. Use of symbols has been standardised throughout so that contributions by different authors are integrated into the monograph and are complementary to each other. Although some repetition may occur between different contributors, this has been kept to a minimum.

Chapter 5, which presents the main computations, models and conclusions from modelling, stands as an independent section, as well as providing the core of the monograph. While the modelling studies provide a considerable amount of material that may be of more interest to glaciologists than climatologists, the studies of temperature profiles at Byrd Station and Camp Century based on the input of 'isotopic' temperatures provide the main justification for the workshop.

For a full assessment of the conclusions one needs to appreciate the results of these modelling studies, as well as the reliability of measurements of individual parameters presented in chapter 3. The summaries of the basic techniques and parameters used in modelling, and the outline of the glacial geology of Greenland and Antarctica, may serve as an introduction to these subjects for those not familiar with this work. Similarly, the final chapter may be understood without mastering all the details of chapter 5. It draws on conclusions of the modelling studies, tests them against other series of data, and suggests a guide to the accuracy with which past climate may be deduced from isotopic profiles. Comparisons between isotopic profiles from different ice cores show that long-term climatic changes can be

similar over wide regions of the Antarctic, but that isotopic 'noise' prevents effective comparison over short periods. The climatic record from ice cores and a global index of past climate is then discussed to show how polar climatic trends compare with world trends.

While this monograph brings together an overall picture of the activities initiated by the workshop in Cambridge, a number of individual studies stimulated by the workshop have been published elsewhere and are referred to at many points in the text. The workshop and subsequent developments have drawn on the help of many people to whom thanks are expressed. Guidance to the editor over the contents of individual chapters has been given by the editorial advisers. This guidance has taken a different form for each chapter and has been a great help. However, responsibility for the contributions in individual sections rests with the named authors, or with the editor where an author is not named.

Production of the monograph would not have been possible without the support of the Royal Society, primarily with travel grants and also with secretarial and other expenses. Cambridge University Press have given patient support and encouragement, while many members of the staff of the Scott Polar Research Institute have assisted in various ways. Thanks are especially due to Miss Elaine Lingham, who has helped first in a secretarial and then in an editorial capacity, from the inception of the monograph to its completion, and her confidence and interest in the project have been of great benefit throughout. Mrs Julie Jones, who provided the secretarial and other services for the workshop sessions, and who later became Director's assistant, provided much help in earlier years. Mrs Alison Wood and Miss Margaret Thomson continued to shoulder the burden of typing and retyping succeeding versions of the manuscript with patience and accuracy over several years, often assisted by other members of the office staff. Our librarian, Mr H. G. R. King, Mrs Ailsa MacQueen of World Data Centre C for Glaciology, and the library staff of the Scott Polar Research Institute have been ever helpful in hunting for obscure references. Other students and staff have helped with compilation and the drawing of many diagrams; these include Miss Anne Swithinbank (now Mrs Anne Howe), Mr Rob Massom and Mrs Sue Jordan. The last named, and Mr Paul Cooper who provided valuable help in computing running means of isotopic data in chapter 6, have been able to contribute through their parallel work on an 'Antarctic Glaciological and Geophysical Folio', which is being produced with a grant from the Natural Environment Research Council under the editorship of Dr David Drewry, whose help and cooperation is acknowledged.

To some, and especially to the early contributors, it must have seemed that this monograph might never appear. I hope that the final production will serve to show both the tremendous value and the limitations of isotopic data as a record of past climate.

G. de Q. Robin

29 November 1981

ICE SHEETS: ISOTOPES AND TEMPERATURES

G. de Q. ROBIN

1

Introduction

The ice sheet of Antarctica covers an area of 13 million km² and is more than 4 km deep in places. The Greenland ice covers 1.8 million km² and exceeds 3 km in depth. Both these ice sheets contain a wonderfully detailed record of the atmospheric history of the Earth that extends back for hundreds of thousands of years.

Each snowfall that adds a layer to the body of an ice sheet deposits also a climatic record, but the oldest parts of this record are being removed ceaselessly by the outward flow of ice to regions where it melts and mixes with ocean waters. The continuing distortion of ice that takes place as the ice sheets are squeezed outward under their own weight also distorts the historical record. If we are to learn to read the history effectively, it is first necessary to understand the way in which a piece of ice has been transported and transformed since it fell as snow many thousands of years ago and perhaps many hundreds of kilometres from the point where it is recovered. The development of the Antarctic ice sheet has taken place over 20 million years or more, but chances of finding pockets of ice dating from its earlier period are very small. However, near bedrock in Greenland and Antarctica we do find ice that was deposited before the last ice age. Since that time vast ice sheets have grown and covered large areas of North America and Eurasia and then retreated. Although present-day polar ice sheets record changes of the Earth's atmosphere during the last ice age, they must themselves have varied in size and flow at the same time. These changes need to be taken into account when interpreting evidence from the lower levels of ice sheets.

Access to the long historical record in polar ice sheets has been made possible by the application and development of deep drilling techniques over the past three decades. Ice cores of up to 15 cm in diameter are brought to the surface, catalogued, and sections are then taken to specialised laboratories in different countries for detailed analysis. The longest records obtained before

1980 were the two cores from the surface to bedrock of ice sheets recovered by the US Army Cold Regions Research and Engineering Laboratories (CRREL) from Camp Century in north west Greenland (1388 m) in 1966, and from Byrd Station, Antarctica (2164 m) in 1968. Each of these cores covers an estimated period of about 100 000 a. Drilling to a depth of nearly 1000 m has been carried out at Vostok Station on the East Antarctic ice sheet by Soviet scientists and engineers, and analyses of the top 950 m cover an estimated period of about 50 000 a.

In 1976–7 a French group, with logistic support from the US National Science Foundation, obtained ice cores at Dome C (74° 40′ S 124° 10′ E, 3240 m elevation) to a depth of 906 m. It is estimated that these cores cover a time span of 32 000 a.

In 1980 and 1981 Soviet scientists obtained an ice core in good condition to about 1500 m depth at Vostok Station, Antarctica. In 1981, the Greenland Ice Sheet Project, run by US, Danish and Swiss scientists, recovered an ice core to bedrock at 2037 m depth at Dye 3 near the southern dome of the Greenland ice sheet. It will be some time before comprehensive results from these two projects are available. In this volume particular attention is given to results from the earlier drilling projects at Camp Century, Byrd Station and Vostok.

The historical record present in the ice takes several forms, not all of which have been studied adequately to date.

Ice from the central regions of Antarctica or Greenland contains only a few parts per million by weight of impurities such as dust, some of which may come from extra-terrestrial sources. However, thanks to the purity of the body of ice, even minute traces of impurity can be determined, as for example the variation of lead content of the Greenland ice sheet over past centuries (Figure 1.1). One can see the increase of atmospheric lead due to man's activities since the early nineteenth century, and particularly since the introduction of leaded petrols at about the middle of this century. Isolated peaks at earlier dates may be due to natural events, such as volcanic eruptions.

A careful study by Boutron & Lorius (1979) of the abundance of twelve elements (impurities) deposited in central Antarctica at Dome C shows evidence of marked fluctuations over the past 60 years. It appears that in Antarctica, industrial pollution has had no significant effect. Impurities of marine origin (Na, Mg) are the most abundant. Another group of impurities (Al, Fe, Mn, K and Ca) come from a continental crustal source, while the heavier elements (Pb, Cd, Cu, Zn and Ag) which show a large variability with time are most likely to be of volcanic origin.

Ice sheets also preserve a continuing sample of the atmospheric gases. Fine snow crystals falling on the surface soon recrystallise into a coarser structure known as firn, a porous solid. In spite of the absence of melting in a cold polar ice sheet, the firn gradually increases in density as it is buried more deeply until it becomes impermeable at a depth of 50–200 m, and is then termed ice by glaciologists. At this point it contains about ten per cent of air by volume, which is then trapped in bubbles and preserved in the ice. This air can be recovered for analysis from ice samples obtained by deep drilling. Composition of air samples from ice of various ages shown in Table 1.1 indicates that only relatively small changes can have taken place in the composition of the Earth's atmosphere over the past 40 000 a.

Isotopes and climate

The most detailed record of climatic change preserved in the ice is provided by the changing proportion

Table 1.1.* *Composition of air trapped in polar ice at Camp Century, north west Greenland.*

Depth, metres	Estimated age, years	N_2, %	O_2, %	Ar, %
108	300	78.2	20.8	0.95
555	2 360	78.3	20.8	0.94
898	4 600	78.1	20.9	0.95
1080	8 000	78.2	20.9	0.95
1201	(14 000)	78.3	20.8	0.94
1300	(20 000)	78.3	20.8	0.95
1345	(40 000)	78.4	20.7	0.95
Atmosphere present day:		78.1	21.0	0.93

* Based on Table 1 of Raynaud & Delmas (1977). Ages shown are based on Figure 1.9 except for the two deepest samples, which is based on curve 4 of Figure 5.22. Figures for CO_2 content in the original table have been omitted and the paper by Delmas, Ascencio & Legrand (1980) should be consulted. These indicate that CO_2 concentrations were almost half the present level at the end of the last ice age.

Figure 1.1. Variation of lead content of snow deposited in Greenland during historical time. From Lorius & Briat (1976). The left-hand scale and open circles show data from Camp Century (77° N) in north west Greenland from Murozumi, Chow & Patterson (1969). The right-hand scale and crosses refer to results from Dye 3 in central Greenland (65° N) (Cragin, Herron & Langway, 1975) which lies closer to the latitude of the industrial zones.

of heavy to light atoms (isotopes) of oxygen and hydrogen which form the ice mass. These ratios are measured by means of mass spectrographs in laboratories organised for the purpose. By regular sampling of the ice along a core, we can determine the way in which isotopic ratios vary with depth, and therefore with time, provided we know the relationship between age and depth. Figure 1.2 shows a very detailed record of the variation of isotopic ratios down the ice core obtained at Byrd Station, Antarctica. This book considers the extent to which a record of past temperature changes can be inferred from such isotopic profiles. In other words, do our isotopic ratios provide us with an isotopic 'thermometer' which records the temperatures at the time the ice fell as snow and, if so, how reliable is this record of temperature?

The relationship between the isotopic ratio and the temperature of snow is due to the tendency of heavier molecules of water to evaporate less rapidly than the normal (lighter) molecules, and for the heavier molecules to condense more readily from the vapour. We can consider these processes as starting from the ocean, the major source of atmospheric water vapour. The oceans, which are of nearly constant isotopic composition, contain 98 per cent of the free water on earth. Of this water, some 99.7 per cent is composed of normal hydrogen (H) and oxygen (^{16}O) atoms in the form of H_2O, as opposed to heavier hydrogen (D, deuterium) or heavier oxygen (^{18}O) atoms. About 0.2 per cent of ocean water consists of $H_2^{18}O$ and about 0.03 per cent of $HD^{16}O$.

Instead of presenting measurements of the proportion of heavier molecules in water, water vapour or ice in absolute terms, it is customary and more convenient to use isotopic ratios defined in relation to an arbitrary standard known as 'Standard Mean Ocean Water' (SMOW).* The ratios are presented as δ values for $\delta^{18}O$ or deuterium δD given in relation to SMOW by

* Standard Mean Ocean Water is defined in terms of a water mass held at the National Bureau of Standards (NBS-1) in Washington, DC, USA, and by secondary reference standards at the International Atomic Energy Agency, Vienna.

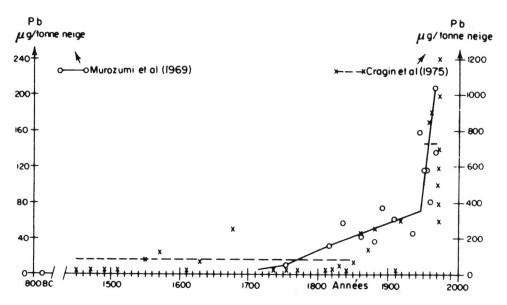

$$\delta^{18}O = \frac{R^{18}O_{Sample} - R^{18}O_{SMOW}}{R^{18}O_{SMOW}} \quad (1.1)$$

where R refers to the ratios of $H_2^{18}O/H_2^{16}O$ in the sample and in SMOW.

In Figure 3.2 we see that, apart from the Arctic Ocean, measured $\delta^{18}O$ values of oceanic surface waters

Figure 1.2. Former climate indicated from measurements of oxygen isotope ratios on the ice core at Byrd Station. Ice at 1000 m depth was deposited around 11 000 a BP, while the lowest ice is of the order of 100 000 a old. From data supplied by S. J. Johnsen.

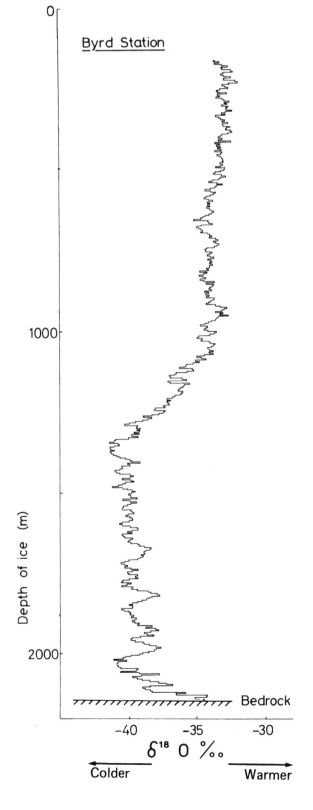

lie between -1.1 and $+1.0$ parts per thousand ($^0/_{00}$), as should be expected if the primary standard SMOW has been carefully collected. Although the present mean δ value of oceanic waters is close to zero, depletion of the lighter isotopic component due to the formation of continental ice sheets means that the average oceanic $\delta^{18}O$ values rose to about $+1.5^0/_{00}$ when Pleistocene ice sheets reached their maximum extent. On the Greenland ice sheet, mean annual $\delta^{18}O$ values of snow now being deposited range from about $-23^0/_{00}$ to approximately $-38^0/_{00}$, while on Antarctica mean values of current deposition run from $-18^0/_{00}$ to about $-60^0/_{00}$. Some laboratories have concentrated on measuring δ values for deuterium. A sufficient number of comparative measurements have been made to show that for ice and snow of polar regions, the relationship between $\delta^{18}O$ and δD values can be considered for practical purposes as linear and given by

$$\delta D = 8.0\delta^{18}O + 10^0/_{00} \quad (1.2)$$

(Lorius & Merlivat, 1977).

We can now consider the physical processes governing the relationship between δ values and temperatures in polar ice sheets. These are shown schematically in Figure 1.3 with figures giving a rough indication of mean annual $\delta^{18}O$ values for water molecules involved at different stages of the cycle ocean–atmosphere–ice sheet–ocean. The relatively uniform composition of ocean waters is due to mixing during their circulation, together with the fact already mentioned that the oceans contain most of the free water on Earth. During evaporation from the oceans, water is depleted of its heavier component molecules by an amount depending on temperature, but of the order of $10^0/_{00}$ in $\delta^{18}O$ content. Greatest evaporation takes place over the tropics, but these areas are largely separated from the circulation patterns in

Figure 1.3. Schematic presentation of isotopic cycle of water vapour through ocean–atmosphere–ice sheet–ocean. The figures show approximate $\delta^{18}O$ values in parts per mil at different stages of the cycle. (Robin, 1977. Reproduced by permission of the Royal Society.)

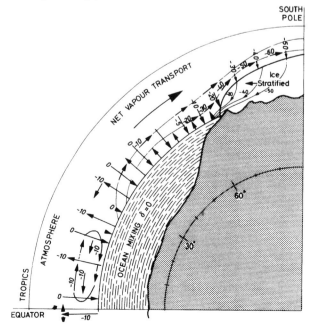

mid-latitudes by the sub-tropical high pressure areas. However, there is considerable transport of air between mid-latitudes and the polar regions. Furthermore, as one moves to higher latitudes, sea surface temperatures are lower, evaporation from the sea surface decreases, and there is a net transport of atmospheric water vapour towards the poles.

In the tropics, condensation from the vapour enriches the water in the heavier isotopes by an amount similar to the depletion during evaporation, hence tropical precipitation has δ values similar to those of the oceanic water, i.e. $\delta^{18}O \approx 0$. As one moves from mid-latitudes towards the poles, however, the original water vapour is not fully replaced by evaporation from the sea, as in the tropics, so the δ value of the atmospheric water vapour falls increasingly as the air mass moves towards the pole. Over the Antarctic ice sheet evaporation is too small to have much effect on the atmospheric composition. Rough mean $\delta^{18}O$ values on the tangential arrows in Figure 1.3 indicate the magnitude of this depletion as one moves over Antarctica.

A more detailed discussion of the distribution of surface δ values in ice sheets and related mean surface temperatures is given in chapter 3, sections 3.2 (Greenland) and 3.3 (Antarctica). While these sections present evidence of a regular variation of δ values with mean ice temperatures, we see also that the relationship between the two varies to some extent between different geographical regions. Nevertheless, because of their regularity of form, which is discussed in chapter 2, polar ice sheets, more than any other topography, favour the development of consistent relationships between δ values and ice temperatures.

In temperate and sub-polar regions, precipitation is mainly produced as the result of frontal activity between different air masses. Instability in individual air masses, and interactions between the surface and the air mass, such as cooling due to upslope winds, may also produce precipitation. Over mountainous regions, snow falling from one system of clouds is likely to have much the same physical characteristics, whether it falls in a valley or on a mountain. Consequently, a regular relationship between isotopic δ values and elevation or mean temperature is less likely to be found in regions of rough topography.

In peripheral regions of polar ice sheets, especially where surface elevations are less than 1000 m, mean δ values of precipitation are less regular in relation to location and elevation than further inland, where upslope winds are an important cause of precipitation. At a distance of more than 100 km from the periphery, ice-sheet slopes rise continuously towards the centre at a gradient of 1:200 or less in Greenland and still more slowly in central Antarctica. This regularity of form produces greater regularity in the pattern of cooling of air masses. This in turn leads to a regularity in the δ values for precipitation which can be related to temperature and climate.

The δ values of an ice sample recovered from an ice core are determined originally by the processes of precipitation. Although direct condensation of rime on

to the surface of ice sheets does take place, the main bulk of precipitation comes from clouds. Studies have been made in Antarctica to measure $\delta^{18}O$ values of falling snow in relation to the cloud temperatures at which the snow was formed. Figure 1.4 shows the data obtained by Picciotto, de Maere & Friedman (1960) at Roi Baudouin Station on the Antarctic coastline (70° 26′ S 23° 19′ E). The results, together with those from a second study made at the South Pole by Aldaz & Deutsch (1967) are shown by lines B and A in Figure 1.5. Also shown on this diagram are dashed curves 1 and 2 which link results

Figure 1.4. Isotopic composition of oxygen in snowfalls with respect to the range of temperature in the corresponding cloud sheet. (From Picciotto *et al.*, 1960. Reproduced by permission of *Nature, London*.)

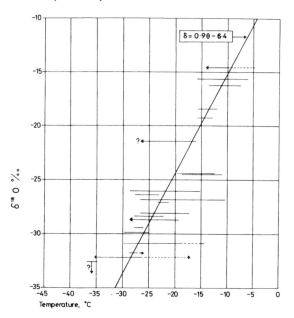

Figure 1.5. Isotopic $\delta^{18}O$–temperature relationships. Line A, clouds at South Pole, from Aldaz & Deutsch (1967); line B, clouds at Roi Baudouin Station on the Antarctic coast, from Figure 1.4. Dashed curves 1 and 2, calculated for Rayleigh condensation with adiabatic cooling, are based on line B; the fractionation factor used to calculate curve 1 is derived from Zhavoronkov *et al.* (1955), that for curve 2 derived from Merlivat & Nief (1967). Line C gives the isotopic–temperature relationship for surface layers of firn of eastern Antarctica (see Figure 3.11). (Based on Aldaz & Deutsch, 1967, Fig. 3. Reproduced by permission of *Earth and Planetary Science Letters*.)

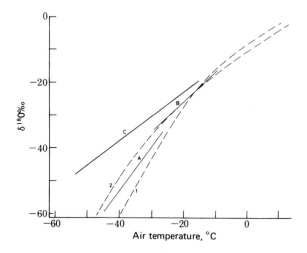

at Roi Baudouin with those at the South Pole by assuming that air from Roi Baudouin undergoes moist adiabatic expansion as it travels up over the Antarctic ice sheet until it reaches the South Pole. Since there is some uncertainty over the correct isotopic fractionation factor to use in the calculations, curves 1 and 2, which use different physical constants, are both shown. These are tied to curve B and are seen to bracket curve A, the δ value–cloud temperature curve for the South Pole; so the physical model appears to be approximately correct. The heavy line C in Figure 1.5 is taken from Figure 3.11. This shows the relationship between δ values of samples of the surface layers of the ice sheet and the mean temperatures of the surface layers. Conversion from δD to $\delta^{18}O$ has been made by equation (1.2). The increasing difference between the δ value of the surface temperature curve and δ value–cloud temperature curves as one moves to lower temperatures is due to increasing differences between the mean surface temperature and the mean temperature of the clouds in which the precipitation is formed. This agrees with meteorological studies of the cold surface inversion over the Antarctic ice sheet, which is caused by intense radiative cooling from the surface throughout the year. As the relatively warm air moves over the continent from the surrounding oceans, strong radiative cooling from the snow surface cools the air above whenever the sky is clear of cloud. Schwerdtfeger (1970) gives mean values of the temperature difference between the surface and the air above the cold inversion (from 450–1000 m above the surface) which have been measured during regular radio-sonde ascents at different stations. At Vostok Station in eastern Antarctica (elevation 3700 m), the air above the inversion is warmer by an average of $+15.7\,^{\circ}C$ than the recorded surface temperature throughout the year; at the South Pole (2800 m), the warming averages $+14.1\,^{\circ}C$, and at Mirny on the coast, although inversions are often present, air at 500 m elevation is on average $0.3\,^{\circ}C$ colder than the surface temperatures. At Byrd Station (1500 m), in western Antarctica, air above the inversion averages $+6.4\,^{\circ}C$ warmer than at the surface. Mean annual temperatures at these stations are Vostok $-55.6\,^{\circ}C$, South Pole $-49.3\,^{\circ}C$, Byrd $-27.9\,^{\circ}C$, and Mirny $-11.5\,^{\circ}C$. The tendency for δ values in western Antarctica to be somewhat higher than in eastern Antarctica (section 3.3) for the same surface temperatures may well be due to heavier cloud cover and hence weaker surface inversions over western Antarctica. Of course, we should not assume that the mean air temperature above the inversion is the same as the mean cloud temperature at which snow is formed. However, the difference is not large at the South Pole where the mean temperature above the inversion is $-35\,^{\circ}C$ compared with the mean condensation temperature of $-39\,^{\circ}C$ (Aldaz & Deutsch, 1967).

The range of temperatures shown in Figure 1.4 is partly due to variations between air masses, but mainly due to seasonal variations as shown in Figure 1.6 where isotopic δ values are plotted as a function of time. We see that these vary from a mean around $-16^0/_{00}$ in summer to about $-29^0/_{00}$ in winter months at Roi

Baudouin Station on the coast of Antarctica. In Figure 1.7 we show similar information from Camp Century, but in this case, the data come from a study of closely spaced samples from various depths below the surface of the ice sheet. The seasonal variation is seen as a variation in $\delta^{18}O$ values with depth, the position deduced for summer and winter deposits being marked by S and W respectively.

The range of δ values between summer and winter firn is approximately $15^0/_{00}$ in the top 0.5 m, similar to the range shown in Figure 1.6 for Antarctic coastal precipitation. However, this range decreases with depth and averages about $5^0/_{00}$ from 100 m to depths approaching 1000 m, when a further decrease occurs. These changes are due to mobility of the water molecules in the firn and ice. At shallow depths larger crystals grow rapidly at the expense of smaller, due to transport of molecules mainly by sublimation. This mobility will tend to smooth out seasonal and shorter-period variations of δ values with depth. However, as ice becomes less porous, this type of mobility decreases and is of little significance for densities greater than $550\,\mathrm{kg\,m^{-3}}$ that are reached at 10–20 m depths. The further decrease at depths around 1000 m is due to molecular diffusion within the main body of ice, which becomes more effective as layer thicknesses are reduced and temperatures increase. These factors are discussed in more detail in section 3.4.

The changes of layer thickness are due primarily to the continuous addition of a mass of snow on the top surface of the ice, and the consequent downward and outward squeezing of the ice mass which causes the flow of the ice sheet. We can follow the mechanics of the process if we consider the history of a vertical column of ice, and assume for the purpose of our model (Figure 1.8) that the column stays vertical as it is pushed outwards by the flow of ice. We treat a two-dimensional case only, but the argument is similar for a three-dimensional case. For the present, we limit our study to the case of an ice

Figure 1.6. Seasonal variation of $\delta^{18}O$ values of surface accumulation at Roi Baudouin Station, Antarctica, April 1958–Feb. 1959. (Gonfiantini & Picciotto, 1959. Reproduced by permission of *Nature, London*.)

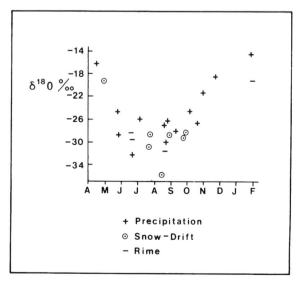

sheet in a steady state, that is one in which the size and shape of the ice sheet does not change with time, so that the flow of ice will just be in balance with the rate of accumulation or ablation (loss of ice) at each point on the surface of the ice sheet. For such discussions we neglect seasonal changes of snowfall and ablation and consider only their annual mean values.

At the centre of the ice sheet, if we are adding ice at the rate A_0 on the top surface, the surface must be sinking at a rate $w_0 = A_0$ if the thickness is to remain constant. To compensate for this downward motion we allow our column to spread out horizontally at a uniform rate at all depths. This corresponds to a vertical velocity that falls uniformly from w_0 at the surface to zero at bedrock, and at any height z above bedrock, the velocity w_0 is given by

Figure 1.7. The $\delta\,^{18}O$ oscillations in firn and ice cores from Camp Century at the depths below surface indicated to the left of the figure. S and W indicate interpretations of summer and winter layers, respectively. As the ice sinks towards the bottom, the thickness (λ m) of the annual layers is reduced due to deformation during flow. (From Johnsen *et al.*, 1972. Reproduced by permission of *Nature, London*.)

$$w_0 = A_0 \frac{z}{Z_0} \qquad (1.3)$$

where Z_0 is the ice thickness at the centre of the ice sheet. This in turn implies that the vertical strain rate is uniform at all depths. Then, if the annual accumulation (A_0) has not changed with time (steady-state climate), the model indicates that the annual layer thickness λ should decrease uniformly with depth from the surface to zero at bedrock at the centre of an ice sheet. At a height z above bedrock we will have

$$\lambda = \bar{A}_0 \frac{z}{Z_0} \qquad (1.4)$$

We show later in Figure 1.9 that this is approximately the case in the ice core from Camp Century.

We can continue to use our vertical column model away from the centre of an ice sheet. The form of ice sheets is discussed in detail in chapter 2, but for the present we need just note that ice flowing under gravity piles up into a vast, relatively flat dome. The surface slope α, which increases slowly from the centre to the periphery of the ice sheet, in approximately inverse proportion to the ice thickness, controls the flow of ice in much the same way as the slope of a river surface

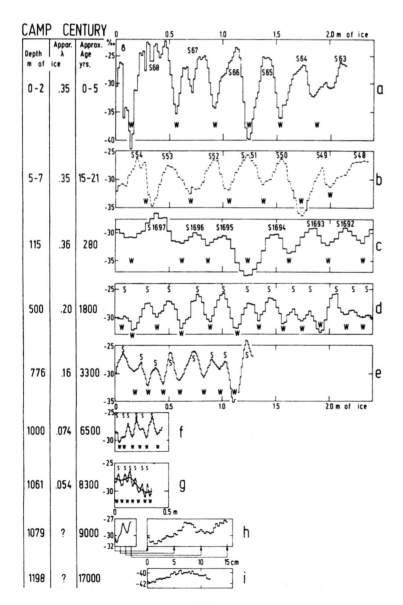

determines the flow of water. Bedrock slopes β have important but relatively local effects on ice deformation and flow.

At a distance x from the centre, the two-dimensional steady state requires that the horizontal velocity of our column (U_x) be given by

$$U_x = \frac{\bar{A}x}{Z_x} \qquad (1.5)$$

where $\bar{A}$ is the mean mass balance along the line x from the centre, and Z_x is the ice thickness at that point. The height of the column will also be changing due to changes of ice thickness at a rate

$$\frac{dZ_x}{dt} = (\alpha - \beta)U_x \qquad (1.6)$$

Figure 1.8. Parameters for a two-dimensional cross-section of an ice sheet in a steady state.

Thus an annual layer thickness A_1 deposited at a distance x_1 from the centre of the ice sheet on ice of total depth Z_1, will have a thickness λ at a distance x from the centre defined by

$$\lambda = A_1 \frac{z}{Z_1} \qquad (1.7)$$

where z is its height relative to bedrock. The vertical velocity w will be

$$w = [A_x + (\alpha - \beta)U_x]\frac{z}{Z_x} \qquad (1.8)$$

Thus the layer thicknesses on this model will still decrease with depth from A_x at the surface to zero at bedrock, but not linearly as at the centre of the ice sheet since A and U_x will vary over the ice sheet.

In Figure 1.9 (from Hammer *et al.*, 1978), layer thicknesses at Camp Century seen in Figure 1.7, together

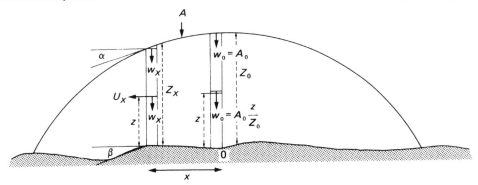

Figure 1.9. Annual layer thickness (λ) in the Camp Century ice core plotted against height above bedrock (z). The figures close to the points are the number of annual layers interpreted in the individual core increments, either directly from $\delta^{18}O$ profiles (open squares), after deconvolution (open circles, see section 3.4), from measured microparticles profiles (filled circles), or by visual stratigraphy (filled squares). The heavy curve is a least-squares fit of a model that assumes a linear variation of layer thickness with depth to 400 m above bedrock, corresponding to a uniform vertical strain rate. Below this level the strain rate decreases linearly to zero at bedrock, which implies $\lambda \propto z^2$ over this interval. The time scale shown on the right is derived from the observed layer thickness. (Hammer *et al.*, 1978. Figure reproduced from the *Journal of Glaciology* by permission of the International Glaciological Society.)

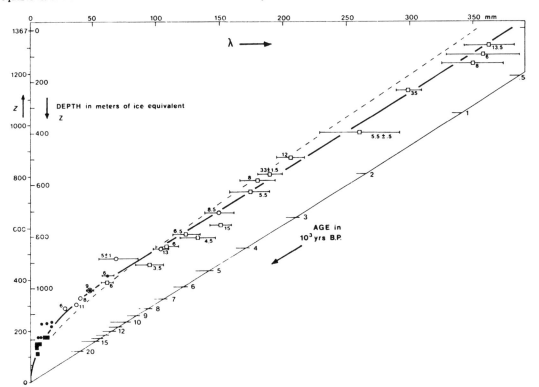

with other observations of layer thicknesses, are plotted against depth in the ice sheet. The heavy line fitted to the observations is identical (apart from a minor thickness change) to that used for the basic model for calculations in section 5.2. The assumption that the vertical strain rate falls to zero over the lowest 400 m takes account of the ice being frozen to bedrock. This would cause ice movement to fall to zero and hence reduce the vertical strain rate to zero also, since there will be no horizontal spreading to permit vertical strain at bedrock. The agreement between the heavy line and the observations is good down to a depth around 1100 m, that is to an age of approximately 9000 a. Below that depth the proportional difference between observed and calculated layer thicknesses increases and the dating appears less reliable.

When we consider the effect of horizontal velocity on vertical movement in equation (1.8) and its consequent effect on the thickness of layers transported from further inland, it is surprising that the calculated thicknesses fit the data in Figure 1.9 so well above 1100 m depth. The simplest explanation is that the accumulation rate recorded in the ice core has varied little over the past 9000 a and that the second term $(\alpha-\beta)U_x$, in equation (1.8) is negligible. Figure 4.6 shows little change in ice thickness for 20 km up the present flowline from Camp Century, that is $(\alpha-\beta)\approx 0$. If the mean velocity of $U_x < 2$ m a^{-1}, which seems possible along an ice divide, the ice at Camp Century down to 1000 m depth will have been deposited in this section of uniform thickness, and would be of relatively local origin.

In general, we use the term 'steady-state ice sheet' to describe an ideal model in which both the size and climate of an ice sheet do not change with time. These assumptions together with our model make it possible to calculate not only the velocity distribution, but also the temperature distribution throughout a large ice sheet. Using this model, we find rough agreement between calculated and observed temperature distributions in the same way as we find agreement between calculated and observed layer thickness (Figure 1.9). Refinements of these calculations to more exact steady-state models, and development of more realistic models that take account of past variations of size of the ice sheet are dealt with in chapter 5. In that chapter we try to determine just what these past variations have been by using these models and the evidence available from isotopic, temperature, and gas-content studies, as well as from glacial geology.

Temperature

A large proportion of the ice of glaciers in lower latitudes is at, or close to, the melting point, and these glaciers are referred to as 'temperate'. In contrast to temperate glaciers, on which there is considerable surface melting during summer months, polar glaciers by the glaciologists' definition are 'cold'. This defines a polar glacier or ice sheet as one on which the surface temperature does not reach the melting point at any time of the year. Temperature of the upper layers of a polar ice sheet are governed by the thermal conductivity of the

firn and ice. Ten metres below the surface the seasonal temperature variations are almost damped out, and temperatures remain close to the mean annual temperature of the surface of the ice sheet. This surface temperature is usually a degree or so colder than the mean annual temperature recorded in a meteorological screen 1 m above the surface.

In Figures 4.1 and 4.3 we show the mean annual (10 m) temperature distribution of the surface layers of the ice sheets of Greenland and Antarctica. We see that temperature is primarily related to surface elevation, but that the latitude and continentality of a location are also factors. The latitude effect is seen more readily in the case of Greenland, while continentality is one of the factors that cause the extremely low temperatures in central Antarctica.

Direct knowledge of bedrock temperatures of Greenland and Antarctica is more limited. At Camp Century the bedrock temperature was -13 °C, while at Byrd Station water rose into the borehole indicating a basal temperature of -1.6 °C, the pressure melting point under 2164 m of ice. Radio-echo sounding of the Antarctic ice sheet has shown the presence of a number of lakes beneath 3–4 km of ice in central areas (Oswald & Robin, 1973), indicating pressure-melting temperatures of about -2 °C to -3 °C. So far, no lakes have been reported beneath the Greenland ice sheet, but the bedrock topography of the ice sheet is not particularly favourable for collection of water. In any case, theoretical predictions which were approximately correct for basal temperatures in the Camp Century and Byrd Station boreholes, indicate that the basal ice over large areas of Antarctica and Greenland will be frozen to the bedrock.

Heat flow, ice flow and temperature distribution in steady-state ice sheets

Ice sheets, like any part of the Earth's crust are affected by the flow of heat from the interior of the Earth. This heat flow does not vary greatly over the Earth's surface. On continents it averages around 28 mW m^{-2}, or about 38 cal cm^{-2} yr^{-1}, enough to melt about $\frac{1}{2}$ cm of ice in a year. It is slightly higher on the ocean bed, and generally higher in the vicinity of mid-ocean ridges. Heat flow is determined by measuring the temperature gradient in rocks, and multiplying this by the thermal conductivity of the rock. As would be expected, therefore, the variation in temperature gradient between different types of rock is much g eater than the variation in heat flow. Typical temperature gradients in soft sediments of the sea bed are 0.05 °C m^{-1}; in limestone, 0.03 °C m^{-1}; in granite, 0.02 °C m^{-1}; and in ice (static), 0.025 °C m^{-1}. The observed temperature gradients in the Earth's crust help us to deduce processes that may be taking place at greater depths in the crust and mantle of the Earth.

In polar ice sheets we have a similar problem, but with certain advantages. We have more knowledge of the deformation (e.g. Figure 1.9) and movement of the ice mass than we have in the case of the Earth's crust and interior. At two locations, Camp Century and Byrd

Station, we have temperature measurements throughout the entire depth of thick ice sheets and many observations at lesser depths. Isotopic data may provide us with a record of past surface temperatures at which the ice at any depth in an ice core was laid down, even though the ice has moved far to reach the borehole site. The motion of the ice itself has such a large influence on the temperature distribution within the ice mass that we can use temperature gradients to determine approximately whether or not an ice sheet is in a steady state, both with regard to size and to temperature distribution. The mathematical and computational methods used in detailed studies are complex and are developed in chapter 5. In this section we aim to give the general reader some understanding of the processes and factors governing temperature distribution, so that he may appreciate the significance of deductions made about the size and motion of ice sheets, present and past, from temperature and isotopic profiles, even if he does not follow through the detailed computations in chapter 5. The simple vertical column model (Figure 1.8) is sufficient to describe the main features. The scale of the various factors involved, such as ice depths, heat flow and time, must be kept in mind.

While the upper-surface temperatures are of great importance in determining the temperature distribution in a polar ice sheet, the problem is best approached by first considering what happens to the geothermal heat entering the base of an ice sheet which is in steady state. If the ice mass is not in motion, then a steady-state condition is only possible if the amount of geothermal heat flowing upward through the ice mass does not change at any level. Otherwise there would be accumulation or loss of heat at levels where the heat flow changed and hence also changes of ice temperature. A uniform geothermal heat flow of $38\,\mathrm{mW\,m^{-1}}$, then, requires a constant temperature gradient throughout the ice mass of $0.024\,^{\circ}\mathrm{C\,m^{-1}}$ provided we regard the density, specific heat and thermal conductivity of ice as constants and neglect their small changes with temperature. Then, if the ice is $1000\,\mathrm{m}$ thick, the temperature at bedrock will be $24\,^{\circ}\mathrm{C}$ warmer than at the surface (Figure 1.10). The latter is, however, determined by the surface climate, since the geothermal heat flux is very small in comparison with solar and atmospheric heating effects at the surface. Then, if the mean annual surface temperature is $-30\,^{\circ}\mathrm{C}$, the bedrock temperature will be $-6\,^{\circ}\mathrm{C}$. If the surface temperature is $-24\,^{\circ}\mathrm{C}$, the bedrock will be at the melting point, $0\,^{\circ}\mathrm{C}$ (neglecting effect of pressure). If the surface is warmer than $-24\,^{\circ}\mathrm{C}$, say $-12\,^{\circ}\mathrm{C}$ for example, the bedrock temperature will be at the melting point and the mean temperature gradient through the ice can then be only $0.012\,^{\circ}\mathrm{C\,m^{-1}}$, which is only half the geothermal gradient. The excess heat reaching the base of the ice will not be conducted upward, and will melt about 2.5 mm per year off the base of the ice. Our ice mass is then no longer of constant thickness and, strictly speaking, the problem changes into one of an ice mass in motion, though in this case the change is small. If the surface temperature is $0\,^{\circ}\mathrm{C}$, the ice mass will all be at the melting point.

We now consider the effect of motion. Under the steady-state assumption, the temperature at any given height above bedrock will not change with time. However, our steady-state model also shows us that the cold, surface ice is moving downwards towards the warmer bedrock with a velocity given in equation (1.3) or (1.8). The steady-state temperature condition can only be met if the ice moving downwards is heated at the correct rate to maintain a constant temperature at each level. The main source of heat deep in the central regions of an ice sheet is the geothermal heat itself, which is therefore used and absorbed in heating the ice. However, after a given layer of ice has absorbed some of the geothermal heat entering its base, the heat flow which penetrates the next layer above will be decreased by the amount absorbed. The temperature gradient will then decrease in each successive layer as we move upward from the base. The effect is shown quantitatively to fit equation (1.3) for the centre of a steady-state ice sheet $3000\,\mathrm{m}$ thick with an accumulation rate of $0.32\,\mathrm{m\,a^{-1}}$ of ice in Figure 1.11. We see here that the upward flow of heat (shown by the length of upward arrows) decreases to zero by $2000\,\mathrm{m}$ above bedrock, so that the ice above this level is isothermal with the surface temperature.

Figure 1.10. Approximate temperature–depth profiles in a static ice mass 1000 m thick, neglecting effects of pressure melting, for the following cases:
(1) basal ice at $0\,^{\circ}\mathrm{C}$ and entire heat flux carried upwards through ice;
(2) same heat flux as (1) with mean surface temperature $-30\,^{\circ}\mathrm{C}$;
(3) mean surface ice temperature at $-12\,^{\circ}\mathrm{C}$;
(4) mean surface ice temperature at $0\,^{\circ}\mathrm{C}$.

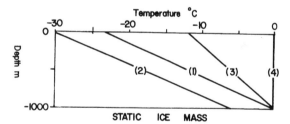

Figure 1.11. Temperature variation with depth in central region of an ice sheet in a steady state with $0.32\,\mathrm{m\,a^{-1}}$ accumulation of ice. The basal temperature gradient (geothermal heat flux) is the same as in Figure 1.10.

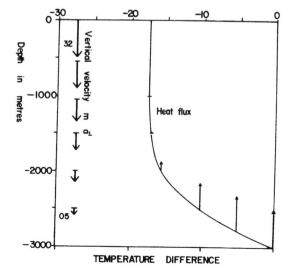

As in the previous example, after calculating the shape of the temperature–height curve from the geothermal gradient and the ice motion, we tie the curve to the surface temperature to predict the actual temperature at any depth. In Figure 1.12 we show the form of the temperature–depth curves for a number of different rates of accumulation, using the same geothermal heat flow (i.e. basal-temperature gradient) in each case. We note that the greater the rate of accumulation, the less the temperature difference between surface and bedrock, and hence the colder will be the basal temperature for a given surface temperature. Although our model does not deal quantitatively with horizontal transport of heat due to the outward movement of our ice column (Figure 1.8), we can readily see what has happened to the geothermal heat reaching the base of the ice. This is best described by the meteorologists' term 'advection', the transport of heat by mass movement of the atmosphere or, in this case, the outward mass movement of ice. The heavier the accumulation, the faster will be the mass movement and the more rapidly will basal heat be carried away by horizontal motion of the ice sheet, so that less heat will flow to the upper layers of the ice sheet.

We have not yet dealt with the question of the heat produced by the work done in deforming an ice sheet under the action of gravity. The rate of heat production in any element of volume is proportional to the product of the effective stress by the effective strain rate (see section 5.4). The amount of heat involved due to the uniform vertical strain of our model is relatively small and does not greatly influence temperatures in the upper part of the ice sheets of Greenland and Antarctica. However, shearing or sliding of the

ice over bedrock produces large strain rates in the lowest 10 or 20 per cent of the ice. One may regard this as frictional heat concentrated at bedrock, in which case the work done is equal to the product of ice movement by basal shear stress. Typical values of basal shear stress for ice sheets involves a heat production equal to the geothermal flux for ice movement of from 20–50 m a^{-1}. Strictly speaking, this heat production will be spread through the lower levels of ice where deformation is considerable, the actual distribution being more concentrated towards bedrock when this is at the melting point so that sliding takes place. This extra heat supply will increase the temperature gradients in the lower levels and will therefore generally increase the temperature difference between the surface and bedrock.

An effect that opposes the increased temperature differences of the previous paragraph comes from the thinning of ice and hence of our 'vertical column' as the ice moves outwards (Figure 1.8). This thinning involves vertical strain in the ice, in addition to that due to accumulation on our steady-state model, and its effect can be estimated from Figure 1.8 by adding the two effects together as indicated by equation (1.8). For example, an ice motion of 50 m a^{-1} over flat bedrock with a surface slope of 1 in 200 will add an extra downward motion of 0.25 m a^{-1} to the surface of a moving column in a steady-state ice sheet. The resultant effect on deep ice temperatures will also depend on the ice thickness and on accumulation rate, so will vary with location. The balance between the opposing effects of increased vertical strain due to ice thinning versus increased frictional heating due to faster ice motion must be calculated for any given situation.

Changing surface temperature with time
Even on our steady-state model in Figure 1.8, the surface temperature at the top of our moving column will not be constant except at the centre of the ice sheet, since as the column moves away from the centre, the surface elevation becomes lower and the mean

Figure 1.12. Calculated temperature–depth curves (as in Figure 1.11) for a surface temperature of −74 °C (upper scale A) and −34 °C (lower scale B) for different values of accumulation of ice shown in cm a^{-1}. The dotted curves relate to scale B only, since ice temperatures cannot be greater than the pressure melting point (−2.2 °C at 3000 m).

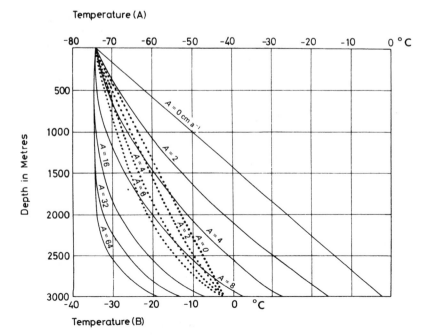

surface temperature correspondingly warmer. In addition, climatic changes cause surface temperatures to change with time over wide regions of the Earth's surface. We need to know how such changes affect the temperature–depth profiles that we observe.

Numerical examples of the effect of harmonically varying surface temperatures in the accumulation zone at the centre of an ice sheet are shown in Figure 1.13. Whereas the amplitude of a one-year cycle decreases to approximately 1 per cent of its surface amplitude when it reaches a depth of about 15 m, a 10 a cycle will fall to 1 per cent at a depth of 60 m, a 200 a cycle by 270 m and a 500 a cycle by 435 m, for an accumulation rate of 0.20 m a^{-1} of ice. Thus, as we penetrate deeper into the ice, the effects of short-term fluctuations, even of centuries' duration, are eliminated and only long-term trends determine the profile.

Another way of illustrating this point is shown in Figure 1.14 which records the temperature change with time in an accumulating ice sheet after a single, but permanent, change of surface temperature. For this case, with an accumulation rate of 0.25 m a^{-1} of ice, we learn that the temperature at 1000 m depth is not affected by any surface changes during the last 1000 a. The temperature at 1000 m depth shows a maximum response to the

Figure 1.13. Response of an accumulating ice sheet to harmonic temperature changes at the surface. The ice is assumed to move downwards at 0.20 m a^{-1} at all depths in response to surface accumulation. The full lines show the envelopes of all possible temperature–depth profiles for surface temperature fluctuations of the periods indicated. The dashed lines show some calculated temperature–depth profiles for harmonic waves of the periods shown (from Robin, 1970).

surface temperatures of approximately 3500 a ago on this model where the vertical ice velocity is uniform. Introducing a uniform vertical strain rate, as in our steady-state model, will increase this age still further from 3500 a (Figure 5.33).

In problems of this type, if we know past temperature changes at the surface, and details of ice flow, we can use the heat-flow equations to calculate the temperature–depth profile. An important part of chapter 5 is devoted to this problem, using δ values to define the past surface temperatures. The close match between the temperature–depth profile calculated in this way for an ice sheet, which in other respects is in a steady state, provides important evidence for our use of δ values as a measure of temperature.

If the surface temperature is changing at a steady rate with time ($d\theta/dt$), and if this is continued over a long period with an accumulation rate A, a temperature–depth gradient ($d\theta/dz$) will tend to be 'buried' in the ice, which is given by

$$\frac{d\theta}{dz} = \left(\frac{d\theta}{dt}\right)/A \qquad (1.9)$$

This has several interesting consequences. Firstly, if our change of surface temperature with time is due only to the movement of ice with velocity U over a region of surface slope α, and Λ is the change of surface temperature with elevation in this region, we then have

$$\frac{d\theta}{dz} = \frac{\alpha U \Lambda}{A} \qquad (1.10)$$

Since surface temperatures normally become warmer with time due to decreasing elevation of our

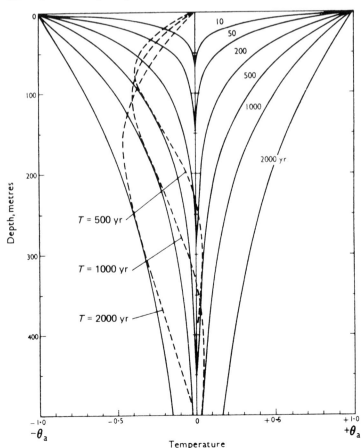

Figure 1.14. Temperature response in an ice sheet to a temperature jump from θ_b to θ_a at time $t = 0$ a. Temperature–depth curves are shown at times ranging from 10–4000 a. The ice is assumed to move downwards at 0.25 m a^{-1} at all depths in response to surface accumulation. The horizontal lines across the temperature profiles indicate the depth of the layer which was at the surface at $t = 0$ a (from Robin, 1970).

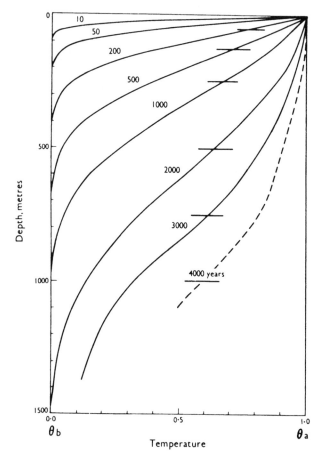

Figure 1.15. Sketch of the temperature distribution in a simplified ice sheet similar to central Greenland. (Robin, 1955. Reproduced from the *Journal of Glaciology* by permission of the International Glaciological Society.)

column, the gradient $d\theta/dz$ defined here is opposite to normal, that is the temperature *decreases* with increasing depth below the surface. This will be opposed by the geothermal flow of heat; but since this upward geothermal flux falls to zero in the upper part of the Greenland ice sheet, we find there that the negative gradients predicted above are present (Figure 4.1). Figure 1.15, from Robin (1955), shows a semi-quantitative estimate of the temperature distribution in a cross-section of the ice sheet of central Greenland based on the hypotheses outlined in this section. The estimates are reasonably consistent with more recent evidence.

The most extensive and detailed application of the moving vertical column model described in this chapter has been made by Budd, Jenssen & Radok (1971a) for a large number of flowlines on the Antarctic ice sheet. In their computations they have used the vast amount of information gathered during various journeys over the Antarctic ice sheet between 1950 and 1970. The programme of the International Geophysical Year was responsible for initiation of many of these studies. The resultant maps provided surface contours that were first used to determine the flowlines of ice along which columns moved. The height of the column of ice was made to change in conformity with the available data on ice thickness along each flowline. Surface accumulation rates were then used for computing ice particle trajectories and the depth to former surface layers of given ages (isochrons). Observed surface ice temperatures and mean geothermal heat flow found in other old continental shields were used in order to calculate temperature–depth profiles. The results are all shown in Figure 1.16 for a flowline running approximately from Vostok to Wilkes Stations. The temperature gradients in the upper 70 m along this profile agreed reasonably well with those measured along this profile in 1962 by an Australian traverse party (Budd, 1969). However, the cold basal ice temperatures between 400 and 700 km on Figure 1.16 were found to be in error by radio-echo

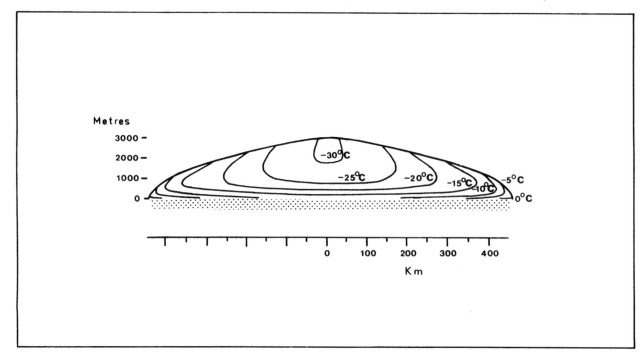

soundings which show the presence of meltwater beneath the ice in this region. More recent field studies indicate that values of the accumulation rate and possibly of the geothermal heat flux used in the preceding calculations need revising, so the discrepancy is probably due more to use of incorrect data than to the use of our simplified model.

Similar studies elsewhere show the approximate validity of the steady-state concept over many areas, including Greenland. The most striking agreement is perhaps that of the Law Dome, shown in Figure 1.17(*a*) and (*b*) where steady-state calculations correctly predict negative temperature gradients along the Cape Poinsett line where accumulation is heavy, and positive temperature gradients along the Cape Folger line where accumulation is much less. The temperatures measured in the Law Dome to a depth of 380 m are controlled mainly by the relatively steady climate during the past few thousand years, whereas further inland we show in chapter 5 that deep ice temperatures will still be affected by surface temperatures from the last ice age.

Form of isotopic–depth profiles

The steady-state concept is also very useful in helping us to understand and interpret isotopic profiles, especially those from the peripheral regions of ice sheets which are presented in section 4.4. The main problem is to relate the depth of a sample in the ice sheet to the location at which it was deposited. Particle trajectories in Figure 1.16 present the answer obtained by detailed computer analysis of one particular case. A simpler appreciation of the problem comes from applying the steady-state concept to sections of an ice column in Figure 1.8 as well as to the entire ice column.

We have already shown in Figure 1.8 that the velocity of a vertical column distant x_1 from the centre

Figure 1.16. Particle paths (arrows), isochrones (dashed lines, ages in 10^3 a) and temperature–depth profiles (full lines, temperatures in °C) along flowline from Vostok to Wilkes Station calculated on a steady-state model by Budd *et al.* (1971*a*).

of a steady-state ice sheet is

$$U_{x_1} Z_{x_1} = \int_0^{x_1} A \, dx \tag{1.11}$$

We want, however, to find the point x_2 (Figure 1.18) at which a particle of ice at distance x_1 and depth $z = (Z_{x_1} - z)$ was originally deposited. Since the surface shape does not change with time, the total mass accumulating on the surface between x_1 and x_2 must be balanced by the outflow of ice down to a depth z, thus

$$U_{x_1} z = \int_{x_1}^{x_2} A \, dx \tag{1.12}$$

Since U_{x_1} is determined from equation (1.11), we combine this with (1.12) to get

$$\frac{z}{Z_{x_1}} = \frac{\int_{x_2}^{x_1} A \, dx}{\int_0^{x_1} A \, dx} = \frac{\bar{A}_2(x_2 - x_1)}{\bar{A}_1 x_1} \tag{1.13}$$

where $\bar{A}_1$ and $\bar{A}_2$ are the respective mean values of the accumulation rate over the intervals 0 to x_1 and x_1 to x_2. If $\bar{A}_1 = \bar{A}_2$, that is the case of uniform accumulation rate over an ice sheet, then we have simply

$$\frac{z}{Z_{x_1}} = \frac{(x_1 - x_2)}{x_1} \tag{1.14}$$

This simple relationship, that the fractional depth of any layer is proportional to the fractional distance of its point of origin to the total length of the flowline, permits a ready appreciation of the form of a steady-state isotopic profile.

A plot of the height of origin of ice in the core against its depth then has the same form as the plot of its height of origin against distance along the flowline from the borehole site. Since isotopic values are approximately proportional to elevation (Figure 3.10) over most of an ice sheet, it follows that the shape of the δ value–depth profile will also be similar to the form of the surface profile inland of the borehole, provided the mass

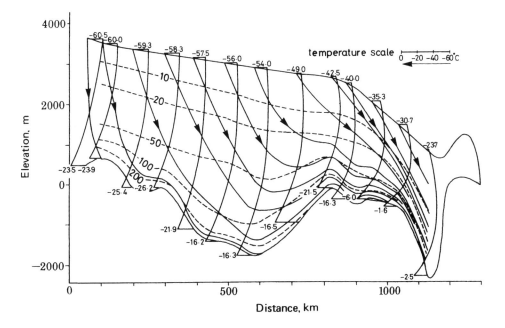

balance does not vary with elevation. Thus, a steady-state ice sheet of parabolic cross-section, in a two-dimensional analysis will tend to generate an isotope-depth profile of parabolic form at the edge of the ice sheet (curve 1, Figure 1.19) or of a portion of a parabola at inland sites. Near the centre of the ice sheet this will be close to a linear profile. If we are dealing with a parabolic surface profile with mass balance inversely proportional to elevation, as is approximately true in East Antarctica above 1000 m elevation, the isotopic-depth profile will be linear (curve 2, Figure 1.19) provided other conditions for equation (1.14) apply.

We can modify equations (1.11)–(1.14), which apply to a moving vertical column, to allow for the variation of horizontal velocity with depth by replacing

Figure 1.17. (*a*) Surface movement (m a^{-1}), surface contours (m) and net accumulation (m a^{-1}) on Law Dome, Antarctica.

(*b*) Calculated (thin lines) and observed (heavy lines) temperature profiles in the Law Dome, Antarctica. The upper figure shows data for slowly moving ice on the line from the summit to Cape Folger. The lower curves show the effect of rapid motion and heavy accumulation on the summit to Cape Poinsett line. (From Budd, Young & Austin, 1976. Reproduced from the *Journal of Glaciology* by permission of the International Glaciological Society.)

the mass flux to any depth $U_x z$ by $\int_0^z u \, \mathrm{d}z$, the z ordinate being taken downwards from the surface. This can be done numerically, either from measured or calculated velocity–depth profiles at the borehole site. Curve 3 in Figure 1.19 shows the effect on curve 2 of applying a velocity–depth curve which falls from a surface value u_s to zero at bedrock, with the value at depth z, u_z being given by

$$u_s - u_z = Cz^4$$

where C is a constant.

For the three-dimensional problem, we assume that the direction of ice flow is the same at all depths. Then the total mass flux Φ between two flowlines separated by a distance Y and distant x from the start of the flowlines is $\Phi = YZ_x U_x$. For a steady-state ice sheet, this must be in balance with the total accumulation falling on the area between the flowlines from the point x_1 to the origin of the flowlines, so that instead of equation (1.13) we have

$$\frac{z}{Z_{x_1}} = \frac{\displaystyle\int_{x_2}^{x_2} (AY) \, \mathrm{d}x}{\displaystyle\int_0^{x_1} (AY) \, \mathrm{d}x} \qquad (1.15)$$

(a)

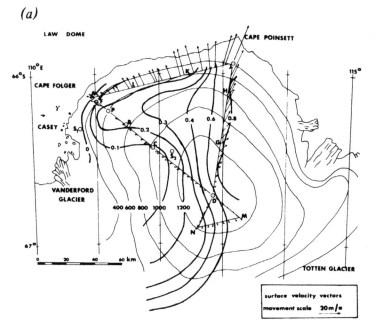

(b)

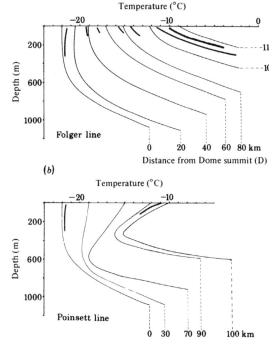

Folger line

Temperature (°C)

Distance from Dome summit (D)

(*b*)

Poinsett line

Temperature (°C)

Distance from Dome summit (D)

Figure 1.18. Position of ice deposited on surface at distance x_2 from centre of ice sheet related to its depth z when it has moved to distance x_1 from centre of a steady-state ice sheet.

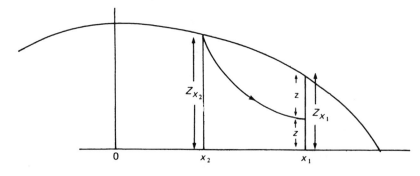

This equation simply tells us that we can estimate the proportional depth of ice that fell at a given elevation inland by taking two areas between given flowlines, one reaching from the borehole site to a given elevation contour and another stretching from the borehole site to the ice divide, and dividing the mass budget of the former by the mass budget of the latter.

In the case of a radial ice sheet, where flow diverges uniformly from the centre to periphery, the effect is to modify a linear profile (curve 2, Figure 1.19) to the shape of curve 4 of Figure 1.19, approximately parabolic in the inverse sense to curve 1 of that figure. If the flow converges radially towards an outlet glacier, the effect is in the inverse sense, so that a linear isotope–depth profile becomes almost parabolic, of similar form to curve 1. We can estimate the form of isotope–depth curves for particular conditions, such as a radial ice

Figure 1.19. Elevation of origin of ice (and hence approximate δ values) as a function of depth near the margin of a continental ice sheet in steady-state flow for the following cases.
Curve 1: two-dimensional vertical column flow for ice sheet of parabolic cross-section and uniform accumulation rate.
Curve 2: as for curve 1 except for an accumulation rate inversely proportional to elevation.
Curve 3: as for curve 2 except for non-vertical ice column in which velocity is proportional to fourth power of height above bedrock.
Curve 4: three-dimensional case for radially divergent flow, otherwise as for curve 2.
Curve 5: radially convergent flow, otherwise as curve 2.

sheet with uniform mass balance and ice movement falling to zero at bedrock by combining the appropriate effects, in this case by use of curves 1, 3 and 4.

Although one can predict the form of isotope-depth profiles for steady-state ice sheets, taking into account mass balance and flowline convergence or divergence, the resultant profile can vary in form so much that one cannot generalise about the results. We need to make individual estimates of form in each case, using the best available data. This has been done in chapter 4 for three boreholes at peripheral sites on the Antarctic ice sheet. The broad agreement of observed and estimated profiles for ice younger than, say, 10 000 a helps our understanding of the more recent flow of the ice sheet, and draws attention to depths at which major changes in ice deposited during the ice age are recorded.

It should be mentioned that although detailed calculations of particle trajectories in Figure 1.16 require information on the velocity of ice motion and on strain rates normal to flowlines, the time dependence of these two factors is eliminated from the calculations when equations (1.13), (1.14) and (1.15) are used. This has certain advantages in principle. If the mass balance over the ice sheet has changed in the same proportion at all points on the ice sheet, then our estimates of the point of origin from equations (1.13) and (1.15) remain valid, provided the form of the ice sheet has been unchanged. While the latter condition is unlikely to apply in peripheral areas during an ice age, changes in ice thickness

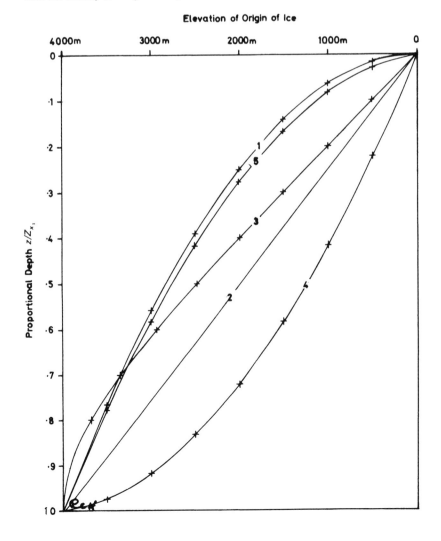

Elevation of Origin of Ice

and form in central regions, especially in eastern Antarctica, were probably relatively small, so calculations of the point of origin of ice should be approximately correct for times up to several hundred thousand years. In peripheral regions the calculations are probably satisfactory for many purposes for ice deposited up to 8000–10 000 a ago.

When we wish to determine the age at any depth in an ice core, as well as the point of deposition, we must introduce ice movement velocity into the calculations. We can readily appreciate that age determinations based only on local strain rates can apply effectively only near the centre of outflow of an ice sheet, and not in peripheral regions. These problems are considered in much more detail in chapter 5.

While the preceding discussion is relevant to the major part of Antarctica, where there is little ablation by melting of the surface of the ice sheet, in Greenland surface ablation accounts for about half of the loss of ice. The equilibrium line separating ablation and accumulation areas lies mostly at elevations ranging from less than 1000 m in the north to about 2000 m in the south. In central Greenland where extensive glaciological studies have been made, the equilibrium line lies around 1500 m. We desire to know how δ values will vary along a flowline across the ablation zone in a steady-state ice sheet.

In Figure 1.20 we show schematic flowlines for a two-dimensional ice sheet, starting from the upstream and downstream limits of a section of the accumulation zone of width δx_A and a mean distance x_A from the equilibrium line (0). The two lines emerge at the surface of the ablation zone separated by a distance of δx_N and at a mean distance x_N from the equilibrium line. If A and N are the accumulation rate and ablation rate on the zones δx_A and δx_N, then for the surface of the ice sheet to be in a steady state, the ablation rate over δx_N must equal the accumulation rate over δx_A, that is $A \delta x_A = N \delta x_N$. Now while a two-dimensional model is reasonably satisfactory in the centre of Greenland, near the coast a substantial fraction of the ice is channelled into outflow glaciers while the remainder is lost by surface ablation along other sections of the margin of the ice sheet. In the three-dimensional case we need, as before, to consider the mass flux between a pair of flowlines. We put the horizontal separation between two

flowlines as Y_A in the zone δx_A, and Y_N in the corresponding zone δx_N. Then, for steady state we now require that

$$A \delta x_A Y_A = N \delta x_N Y_N \qquad (1.16)$$

Our aim is to determine the ratio $\delta x_N/\delta x_A$ and hence the distribution of height of origin of ice across the ablation zone. Values of A and N can be estimated from available surface measurements and we need to know the ratio Y_N/Y_A. Direct measurement of flowline direction along two entire flowlines would be a vast undertaking (see section 3.7), but if surface velocity measurements are available, as is the case across central Greenland (see Figure 2.17), we can derive the ratio of Y_N/Y_A for the steady state. For steady state to apply between two flowlines, the mass transport of ice from upstream at x_A must equal the mass transport at x_N in the ablation zone if there is to be no build up or deficit of ice between x_A and x_N. If U_A and U_N are the mean ice column velocities, and Z_A and Z_N the ice depths, we can equate the mass fluxes as

$$U_A Y_A Z_A = U_N Y_N Z_N \qquad (1.17)$$

We can then obtain a value for Y_N/Y_A from measurements of velocity and thickness. Substituting these values in (1.16) we get

$$\frac{\delta x_A}{\delta x_N} = \frac{N}{A} \frac{U_A}{U_N} \frac{Z_A}{Z_N} \qquad (1.18)$$

The quantities on the right-hand side can all be measured, the main uncertainty being the difference between the measured surface velocities and the column velocities U_A and U_N.

In general Z_A and N are considerably greater than Z_N and A except at elevations around the equilibrium line. This is to be expected, since, in general, ablation areas are much smaller than accumulation areas, the ratio being of the order of 1:6 in central west Greenland. We present relevant data for this area in section 4.2. Over a considerable area near the equilibrium line $\delta x_A/\delta x_N$ lies between 1/1 and 1/2. While at altitudes well below the equilibrium line, where flow has been channelled off towards an outlet glacier, $\delta x_A/\delta x_N$ is of the order of 50:1 for ice that originates from a large part of the ice sheet. However, this refers to one particular case and the extent to which this is typical of the marginal ablation zone is not known.

Comparison of isotopic and temperature profiles
While the previous section helps our direct interpretation of the historical record of isotopic values

Figure 1.20. Two-dimensional cross-section of ice sheet showing symbols used for analysis of ice transport between pair of flowlines.

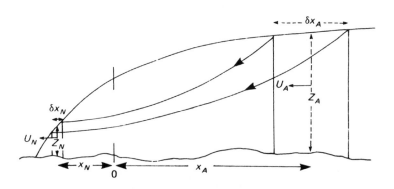

obtained in an ice core, the temperature at any depth depends on the integrated effect of the past surface temperature and flow of our ice column. The greater the depth, the longer the time period over which effects are to be integrated (Figures 1.13, 1.14). However, if we start from an observed temperature–depth profile we can determine which boundary conditions give a best fit to the observed profile when using simple steady-state equations. With this approach, the geothermal heat flow is given by the temperature gradient immediately above bedrock, and we find best-fit values for an assumed steady rate of change of surface temperature with time and for the vertical velocity of ice at the surface. Rapid computing techniques make this approach practicable. Application of this method to the temperature profiles at Camp Century, Byrd Station and Vostok is described in chapter 5. In each case the best-fit vertical velocity matches the observed net accumulation rate during recent decades to within the error of observation (10–25 per cent), which implies that the ice sheet is in a steady state with regard to thickness within these limits. The temperature trends calculated in this way show that, whereas surface temperatures at Byrd Station have been getting warmer with time, at Camp Century the surface has been getting colder. Figure 1.21 shows that, although opposite in sense, both trends agree with changes of

Figure 1.21. Comparison of temperature change with time (straight lines) derived from analysis of temperature–depth profiles at Byrd Station and Camp Century with δ ^{18}O profiles on a time scale shown by Johnsen *et al.* (1972). Conversion factors from temperature to isotopic scales are δ ^{18}O ∝ 0.76θ for Byrd Station and δ ^{18}O ∝ 0.62θ for Camp Century. (From Robin, 1976. Reproduced from the *Journal of Glaciology* by permission of the International Glaciological Society.)

δ values with time over the past 2000 a (Camp Century) and 4000 a (Byrd Station). This comparison supports the interpretation of isotopic values as indicators of past temperatures. That this is a simplified interpretation of a complex problem should be apparent from this chapter. The complexities of such comparisons and the additional conclusions that can be obtained from further analysis of the data available are set out in some detail in chapter 5.

Conclusion

Our present knowledge of polar ice sheets can be summarised briefly. Such ice sheets are of continental dimensions, they are continually nourished by falling snow, and over long periods this snow builds the ice into vast domes which slope, and therefore flow, outward from central areas. Analysis of the limited number of temperature profiles available from Greenland and Antarctica indicates that the ice sheets are approximately in a steady state as regards temperature distribution, and hence also as regards their size and climate. Recent evidence from measurements of surface movement by satellite fixes, and analysis of surface strains along a line on the inland ice also seems to confirm this picture. The agreement between these two methods gives confidence in the approximate validity of our model of ice flow. We can therefore use this model to interpret isotopic values from ice cores back to still earlier ages than the 10 000 a or so covered by our temperature profiles. How much further back we can go with confidence is considered in later chapters.

In studying the earlier history of ice sheets, evidence from glacial geology and allied studies is of value in helping us to determine past variations in their

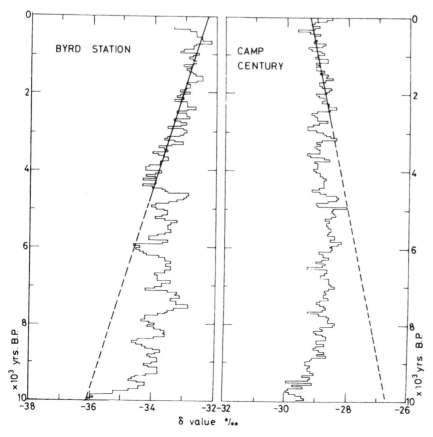

size and flow. Chapter 2 therefore summarises and discusses such evidence, especially in relation to Greenland and Antarctica. In chapter 3 we consider the accuracy with which we can measure and interpret different factors which govern the behaviour of ice sheets. Chapter 4 presents observations of the principal isotopic and temperature profiles that are available, along with associated glaciological data that can help our interpretation. Chapter 5 presents some theories by which we can interpret the observations; it describes the basic assumptions and methods used in calculating the various models of the ice sheet, and gives the results for comparison with observed profiles. Finally, in chapter 6 we survey evidence on the numerical relationship between mean δ values and mean temperatures and discuss the magnitude and cause of errors involved in the relationship. We then briefly relate evidence from isotopic profiles to other evidence on climatic changes over the past 100 000 a.

Throughout the book we have kept in mind those working in disciplines related to glaciology, such as oceanographers, climatologists, glacial geologists and botanists who need to understand the basis on which δ profiles have been interpreted by glaciologists. We have therefore endeavoured to produce a work that is broad in scope, informative about glaciological details, and outlines relevant knowledge from related fields. We hope the result will interest non-specialists and be of value to glaciologists.

ICE SHEETS: GLACIAL GEOLOGY AND GLACIOLOGY

2

2.1

Data on the size of former ice sheets

W.S.B.PATERSON

In order to estimate the dimensions of former ice sheets, and their pattern of flow, we may use both observational data and theoretical models, provided the latter have been adequately tested by comparison with present-day ice sheets. We discuss both approaches in this section, and outline briefly the types of information from which the size of earlier ice sheets can be deduced. A textbook of glacial geology (e.g. Flint, 1971) should be consulted for further details.

Glacial deposits

The history of glaciation in a region can be reconstructed by studying the sequence of deposits and eroded landforms that the ice leaves behind. Moving ice may, under certain conditions, pick up stones and other debris from its bed. When this ice reaches a place where the basal temperature corresponds to the melting point, geothermal heat, and the heat produced by friction of ice on bedrock, will slowly melt basal ice and release the material in it. Again, during recession and downwasting of a retreating ice sheet, the debris contained in the ice will be deposited. These processes form layers of 'till', unsorted material of a wide range of shapes and sizes. The extent of sheets of till and the locations of end moraines, formed by deposition of debris at the terminus of a glacier during periods when its position is stationary, show the extent of former ice sheets in any region. Additional signs of glaciation are erratics (boulders, of different material from the underlying bedrock, that have been carried by ice) and bedrock that has been grooved, scratched or polished by ice movement.

To determine the extent of a former ice sheet at different times, deposits in different regions must be correlated. While this can sometimes be done on the basis of their appearance, dating of the deposits is more satisfactory. ^{14}C dating of materials such as wood, peat, or shells is commonly used. For example, if an advancing ice sheet overrides a forest, wood will be incorporated into the glacial deposits. Or peat may start to accumulate in an area after the ice has retreated. Another dating method is to count 'varves'; these are sediments, stratified in annual layers, which were deposited in glacial lakes.

Other necessary information about a former ice sheet is its thickness and the direction of ice movement. In mountain regions, the ice thickness at the glacial maximum may be determined from the upper limit of strong erosion on valley walls or from the upper limit of erratics. Erratics also show the direction of ice movement if their source area is known. 'Striations', scratches on bedrock surfaces made by small pieces of rock embedded in the moving ice, show the direction of ice

Figure 2.1. Sea-level curve for Ottawa Islands, Hudson Bay, Canada drawn on the basis of the elevation and age of radiocarbon-dated samples. The vertical and horizontal dimensions of the rectangles correspond to the probable errors in elevation and age. The solid curve is the probable maximum sea-level position, the dotted curve the minimum. From Walcott (1972) based on data of Andrews & Falconer (1969).

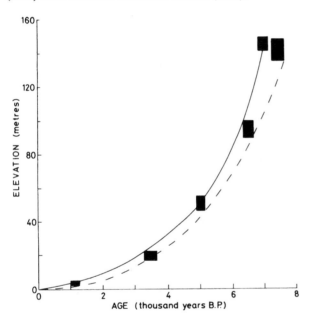

movement, but, in most cases, the two directions 180° apart cannot be distinguished. Statistical analysis of the orientations of pebbles in till has also been used for this purpose. In many glaciated areas, the bedrock has been eroded into asymmetrical bosses and hillocks; the upstream side has been smoothed by abrasion, while the downstream side is steep and irregular as a result of quarrying action by the ice. This is another indication of direction of ice movement, as are asymmetrical depositional features such as drumlins and eskers. The drawback of most of these methods is that they show the direction of ice movement during the final stages of glaciation; signs of earlier movements have usually been obliterated. For reconstructions of past ice sheets, on the other hand, the direction of ice flow at the glacial maximum is usually of most interest and this may be quite different from the final direction.

By such means the extent of the ice sheets in North America and Eurasia at various times in the last glaciation has been mapped. The uncertainties in the results can be illustrated by comparing two North American studies, that of Prest (1969) and of Bryson, Wendland, Ives & Andrews (1969). These authors agree closely on the maximum extent of the ice sheet; the main discrepancies are in the timing of its retreat and eventual disappearance. For example, the area of the Laurentide ice sheet at 10 000 BP on the map of Bryson *et al.* is 1.25 times that on Prest's map. Again, according to Prest, Laurentide ice had disappeared from the Canadian mainland by 6000 BP, whereas Bryson and others consider that remnants remained until about 4500 BP.

Sea level changes and uplift data

The weight of a large ice sheet depresses the land underneath it by an amount proportional to the ice thickness. For example, because the density of the earth's mantle is 3.7 times that of ice, 3700 m of ice would cause a depression of about 1000 m. As the ice melts, the land starts to recover and rises slowly, a process that continues for thousands of years. Uplift following the removal of the ice sheets of the last glaciation is still continuing. Measurements of the amount of uplift in

Figure 2.2. Contours of elevation above present sea level of shoreline in eastern North America at 6000 BP. The contours are based on values, interpolated to 6000 BP, on curves of sea level versus time at the points shown. After Paterson (1977, Fig. 2) and Walcott (1972, Fig. 16).

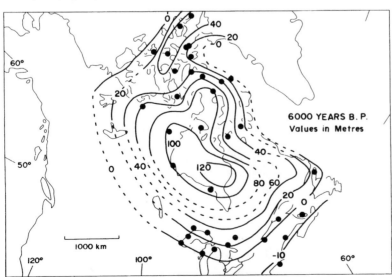

different places indicate the extent and relative thickness of the former ice sheet. Uplift can be measured in certain regions because the ice sheets in North America and Europe depressed low-lying areas, such as the lands surrounding the present Hudson Bay and Baltic Sea, below sea level. Thus these areas were flooded by the sea when the ice sheets melted. Thereafter the land started to rise, so that beaches and marine deposits appeared above sea level. Materials such as marine shells or driftwood buried in beach gravels can be dated by the ^{14}C method. In any given locality, such dated specimens at different elevations provide a 'sea-level curve', that is, a record of the emergence of the land relative to sea level. Figure 2.1 is an example.

A complication arises because world sea level changes with time; as the ice sheets melt, the amount of water in the oceans increases and sea level rises. Simultaneous world-wide changes of sea level are called eustatic changes. They are determined by combining emergence and submergence data from all parts of the world. In this way it is hoped that anomalous changes in certain areas, such as tectonic effects or movements resulting from the removal of ice sheets, will be averaged out. Between 18 000 BP and 6000 BP world sea level rose by roughly 120 m; it has been stable to within a few metres since then (Walcott, 1972). Thus, before a sea-level curve for a particular locality will represent the emergence of the land relative to present sea level, it has to be corrected for eustatic changes.

Sea-level curves from different places provide the data for drawing contour maps of the elevation, above present sea level, of old shorelines of different ages. Figure 2.2 shows the present elevation of the 6000 BP shoreline in eastern North America. The region of uplift corresponds closely to the area covered by ice sheets at the maximum of the last glaciation as deduced from glacial-geological data. See, for example, the map of Prest (1969). The region of greatest uplift is centred over Hudson Bay; this suggests that the ice was thickest there. The separate centre of uplift in the northernmost islands suggests that the Laurentide ice sheet, centred over Hudson Bay, did not extend that far north; geological evidence confirms this.

It is important to note that such maps may be interpreted only in terms of relative ice thickness in different areas, not absolute thickness. This is because measured post-glacial uplift is only a fraction of the total uplift. Uplift begins as soon as the ice thickness starts to decrease. The uplift measured, on the other hand, is restricted to that which has occurred since the ice disappeared. Moreover, uplift is still continuing in most areas. Walcott (1972), for example, has estimated, on the basis of the negative free-air gravity anomaly, that another 300 m of uplift will occur in the Hudson Bay area.

Clark & Lingle (1977) have shown that sea-level changes as the ice/water load is redistributed over the earth result from changes in the gravity field and elastic rebound as well as from the previously mentioned slow recovery due to the flow of material beneath the earth's crust. Uplift measured by raised beaches in regions

formerly covered by ice sheets is dominated by the last factor. Elsewhere, however, the pattern of sea-level changes is more complex than the uniform rise or fall assumed by most students of eustatic changes.

However the study of glacial deposits, combined with uplift measurements, can give a good indication of the extent of former ice sheets and some idea of where the thickest ice, and thus the centres of outflow, were. We next consider how to reconstruct the ice sheet surface.

2.2

Surface profiles of ice sheets

W.S.B.PATERSON

In this section we treat the problem of deducing the shape of the surface of an ice sheet from its horizontal dimensions. The problem arises in ice core studies because, to interpret oxygen isotope data, one must know the surface elevations at which the different layers in the core originated. Moraines and other glacial deposits, however, usually indicate only the former extent of the ice sheet.

We first consider the problem from the theoretical aspect, based on analyses by Nye, Weertman, and others. Paterson (1969, Chp. 9; 1972) has made summaries of this work to which the reader is referred for further details of the mathematics and for references to the original papers. We consider an ice sheet in a steady state (that is, the thickness does not change with time), on a horizontal bed. Figure 2.3 represents a cross-section of the ice sheet and shows the coordinate system and the symbols used.

The surface profile might be expected to depend on the accumulation and ablation, and on the way in which ice deforms in response to an applied stress (the

'flow law'). A reasonable approximation is to regard ice as a rigid/perfectly plastic material. Such a material does not deform as long as the applied stress is less than a certain critical value; when the stress reaches that value the deformation rate becomes very large. The critical value is called the yield stress, τ_c. In a perfectly plastic ice sheet, the thickness at each point is such that the shear stress at the base is equal to the yield stress. A value $\tau_c = 1$ bar is often used for ice.

In the central plane of the ice sheet, OB in Figure 2.3, there is no shear stress; the normal pressure at depth $(H-z)$ equals the hydrostatic pressure $\rho g\,(H-z)$. Thus the total horizontal force on OB, per unit length perpendicular to the plane of the diagram, is

$$\int_0^H \rho g(H-z)\,\mathrm{d}z = \tfrac{1}{2}\rho g H^2$$

Here ρ is density (assumed constant) and g is acceleration due to gravity. The only other horizontal force on OB is the force due to the shear stress τ_c on the base OA. Thus, for equilibrium,

$$\tfrac{1}{2}\rho g H^2 = \tau_c L$$
$$H = K_0 L^{1/2} \tag{2.1}$$

where $K_0 = (2\tau_c/\rho g)^{1/2}$, a constant. If $\tau_c = 1$ bar, and H and L are measured in metres, $K_0 = 4.7\,\mathrm{m}^{1/2}$. This important formula relates the thickness at the centre of an ice sheet to its radius or half-width. For example, in Greenland at latitude $72°$ N, $L = 450$ km. The formula gives $H = 3150$ m, close to the true value of about 3200 m.

By considering the equilibrium of a section of ice sheet, of height h and length δx, it can be shown that

$$\frac{\mathrm{d}}{\mathrm{d}x}(\tfrac{1}{2}\rho g h^2)\delta x = \tau_c \delta x$$

which integrates to give the equation of the surface profile

$$h = K_0(L-x)^{1/2} \tag{2.2}$$

This can also be written

$$h = K_0(x')^{1/2} \tag{2.3}$$

where x' is the distance from the edge.

Two points should be noted:

(1) the shape of the profile depends only on the plastic properties of ice (the yield stress) and not on the accumulation and ablation rates;

Figure 2.3. Coordinate system and symbols used in theoretical analysis.

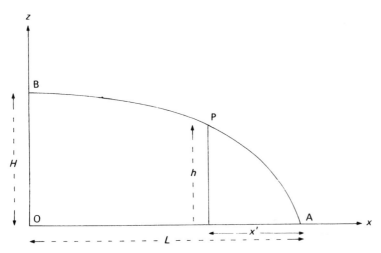

(2)   the shape of the profile is the same whether the ice sheet is visualized as a long ridge perpendicular to the plane of the diagram (Figure 2.3) or whether, in plan view, it has a circular or even an irregular outline; in the last case, x' in (2.3) has to be interpreted as distance from the edge measured along a flowline, that is, a line of maximum slope.

To treat ice as a rigid/perfectly plastic material is only a first approximation. For an ice sheet deforming in simple shear, the actual flow law is

$$du/dz = 2(\tau/C)^n \qquad (2.4)$$

Here u is the x component of velocity, C is a parameter that depends on the crystal structure, impurity content, and temperature of the ice but not on the applied stress, n has a value of about 3, and τ is the shear stress given by

$$\tau = \rho g(h - z)\alpha \qquad (2.5)$$

if α, the surface slope, is small. This formula is easily derived by considering the balance of forces in a parallel-sided slab of slope α. It is a good approximation in the present instance if α is small (Nye, 1969). Because α is not small near the edge of the ice sheet, the theoretical profile may be unrealistic there. In the present case the bed of the ice sheet is assumed to be horizontal, and so $\alpha = -(dh/dx)$.

To determine the surface profile we also use the fact that, in a steady state, the amount of ice that accumulates on the surface between the centre and any point P, must be equal to the amount of ice flowing through a vertical section at P. Thus,

$$Ax = \int_0^h u \, dz \qquad (2.6)$$

where A is the accumulation rate, assumed constant over the ice cap. This relation holds when the ice cap is a long ridge with parallel flowlines.

The equation of the profile is derived as follows:
substitution of (2.5), with $-dh/dx$ written for α, in (2.4);
integration of (2.4) to determine u;
substitution of this value of u in (2.6), and integration.
This gives a differential equation relating h and x. The solution, for $n = 3$ is

$$h = K_1(L^{4/3} - x^{4/3})^{3/8} \qquad (2.7)$$

where

$$K_1^2 = H^2/L = 2.1(\bar{C}/\rho g)^{3/4} A^{1/4} \qquad (2.8)$$

The flow law multiplier $\bar{C}$ depends on ice temperature and other quantities that vary with x and z. Thus the value in (2.8) is an average. Equations (2.7) and (2.8) can be compared with (2.2) and (2.1). For radial flow in an ice cap of circular plan, (2.6) is replaced by

$$\tfrac{1}{2}Ax^2 = x\int_0^h u \, dz$$

and the numerical factor in (2.8) becomes 1.7 instead of 2.1.

In contrast to the case of perfect plasticity, the profile calculated from the flow law depends on accumulation rate, temperature, and the shape of the

ice cap in plan. The dependence is not very sensitive however, particularly as regards accumulation; maximum thickness H is proportional to the eighth root of the accumulation rate A. The relation between the two cases is that perfect plasticity is the limit where the flow law index n tends to infinity. The strain rate du/dz is then zero if $\tau < C$ and infinity if $\tau > C$.

So far, we have considered an ice sheet, such as the one in Antarctica, in which there is accumulation all over the surface; ice is lost only by calving of icebergs at the edge. An alternative is an ice sheet with accumulation at rate A at the higher elevations and ablation at rate N elsewhere. The accumulation area profile can be shown to be

$$h = K_2[L^{4/3} - (1 + A/N)^{1/3}x^{4/3}]^{3/8} \qquad (2.9)$$

and, in the ablation area,

$$h = K_2(1 + A/N)^{1/8}(L - x)^{1/2} \qquad (2.10)$$

which is a parabola. In (2.9) and (2.10)

$$K_2 = H^2/L = 2.1(\bar{C}/\rho g)^{3/4}(AN)^{1/4}(A + N)^{-1/4} \qquad (2.11)$$

Equation (2.11) applies to an ice sheet in which the crest is a long ridge. As before, the numerical factor is changed to 1.7 for an ice sheet of circular plan.

In some ice sheets, motion may take place largely by ice sliding over the bedrock rather than by deformation within the ice itself. In such cases the theory starts, not from the flow law, but from the equation for sliding velocity U_{sl}.

$$U_{sl} = B\tau_b^m$$

Here τ_b is the shear stress at the base and B is a parameter that depends on the bedrock roughness, the thermal properties of ice and rock, and the plastic properties of ice; m is about 2 (Weertman, 1957). Equations that differ only slightly from (2.7) to (2.11) can be derived.

The ellipse

$$h^2 = (H/L)^2(L^2 - x^2) \qquad (2.12)$$

has been used as an empirical fit to portions of the Antarctic ice sheet.

Other theoretical work on profiles includes;
analysis of the minor modification needed to the standard profile near the centre where the shear stress is zero (Weertman, 1961a);
profile calculations for the terminus of a perfectly plastic glacier (Nye, 1967);
allowance for the isostatic depression of bedrock under the ice (Weertman, 1961a);
analysis of the effect of water at the base of the ice (Weertman, 1966).

Two questions may be asked:
(1)   how much difference is there between these various profiles?
(2)   how well do they fit profiles of real ice sheets?

In Figure 2.4 the parabola (equations (2.2) and (2.10)), the ellipse (2.12), and equation (2.7) are compared. The parabolic profile is lower than the others; however, the differences are relatively minor. Figure 2.5 shows that (2.7) fits the profile of part of the Antarctic ice sheet very well; other similar examples could be quoted. The theoretical profiles are good representations of many real ice sheets, in spite of the drastic simplifying assumptions in the analysis. Again, Figure 2.6 shows surveyed profiles along various ice sheets. There is a noticeable similarity between them although they cover a wide variety of sizes, ice temperatures and accumulation or ablation rates. This supports the basic theoretical prediction that the shape of an ice sheet is mainly determined by the plastic properties of ice; variations in accumulation rate, ice temperature, bedrock characteristics and other factors are relatively unimportant.

Not all ice sheet profiles conform to theoretical ones however. This may arise from failure of one or both of the main assumptions; namely, that the ice sheet is in a steady state and that its bed is horizontal. Paterson (1969, p. 154) shows examples. No real ice sheet is ever, strictly speaking, in a steady state; its dimensions are always changing in response to changes in climate. In most ice sheets, however, ice flow more or less counterbalances the effects of accumulation and ablation. Thus the rate of thickness change with time is small compared with the accumulation rate. In such cases, a steady-state theory should be a good approximation. The effect of irregularities in the bedrock is to introduce longitudinal stresses in the ice so that it no longer deforms in simple shear as assumed in the theory. The resulting changes in

the surface profile can be calculated (Robin, 1967; Collins, 1968; Nye, 1969; Budd & Carter, 1971) if additional data, such as surface strain rates, are available.

In some cases, it is not clear why certain profiles differ from theoretical ones. For example, Hughes (1973) has pointed out that part of the West Antarctic ice sheet has a concave profile. He interprets this as evidence that the ice sheet is unstable and may be disintegrating at present. The shape of the profile certainly suggests that the ice sheet is not in a steady state, and it is not clear why fast-moving ice streams form near the edge of the ice sheet where it joins the Ross Ice Shelf. The step from this to instability seems more doubtful. Again Mathews (1974), by studying elevations of moraines and upper limits of erratics on valley walls, has reconstructed the profiles of certain lobes of the Laurentide ice sheet that covered most of Canada during the last glaciation. While some lobes had normal profiles, others were much flatter than expected. Mathews suggested that the abnormal profiles might result from water at the base of the ice.

Which profile to use in any particular case depends on the amount of data available. For a detailed study of an existing ice sheet, for which accumulation and ablation rates, ice temperature and bedrock characteristics are known, profiles such as (2.7), (2.9) or (2.10) could be used. For reconstructions of past ice sheets, which are at

Figure 2.4. Theoretical profiles of ice sheets. From Paterson (1969, Figure 9.2).

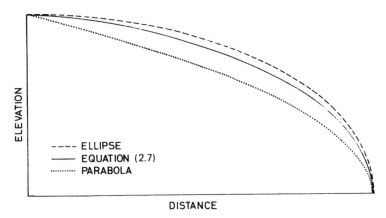

Figure 2.5. Profile of Antarctic ice sheet between Mirny and Komsomolskaya, compared with theoretical profile. Data from Vyalov (1958), diagram after Paterson (1969, Figure 9.3).

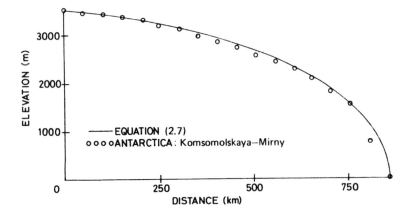

best only rough approximations, the parabola, equation (2.2), may well be adequate. A value of the average basal shear stress, which determines the value of K_0, has to be chosen. A value $K_0 = 4.7 \, \text{m}^{1/2}$, which corresponds to a shear stress of 1 bar, gives a good estimate of ice thickness in central Greenland, and is a good average value for valley glaciers in temperate regions (Nye, 1952).

Figure 2.6. (*a*) Surface profiles over sections of the ice sheets of Greenland and Antarctica, where the ice lies on rock generally within ±500 m of sea level. From Robin (1964).

(*b*) Surface profiles of smaller ice sheets or ice caps over sections that have an approximately horizontal bed. From Robin (1964).

On the other hand, it seems to be too high for most ice sheets (Orvig, 1953; Bull, 1957; Haefeli, 1961). Budd, Jenssen & Radok (1971*a*) calculated that, in Antarctica, the basal shear stress varies from about 0.2 bar in the centre to 1 bar near the coast. A mean of 0.6 bar corresponds to $K_0 = 3.6 \, \text{m}^{1/2}$. Again, the ice lobes with normal profiles in Mathews' (1974) study, mentioned above, gave values of K_0 between 2.9 and 4.1 $\text{m}^{1/2}$. Thus, in the absence of any information about basal shear stress, accumulation and ablation rates, or other parameters, an appropriate profile would be

$$h = 3.6(L - x)^{1/2} \qquad (2.13)$$

where all quantities are in metres.

(*a*)

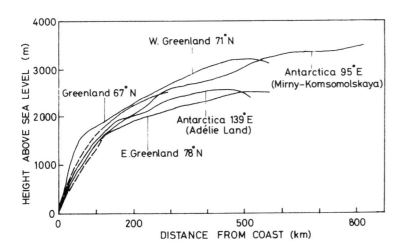

(*b*)

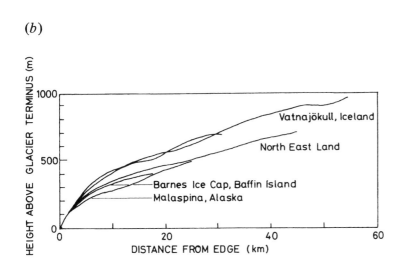

2.3

Migration of centres of ice domes and of ice-shelf–inland-ice boundaries

J.WEERTMAN

Ice domes of the Greenland and the Antarctic ice sheets undoubtedly have migrated over the past in response to changes of the climate. However, model calculations (Weertman, 1973) show that the magnitude of the shift of the position of the centre of an ice dome is relatively small, even for rather large changes in the accumulation pattern, *if* the edges of an ice sheet remain in about the same position. As the edges of the East Antarctic ice sheet and the southern part of the Greenland ice sheet probably have not migrated over great distances since the time these ice sheets were formed (sections 2.4 and 2.5); the centres of their ice domes are not likely to have shifted very far from their present-day positions.

It is much easier to analyse the flow and deformation of ice under the centre of an ice dome than at other parts of an ice sheet. Therefore it is easier to obtain information from ice cores obtained from holes drilled at the centres provided that the centres have not migrated over appreciable distances in the past. For this reason it is of interest to estimate the effect that a change in the accumulation pattern on an ice sheet has on the position of the centre.

The physical basis for the relative stability of the centre of an ice dome whose edges are fixed is the fact that the flow law of ice approximates to that of a perfectly plastic solid. Consider the simple example of a two-dimensional ice sheet that rests on a flat rigid bed (see Figure 2.7). Let the origin be at the bed directly

Figure 2.7. Cross-section of a two-dimensional ice sheet that rests on a flat bed.

under the ice divide; let x be distance measured in the horizontal direction; let $h_1(x)$ and $h_r(x)$ be the ice thickness at x for the left-hand and the right-hand sides, respectively, of the ice sheet; and let L_1 and L_r be the distance from $x = 0$ to the left-hand and right-hand edges of the ice sheet.

If ice obeys the equation of flow of a perfectly plastic solid, the ice profiles from equation (2.2) are

$$h_1(x) = K_0(L_1 + x)^{1/2} \qquad (2.14a)$$

and

$$h_r(x) = K_0(L_r - x)^{1/2} \qquad (2.14b)$$

The profiles given by equations (2.14) are independent of the accumulation rate. If the accumulation rate is increased, the ice flows faster but the yield stress is not exceeded.

At the ice divide ($x = 0$) the right- and left-hand profiles must have the same thickness ($h_1(0) = h_r(0)$). Therefore $L_1 = L_r$ and the ice divide is equidistant from the edges of the ice sheet. The position of the ice divide is not shifted if the accumulation rates on the two sides of the ice sheet are not equal or if the rates are changed. If the bed is not flat and the average elevation under the left-hand side differs from that under the right, the position of the ice divide will not be symmetric with respect to the edges. (John Hollin (private communication) has pointed out that Fig. 3 of Weertman (1973) is in error. The side of the ice sheet with the higher bed should have the smaller width.) However the position of the divide for this situation likewise cannot be changed by changing the accumulation rate. Similar results hold for a three-dimensional ice sheet.

For the more realistic case in which it is assumed that the flow of ice is described by Glen's power law creep equation (that is creep rate is proportional to stress raised to approximately a third power) the average ice velocity $U_i(x)$, where $i = 1$ or r, in the horizontal direction for the ice sheet shown in Figure 2.7 is given by

$$U_i(x) = -C'h_i^m (dh_i/dx)| dh_i/dx|^{n-1}$$

$$= h_i^{-1} \int_0^x A_i \, dx \qquad (2.15)$$

where C' is a positive constant, A_i is the accumulation rate at x, $n \approx 3$, and $m = n + 1$ for laminar flow if no sliding at the bed occurs and $n \approx m \approx 2$ if almost all the motion is produced by sliding (Paterson, 1969).

The left- and right-hand profiles of the ice sheet are obtained by solving equation (2.15) for h_i and setting $h_1(0) = h_r(0)$. For the approximation in which the right-hand side of equation (2.15) is set equal to $\bar{A}_i x$, where $\bar{A}_i = L_i^{-1} \int_0^{L_i} A_i \, dx$, the position of the ice

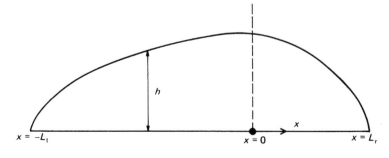

divide is given by the equation

$$\Delta L/L = \Delta A/2(n+1)\bar{A}_r \qquad (2.16)$$

where $L = \frac{1}{2}(L_1 + L_r)$, $\Delta L = L_r - L$, and $\Delta A = \bar{A}_1 - \bar{A}_r$. The term ΔL measures the shift of the ice divide from the symmetry position when the accumulation rates on the two sides of the ice sheet are not equal. For the case of $n = 3$ the ratio $\Delta L/L = \frac{1}{8}\Delta A/\bar{A}_r$. In other words, the amount of the fractional shift is almost an order of magnitude smaller than the fractional difference in the accumulation rate. (It should be noted that the side of the ice sheet with the heavier accumulation rate has the smaller width.)

Ice-shelf–inland-ice boundary

The thickness of ice sheet shown in Figure 2.7 goes to zero at the edges. For all practical purposes this model is approximately correct for the Greenland and East Antarctic ice sheets. The beds of these ice sheets are above sea level (or would be above sea level if the ice sheets were removed and isostatic rebound took place). The ice from these ice sheets flows into the sea at outlet glaciers, and the ocean–land boundary is approximately the boundary of the edge of the ice sheets. The West Antarctic ice sheet has a bed that is below sea level over a major fraction of its area (even if the ice sheet were removed and isostatic rebound occurred). Most of the ice flow from this sheet goes into two ice shelves, the Ross and the Filchner Ice Shelves. The boundary of the ice sheet occurs where the ice mass loses contact with the bed and floats in sea water.

Consider the simple case of a two-dimensional ice sheet that flows into a two-dimensional ice shelf. Figure 2.8 shows the right-hand side of such an ice sheet and ice shelf. The ice sheet is considered in this figure to rest on a rigid bed of constant slope β. The ice sheet joins the ice shelf at $x = L$. The depth D of the bed below sea level is equal to D_0 at the centre ($x = 0$) and is equal to D_L at $x = L$ ($D_L \geqslant D_0$). The ice thickness again is equal to $h(x)$ at any point x. It is assumed that there is a positive rate of accumulation $A(x)$ at any point x on the ice sheet or on the ice shelf.

If the flow law of ice were that of a perfectly plastic solid, the thickness h of the ice shelf of Figure 2.8 would have a unique value that is given by the expression

$$h = 4\tau_C/\Delta\rho g \qquad (2.17)$$

where $\Delta\rho = (\rho_w - \rho_i)$ and ρ_w and ρ_i are the densities of sea water and ice respectively (Weertman, 1974). The ice shelf of Figure 2.8 joins the ice sheet at the distance $x = L$ from the centre at which the ice sheet touches the bed. The depth D_L from sea level to bedrock at $x = L$ is given by

$$D_L = D_0 + \beta L \qquad (2.18)$$

and the distance from sea level to the bottom of a floating ice mass of thickness h is equal to $(\rho_i/\rho_w)h$. Thus the position of the ice-shelf–inland-ice boundary is given by

$$L = \beta^{-1}[(\rho_i/\rho_w)(4\tau_0/\Delta\rho g) - D_0] \qquad (2.19)$$

It should be noted that if D_0 is greater than $(\rho/\rho_w)h$, where h is given by equation (2.17), no ice sheet can exist.

For a flat bed ($\beta = 0$) equation (2.19) predicts that $L = \infty$ if D_0 is smaller than $(\rho/\rho_w)h$. In other words, in this situation the ice sheet will extend out to the edge of the continental shelf.

The distance L to the ice-shelf–inland-ice boundary can be calculated for the realistic case in which the flow law of ice is taken to be Glen's power law creep equation (Weertman, 1974). The analysis is too long to be repeated here. For the realistic flow law the ice shelf thickness is not unique but depends weakly on the accumulation rate and distance from the ice-shelf–inland-ice boundary. The result that is obtained (Weertman, 1974) is that the ice thickness $h(L)$ at the junction of the ice sheet and the ice shelf is given by the expression

$$h(L) = G(\bar{A}L)^{2/9} \qquad (2.20)$$

where $\bar{A}$ is the average accumulation rate over the ice sheet and G is a constant equal to $37.9\ \mathrm{m}^{5/9}\bar{A}^{2/9}$. (This value of the constant G is calculated for the case in which isostatic sinking of the ice sheet into the bedrock can occur. If the bed is considered to be rigid the value of G should be reduced by a factor approximately equal to $\frac{2}{3}$.) Combining equations (2.18) and (2.20) gives the following equation for the position $x = L$ of the boundary of the ice shelf and the inland ice:

$$G(\bar{A}L)^{2/9} = D_0 + \beta L \qquad (2.21)$$

Depending upon the values of β and D_0 there may be one, two, or no real, positive values of L that satisfy this equation. (These solutions are easily found by graphical means.) If $\beta = 0$ there is only one solution. This solution is one of unstable equilibrium. That is, if the actual value of L is larger than the value given by equation (2.21), the ice sheet will continue to grow and increase its width, and if the actual value of L is smaller the ice sheet will shrink until it disappears. (The reader should

Figure 2.8. Cross-section of the right-hand side of an ice sheet that flows into an ice shelf and which rests on a bed of constant slope β.

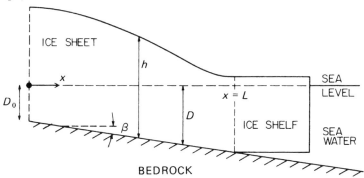

note that for $\beta = 0$ the equilibrium value of L decreases in value as $\bar{A}$ increases in magnitude. This result is a symptom of unstable equilibrium.) For the situation in which there are two solutions of equation (2.21) the solution with the larger value of L corresponds to stable equilibrium and the smaller value to unstable equilibrium.

Hughes (1973) has suggested that the boundary between the West Antarctic ice sheet and the Ross Ice Shelf is retreating inland at the rate of about $70 \, \mathrm{m \, a^{-1}}$. The value of L predicted from equation (2.21) with estimated values of the constants β and D_0 for the West Antarctic ice sheet is rather uncertain (Weertman, 1974). With estimated values of $\beta \approx 4 \times 10^{-4}$ and $D_0 \approx 200 \, \mathrm{m}$ the value of $L \approx 700 \, \mathrm{km}$, which is approximately the observed value. However, for an estimated value of $\beta \approx 6 \times 10^{-4}$ and $D_0 \approx 250 \, \mathrm{m}$, values well within any reasonable error limit, no steady-state ice sheet can exist!

A complete treatment of this problem must take into account the fact that fast moving ice streams exist where the West Antarctic ice sheet flows into the Ross Ice Shelf. More recent studies of strain rates and movement in the southeast corner of the ice shelf led Thomas (1976) to conclude that the Ross Ice Shelf is thickening by around $1 \, \mathrm{m \, a^{-1}}$ near the grounding line in this area. Use of this strain data together with a borehole temperature profile further out at site J9 by Thomas & MacAyeal (1977) indicates a mean thickening of $0.3 \, \mathrm{m \, a^{-1}}$ in the same region.

The problem involves such a complex combination of glaciological, tectonic and oceanographic factors that we cannot answer the question of the stability of the West Antarctic ice sheet with confidence on the basis of present knowledge, although this theory indicates a possible basis for assessing the potential instability of the ice sheet. Other observations, such as isotopic profiles may in time give a more direct indication of how long a period has elapsed since unstable flow of the ice sheet last took place, if at all.

2.4

Form and flow of the Antarctic ice sheet during the last million years

D.J.DREWRY & G. de Q. ROBIN

This section discusses evidence of variations of the Antarctic ice sheet that is available from glacial geology and branches of geophysics and its relation to direct glaciological observations. We deal primarily with the last million years, which covers the age of the oldest ice we may expect to study in ice cores during the next decade.

For more purposes, we may consider the Antarctic ice sheet as consisting of two parts, the larger East Antarctic ice sheet ($10.15 \times 10^6 \, \mathrm{km^2}$), which is divided by the Transantarctic Mountains from the smaller 'marine' West Antarctic ice sheet ($2.99 \times 10^6 \, \mathrm{km}$) with which we include the floating Ross and Ronne–Filchner Ice Shelves. Bedrock beneath the ice sheet of East Antarctica lies mainly above present sea level. The history of the two sections differs considerably.

Today, in the warmer northern hemisphere, there is very little ice aground on rock below sea level in coastal regions, whereas ice is aground to at least 200 m below sea level along 60 per cent or more of the seaward boundary of the cold inland ice of Antarctica. The remaining 40 per cent of the Antarctic coastline is in warmer areas such as the west coast of the Antarctic Peninsula, to the south of the Indian Ocean or coastal areas with low precipitation such as the western Ross Sea which are protected from the outflow of inland ice by mountain ranges.

The antarctic crustal plates, particularly East Antarctica, had drifted to approximately their present position by 40 Ma BP (Drewry, 1976). Mountain glaciation was followed first by development of a continental ice sheet on East Antarctica some 10–15 Ma BP (Drewry, 1975a). Colder conditions allowed this ice to expand over the shallower seas around Antarctica and led to the formation of the West Antarctic ice sheet around 4–5 Ma BP (Drewry, 1978; Hayes & Frakes, 1975; Shackleton & Kennett, 1975). This younger ice sheet flowed across some glaciated valleys previously eroded by discharge from the older East Antarctic ice sheet (Rose, 1982), while in other areas changed atmospheric circulation, perhaps combined with tectonic movements, led to the later wastage of glaciers, such as those flowing from and through the Transantarctic Mountains into McMurdo Sound. These changes all took place well before the last million years which we consider here and have been critically reviewed by Mercer (1978).

Present configuration and flow

Figures 2.5 and 2.6 have already indicated that the form of the ice sheet along certain profiles in East Antarctica conforms reasonably well with theoretical equations related to the flow properties of ice. Although it is clear that we are dealing with a mature ice sheet in East Antarctica, it is only recently that sufficiently detailed maps of some areas have made it possible to check the validity of our models of flow. We will see in section 3.7 that ice flows in the direction of maximum surface slope, using mean slopes over distances that are an order of magnitude greater than the ice thickness in order to eliminate the relatively local effects of bedrock irregularities on surface slopes (Robin, 1967; Budd, 1971).

In the classic computer modelling studies of Budd, Jenssen & Radok (1971a), in *The derived physical characteristics of the Antarctic ice sheet*, all lines of ice flow in East Antarctica are shown to originate from the highest point of the ice sheet, a near-central 'dome' at approximately $81°$ S $75°$ E. Another concept of Antarctic ice flow, used by Gould (1940) refers to flow diverging from an 'ice divide'. Due to lack of evidence, he postulated

Figure 2.9. Surface contours (light lines), flowlines (heavy lines) and major ice divides (dashed lines and circles) of the Antarctic ice sheet. Lines (some dotted) of the longer profiles shown in Figure 2.10 are lettered from A to G, while the position of shorter cross-profiles normal to the divide A B C D are shown by letters a, b, c, d, e. Ice divides along which movement may be less than 1 m a^{-1} are shown by circles, and where ice motion is probably faster heavy dashed lines are used.

a single divide close to the Transantarctic Mountains. Factors governing the position of ice divides have been discussed on a two-dimensional basis in section 2.3.

Now that more detailed studies of the surface form of Antarctica are becoming available we see that the answer is intermediate between the two concepts outlined above. A new sketch of flowlines based on radio-echo profiling studies of the ice sheet of East Antarctica and other sources is shown in Figure 2.9. Ridges of 'ice divides' along which slopes are so low that ice movement is unlikely to exceed 1 m a^{-1} are shown by open circles. These ridges define part of the boundary between major discharge systems of the ice sheet discussed in Giovinetto (1964), but elsewhere these boundaries are defined by flowlines (dashes). Profiles starting from the central dome (A) of the ice sheet along lines lettered in Figure 2.9 are presented in Figure 2.10. Along the flowline AF to Byrd Glacier and the Ross Ice Shelf (1), the profile approximately fits that defined by equation (2.13), shown by the dashed curve (3). The profile AG to Lambert Glacier (2) falls more steeply from the Dome (A), possibly due in part to higher bedrock beneath the ice, but also because of strong convergence of ice flow in this sector towards the head of Lambert Glacier. Curve 4 shows the profile along the line ABCD in Figure 2.9, which has formerly been presented as a flowline from the central dome. The section ABC follows the boundary between ice flowing towards the Ross Sea on one side and the Indian Ocean on the other. Along two sections of this profile, one 200 km length, the other 400 km, the surface slope is zero or possibly reversed. However, cross-profiles (a) and (c) normal to these sections show that the ice-surface slopes off rapidly to either side to produce the stresses necessary for ice to flow. The term 'ice divide' describes these sections well, as the diverging flow to either side will prevent flow along the long horizontal sections of the ridge. On other sections of ABCD, the profiles (b), (d) and (e) show that a sloping ridge forms

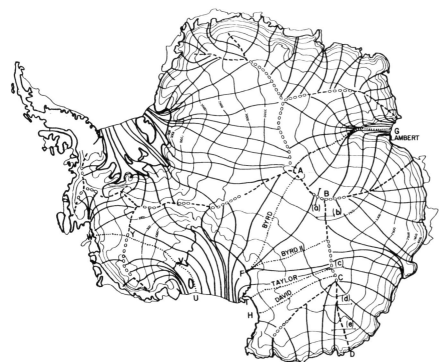

to a varying extent along the boundary between the two regions. The ratio of slope to either side of the boundary to that along the boundary is variable and is much less across profiles (b), (d) and (e) than for (a) and (c). Since surface slope drives the ice flow, we conclude that the more sharply a ridge is defined by the slope off to either side, the less will be the flow along the crest of the ridge. When the slope along the line of the ridge itself equals the slope off to either side, the problem becomes that of radial flow. This seems from Figure 2.9 to describe the approximate situation at the end of many major ridges.

The effect of the Transantarctic Mountains as a barrier to ice flow is shown in Figure 2.11 by three

Figure 2.10. Elevation of surface profiles on the Antarctic ice sheet from radio-echo profiling studies. Location of lines are shown on Figure 2.9. The dashed curve (3) shows a theoretical mean profile defined by equation (2.13). Other profiles are (1) line AF of Figure 2.9, (2) AG, (4) ABCD; while profiles a, b, c, d and e are normal to (4).

profiles along flowlines from the ridge near C to Byrd Glacier by line CF(1), to David Glacier (2) by CH, and to Taylor Glacier by CT(4). Curve (3) shows the standard parabolic profile from equation (2.13) for comparison. Whereas the profiles along CF and CH to the two major glaciers are broadly similar in character to curve (3), the line CT which follows an ice divide to Taylor Glacier lacks the slope to maintain ice flow as it approaches the mountains. Drewry (1980) has shown that the ice drains off from this section of the profile towards Mackay, Mawson and David Glaciers to the north and towards Mulock and Byrd Glaciers to the south, so that Taylor Glacier discharges only local ice. The relatively steep slope down Taylor Glacier is typical of a thin mountain glacier, as opposed to the moderate slopes of thick ice shown by Byrd and David Glaciers. Near C, slopes lower than those in the vicinity of the central dome A in Figure 2.9 should be noted. Calculated basal ice temperatures and the observed presence of sub-ice lakes near C (Oswald & Robin, 1973) suggest that the lower slopes

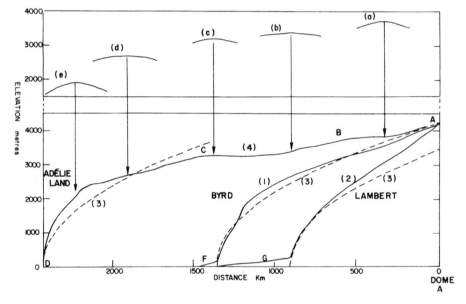

Figure 2.11. Surface profiles along lines CF, CT and CH of Figure 2.9 together with theoretical mean profile (3) as in Figure 2.10. (1) Runs from Dome C to Byrd Glacier (CF).

(2) From Dome C to David Glacier (CH) and (4) from Dome C to Taylor Glacier (CT).

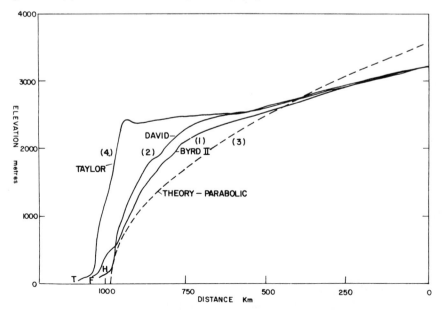

in the region of C are due to warmer basal ice and the effect of water on basal sliding.

Flow in parts of East Antarctica where no prominent mountain ranges block the flow to the coast is described by Shumskiy (1970, p. 329) as follows.

'Characteristic features of the peripheral zone are the so-called outflow glaciers which are ice rivers embedded in the ice sheet. They occupy about 15 per cent (4500 km) of the length of the outer boundary. Most of them differ very little from the surrounding ice sheet as regards surface and bottom relief (which contain shallow depressions only), but they move at much greater velocities and are bordered by systems of crevasses. Short glaciers of this type form crevassed amphitheatres at the edge of the ice sheet, but the longest extend over hundreds of kilometres ... The depressions whose sides approach one another in the upper reaches of outflow glaciers separate numerous small ice domes by sloping saddles at the edge of the ice sheet.'

Table 2.1, from Suyetova (1966), presents an estimate of discharge along different types of coastline. The reduction in coastline occupied by outlet glaciers compared to Shumskiy's figure of 4500 km is presumably due to omission of some outlet glaciers that enter ice shelves rather than discharge into the open sea. When one considers that about half the accumulation of snow on Antarctica falls within 200 km of the coast (see Figure 4.4) and that outlet glaciers are estimated to discharge more ice from 9.5 per cent of the coastline than crosses the remaining 45 per cent of the coastline not fringed by ice shelves, it appears that the major fraction of snow falling more than 200 km inland is likely to be discharged through outlet glaciers.

Shumskiy's description applies in many respects to the flow of the West Antarctic ice sheet from Byrd Land into the Ross Ice Shelf. Surface contours in Figure 2.12, from Rose (1979), show the existence of ice streams of low surface slope for at least 200 km inland from the ice shelf, while their boundary crevasse systems remain visible or are detected by radio-echo sounding for a further 200–300 km inland in areas where surface contours do not indicate any channelling of flow. Between the ice streams, and further inland, surface slopes are similar to those of other ice sheets of the

same thickness. Relatively rapid movement of ice streams of low surface slope is attributed to the effects of sliding, especially at higher velocities when frictional melting will lubricate the base of the ice. Radar returns from the bottom of ice streams suggest that parts are underlain by a water layer. The steeper sloping ice surfaces to either side of the low gradient ice streams in Figure 2.12 are attributed to basal ice temperatures below freezing point which inhibits sliding (Rose, 1979).

Studies of sub-glacial relief over the area covered in Figure 2.12 show that ice streams in Byrd Land occupy relatively shallow bedrock depressions of a few hundred metres in depth on relatively flat terrain. The typical bedrock profile along the flowline of ice stream (e) of Figure 2.12 is seen in Figure 2.13(a). It contrasts strongly with Lambert Glacier (Figure 2.13(b)) and some outlet glaciers in Greenland which occupy deeply eroded glacial valleys which deepen inland in typical fjord style, as explained by Crary (1966). Over the section of Lambert Glacier from 300–500 km from the ice front, bottom slopes around ten times the surface slope and of opposite sense suggest that this portion of the glacier may be mainly afloat. Figure 2.13(b) from Morgan & Budd (1975) shows this portion is aground although they report the presence of sub-glacial water at 500 km. In either case, it appears that the bedrock profile beneath Lambert Glacier has reached erosional stability in relation to present sea level. This is not true of the ice streams of Byrd Land (see Figure 2.13(a) for example), which again suggests a much younger age for these features in contrast to the large discharge glaciers of East Antarctica.

We may summarise our conclusions on the present configuration and flow of the Antarctic ice sheet shown in Figure 2.9 as follows. Ice sheet contours at any location are primarily governed by distance from the outer boundary of the inland ice, this boundary being either the mapped coastline or the inner boundary of a floating ice shelf, if present. Low surface slopes of rapidly moving ice in ice streams and glaciers also influence surface form, as do peripheral mountain ranges and sub-ice topography to a limited extent. Temperature and accumulation rate have only a second-order effect on the form of the ice sheet, except where they permit ice sheets to extend over areas below sea level.

Table 2.1. *Ice discharge – all Antarctica*

			Annual discharge into the sea		
Type of coastline	Length, km	Length, % of total	km^3 water	% of total	per km of coast (km^3 water)
Outlet glaciers	2 860	9.5	260	22	0.091
Ice shelves	13 660	45.5	730	62	0.053
Ice wall	11 090	37.0	190	16	0.017
Rock outcrops	2 420	8.0	–	–	–
	30 030	100.0	1180	100	0.043*

* Per km of total ice coastline.

Figure 2.12. Surface contours of Byrd Land from Rose (1979, Fig. 1) showing boundary of the Ross Ice Shelf and the location of ice streams A–E flowing into the ice shelf. The square grid shows the location of flight lines while the dashed lines show surface traverses for which elevation data was available. PQ shows the location of the profile on Figure 2.13(*a*). (Reproduced from the *Journal of Glaciology* by permission of the International Glaciological Society.)

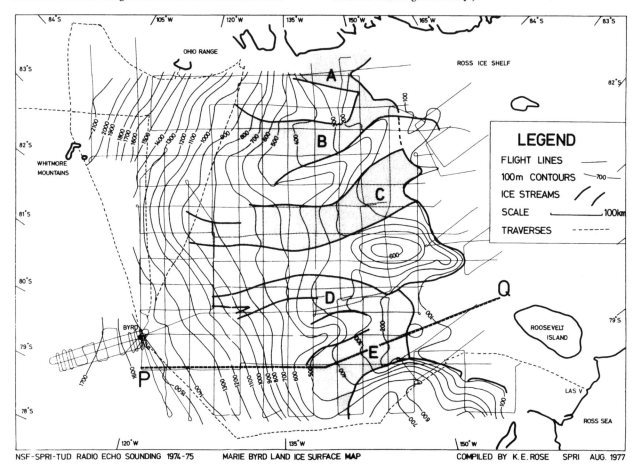

Figure 2.13. Ice stream profiles showing surface and bedrock elevations at the same scale: (*a*) along flowline PQ shown in Figure 2.12 in Byrd Land (from SPRI data); (*b*) from the inland ice sheet along Lambert Glacier to the Amery Ice Shelf from Morgan & Budd (1975).

(*a*)

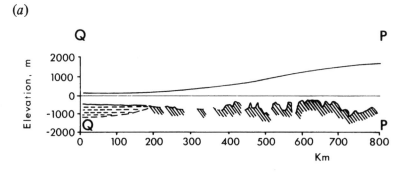

(*b*)

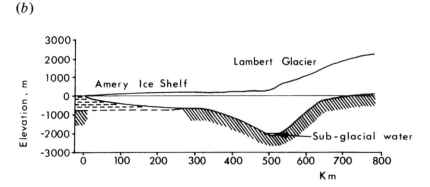

Past changes of configuration and flow

In contrast to the growth and disappearance of the vast ice sheets of the northern hemisphere during the Pleistocene period, the Antarctic ice sheet appears to have undergone relatively small changes in size during the last million years. The volume changes over this period have been estimated at around 10 per cent of the total volume, with larger proportional changes being more likely in the West Antarctic ice sheet than in East Antarctica (Voronov, 1960; Hollin, 1962; Mayewski, 1975).

Both terrestrial and marine evidence indicates that a considerable but temporary expansion of the ice sheet covering East Antarctica took place between 4.7 and 3.6 Ma BP, when ice may have extended to the edge of the continental shelf (Shackleton & Kennett, 1975; Drewry, 1978). Since then, in spite of some fluctuations, especially around 2.7 Ma BP, the ice of East Antarctica appears to have been relatively stable.

Formation of the ice mass in West Antarctica took place around the same time as the maximum spread of the ice of East Antarctica, the most likely period being from 4.0–5.0 Ma BP (Drewry, 1978; Mercer, 1978). The pattern and characteristics of marine sediments around Antarctica imply the continued presence of a substantial ice sheet over West Antarctica since this time, even though its volume must have changed significantly in response to sea-level fluctuations, as will be discussed later. We do not accept the hypothesis that higher sea levels in the past as reported from the Caribbean and other sites, are due to the collapse of the West Antarctic ice sheet for reasons we discuss later.

Geological evidence described in section 2.1 on the former extent and thickness of ice sheets is relatively scarce in the Antarctic owing to the limited areas of bare rock and to the scarcity of suitable material for dating deposits. A large proportion of all glacial debris from Antarctica must be deposited on the sea floor, hence the study of deposits of the sea bed by marine coring is of special importance in this region.

The relative stability of the ocean environment compared to that on glacierised lands means that sea-floor sediments often contain an undisturbed and relatively detailed history of glacial events. Nevertheless, the interactions between climatic changes, ice volume and marine sediments around Antarctica are complex and our knowledge is to some extent speculative. Drewry & Cooper (1981) suggest that the dominant source of ice-rafted debris (IRD) is the fast-flowing sediment-rich outlet glaciers, and that ice shelves and other sources make only a minor contribution to sedimentation in the open ocean. The debris from these sources is mainly confined to the lowest one to five per cent of the ice thickness, and hence is usually released through melting relatively close to the continent. Mixed with this is marine debris which comes primarily from oceanic microfauna. Its quantity depends on oceanographic conditions such as current patterns, and hence the available nutrients, temperature, ice cover and light. Zones of maximum marine productivity shift with climatic changes. Furthermore, deposition is affected by bottom water velocities, and can also be disturbed by icebergs. Sedimentation must be greatly reduced beneath ice shelves and halted and/or disturbed by expansion of grounded ice during times of lower sea level.

In spite of the difficulty of deducing terrestrial glaciological variations from marine sediments, the latter offer one major advantage over land-based evidence. It is somewhat easier to date marine sediments since uranium-series derivatives can be used for the period up to 400 000 a. In addition, reversals of the Earth's magnetic field, recorded in the remnant magnetisation of marine cores, include the Matuyama–Brunhes magnetic reversal and several other minor polarity excursions, but of regional extent, and studies of the fossil content of marine cores provide more plentiful evidence for dating material than is the case for terrestrial deposits in Antarctica.

In Figure 2.14 we summarise some information on climatic periods in and related to Antarctica. The four top rows show cold periods (lines) and warm periods

Figure 2.14. Dates of climatic periods in the Antarctic and southern hemisphere. On curves (1) to (4) cold or glacial periods determined from a study of marine cores are shown by heavy lines based on the following.
(1) equatorial cores described in Shackleton & Opdyke (1973, Table 3);
(2) Weddell sea cores from Anderson (1972, Figs 30 and 31);
(3) sub-Antarctic cores, data modified from Kennett (1970, Fig. 8);
(4) Southern Ocean core south west of Australia from Vella, Ellwood & Watkins (1975).
On curve 5, heavy lines show periods of flooding (warm) of Taylor Valley, Antarctica from Hendy *et al.* (1979, Fig. 4).

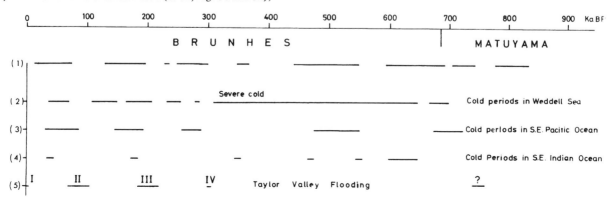

(gaps) determined from analysis of microfaunal remains and isotopic ratios in dated marine cores from the equatorial Pacific and different parts of the Southern Ocean. Lines in the fifth row indicate warmer periods when lakes in Taylor Valley were of large extent. The dates of warmer periods from these lakes agree reasonably well with those from marine sediments.

Evidence of the form of the Antarctic ice sheet over the past few million years is very limited due to lack of dates of terrestrial deposits. We can learn from glacial geology the maximum height of ice in any area and use this evidence to deduce the form of the ice sheet at its maximum extent. This appears to have occurred from 3–5 Ma BP, well before the period which we wish to study. Since then, the form of the ice sheet may have changed to a limited degree due to tectonic movements, especially in coastal regions such as the Transantarctic Mountains in Victoria Land (Bull & Webb, 1973; Drewry, 1975a). In addition, erosion beneath outlet glaciers may have lowered the whole ice sheet over a wide region, as appears likely in the Prince Charles Mountains, where old morainic deposits on Mount Menzies are found 1000–1500 m above present ice levels (Trail, 1964; Wellman & Tingey, 1981).

Broad reviews of the evidence of the former extent of the ice sheet have been given by Grindley (1967), Hollin (1969, 1970), Denton, Armstrong & Stuiver (1971), Mercer (1973) and Denton & Hughes (1981). Denton's conclusions are that, around much of the coast of East Antarctica, morainic deposits and evidence of erosion on many exposed nunataks and mountains show that the ice sheet underwent peripheral thickening of 300–800 m. In West Antarctica, increases of thickness of the ice sheet have ranged from 600 m near the coast to 200 m at Whitmore Mountains, 500 km inland from the ice shelves. However, LeMasurier (1976) reports moraine 820 m above present ice level on pumice dated at 1.6 Ma BP on Chang Peak (77.1° S 127° W), some 300 km from the nearest ice shelf, although it appears difficult to separate out possible tectonic effects in this region.

The most detailed terrestrial evidence from glacial geology for which dates are available comes from the McMurdo region, especially the Dry Valley system.

Unfortunately, we believe this area does not consistently reflect changes of height of the main East Antarctic ice sheet, since ice entering this region from the west is of local origin only (Drewry, 1980). It appears that the ice sheet behind Taylor Valley would have to thicken by at least 500 m before the main body of inland ice entered the valley, but this is prevented by effective drainage through the Byrd, David, and other, Glaciers.

We can estimate the varying form of the ice sheet over the past million years by assuming that during interglacial periods it was of similar form to the present, and deduce its configuration during the last glacial period as a guide to its form during other glacial periods. Extensive studies of the last glaciation in the northern hemisphere under CLIMAP, an international project on 'Climate/ Long-range Investigation, Mapping and Prediction', have concentrated on conditions at 18 000 a BP, the coldest stage. This corresponds approximately with the date of coldest conditions in Antarctica during the last glaciation (Robin, 1977), so we will discuss the form of the Antarctic ice sheet at this time.

The main change in the form of the Antarctic ice sheet during colder periods will be due to the larger extent of inland ice in areas where outflow is not blocked by mountains. This is due primarily to changes of sea level. If we know the areal changes of ice extent, we can use our knowledge of the form of ice sheets (see sections 2.2 and 2.3) to make general estimates of changes in form during ice ages. Table 2.2 shows the present form of the ice sheet as the mean of profiles 1, 2 and 4 (to point C) in Figure 2.10 and the profile of equation (2.13) for comparison. Columns (4)–(6) show the increase of ice thickness at given points on the present profile caused by a seaward advance of 50, 100 and 200 km of the outer boundary, assuming the form (column 2) is maintained. We see that the greatest effects of an advance of the outer boundary of an ice sheet take place near the boundary. Further inland, the increase of height is more limited and could be counterbalanced to some extent by a decrease of accumulation during an ice age.

The time lag between a change of climate and of sea level may also be significant. With present rates of accumulation it would take from 1000–5000 a of snow-

Table 2.2. *Elevation profiles of ice sheets*

(1) Distance from present ice front, km	(2) Present surface elevation, m	(3) Surface elevation by equation (2.13), m	(4) Increase in elevation for advance of inland ice boundary by: 50 km, m	(5) 100 km, m	(6) 200 km, m
0	0	0	850	1270	1910
50	850	805	420	810	1240
100	1270	1138	390	640	980
200	1910	1610	180	340	590
500	2700	2546	110	220	450
1000	–	3600	(89)	(176)	(343)

Values in brackets calculated from equation (2.13).

fall to accumulate the increased ice volumes shown in Table 2.2, and since outflow of ice will continue, the response time will be appreciably greater, say from 3000–10 000 a. In a time of rising sea level and increasing accumulation, we can use Nye's (1960) results from kinematic wave theory to estimate that the response time will be of the order of 5000 a. If these estimates are correct we expect the present day ice volume to be in approximate equilibrium with the present-day climate, since the climate has changed little during the past 10 000 a.

A lowering of sea level around Antarctica of about 120 m at around 18 000 a BP has been well established (Lingle & Clark, 1979). At this period we expect that the inland ice would have extended at least to the present −200 m isobath in areas where ice now terminates at sea level. Where ice shelves are now present, we expect inland ice to have covered areas where there is now less than 120 m of water between the bottom of the ice shelf and the sea bed. Cooling of 5 °C or more at this time, shown by isotopic profiles at Byrd and Vostok Stations, will also have caused more extensive formation of ice shelves around more northerly sections of Antarctica. Since the present-day grounding line between ice shelves and inland ice mostly lies from 200–400 m below sea level, the seaward limit of the ice sheet around East Antarctica is estimated to have extended to the present 300–500 m isobaths at 18 000 a BP. In general, the 500 m isobath lies from 50–200 km off the coast of East Antarctica, which suggests that thickening over coastal areas averaging 500–1000 m could have occurred at 18 000 a BP.

Dated evidence in favour of this model comes from US and Australian studies around Law Dome (112° E). A radio-carbon date on a raised beach deposit on Windmill Islands indicates that ice retreat took place more than 6000 a BP. Ice core studies on Law Dome by Budd & Morgan (1973) show that glaciological conditions there have been stable since 8000 a BP. Prior to that, studies of isotopic $\delta^{18}O$ levels and total gas content (see section 4.3) show much colder conditions and suggest that ice thicknesses may have been 500 m greater than at present at the summit of Law Dome, some 100 km from the ice margin.

In addition, in the Rennick Glacier area the last major ice expansion (termed the 'Rennick Glaciation') possesses a minimum retreat data of 1265 a BP (Mayewski, Attig & Drewry, 1979) and is here inferred to be Wisconsin. Up to 600 m of ice thickening is observed in the coastal region around Rennick Glacier at this time. Bardin (1982) reports between 40 and 200 m of former ice thickening in the Prince Charles Mountains at Mount Collin, as given by his youngest identified moraines which he has dated by degree of chemical alteration to the Late Pleistocene.

In Terre Adélie, measurements of isotopic ratios and total gas content at site D10, 5 km from the coast, have been modelled in Raynaud, Lorius, Budd & Young (1979) along lines presented in section 5.3. They tentatively conclude that about 400 m of general thinning took place some 250 km from the coast, but that since around 5000 a BP the ice sheet has been close to balance.

The results also suggest that the time of maximum cold preceded the maximum ice thickness. Table 2.2 indicates that the thinning reported by Raynaud *et al.* can be explained by a change of seaward extent of the ice sheet in this area of a little over 100 km, which is consistent with bathymetric contours in this area.

The existence of local wind ablation areas on the ice sheet inland of the Transantarctic Mountains, on which meteorites carried by ice motion have accumulated, suggests that there was little, if any, increase of ice thickness during the last glacial period in these areas. On one such area, of approximately 100 km² at the Allan Hills, some 300 meteorites but no terrestrial rocks have been reported by Fireman, Rancitelli & Kirsten (1979). A number have been dated by several techniques and show ages between 30 000 and 750 000 a BP, and one apparently is about 1.5 Ma old (Fireman *et al.*, 1979; Nishiizumi *et al.*, 1979). Meteorites covering the site have been collected more efficiently over the past 800 000 a, and any which accumulated earlier had been removed. Since the present ice level is only about 100 m below the height of the nunataks, a general rise of surface level of say 200 m would be sufficient to start flow and remove the meteorites, but this has not taken place. The area lies inland of the Transantarctic Mountains where the ice sheet drains through trunk glaciers. Although lowering of sea level may result in grounding of the lower part of these glaciers, it appears that this has not been sufficient to cause any considerable thickening of the inland ice sheet behind the mountains. Although this conclusion agrees with those based on evidence from the McMurdo Dry Valleys, it differs from our estimate of the effects of a lower sea level in other parts of Antarctica.

Our tentative glaciological evidence from ice cores from margins of East Antarctica suggests that at the culmination of the last ice age, around 18 000 a BP, the ice was substantially thicker on Law Dome and inland of the present coast of Terre Adélie. In addition, we have a moderate amount of geological evidence that the ice was from 300–800 m thicker within 100–200 km from the coast, but this evidence is generally undated and some deposits relate to much earlier glaciations when sea level lowering may have been 25 per cent greater.

The continental shelf off much of West Antarctica is considerably wider than that off East Antarctica, especially in the Ross and Weddell Seas, although deep troughs in front of the Transantarctic and Ellsworth Mountains lie beneath the present Ross and Ronne-Filchner Ice Shelves. A simple lowering of sea level by 120 m would not be sufficient to ground the ice shelves in these troughs, but extensive grounding of ice shelves on outer parts of the continental shelf could restrain ice flow sufficiently to cause thickening that would fill these troughs. Thomas & Bentley (1978*b*) have discussed the dynamics of the advance and retreat of such marine ice shelves, but have reached no firm conclusion on the extent of the Ross Ice Shelf at 18 000 a BP.

According to Hughes (1973, 1975), Denton *et al.* (1975) and Denton & Hughes (1981), during the last ice age around 18 000 a BP the present area of the Ross Ice

Shelf was fully grounded and grounding may have extended to the edge of the continental shelf. Drewry (1979) proposes a model in which grounded ice at 18 000 a BP is based on a lowering of sea level of 120 m, as shown in Figure 2.15, but is sustained for less than 5000 a. Extensive areas of sea floor now 400 m below sea level or less are shown as ice rises, equivalent to present Roosevelt Island. The front of the ice shelf is shown as extending further northwards as a result of these additional anchoring points. Ice piedmonts by the Transantarctic Mountains will become larger also. While extensive areas of grounded ice shown in Figure 2.15 will produce additional thickening of the inner part of the ice shelf, we consider that three lines of evidence favour the model in Figure 2.15.

Since the form of the West Antarctic ice sheet is of importance for interpretation of the ice core from Byrd Station, we will summarise this evidence. In Figure 2.16 we show the latest profile of surface and bedrock form up the flowline from the Ross Ice Shelf that passes through Byrd Station Strain Network to the ice divide (line UVY in Figure 2.9). We continue the profiles across the ice divide down the flowline to Thwaites Glacier Tongue and the Amundsen Sea (line YW). Although data are more limited on the latter section, it appears that beneath much of the thicker inland ice, especially near the ice divide, bedrock elevations are around 1000 m below sea level. Comparison of surface form with profiles in Figure 2.10 and Table 2.2, allowing for bedrock 1000 m lower, shows that the surface slope from the ice divide is slightly steeper than that indicated by Table 2.2, column 2 towards the Amundsen Sea but less steep

Figure 2.15. Reconstruction of Ross Sea embayment during late Wisconsin times (about 18 000 a BP) showing seaward extension of Ross Ice Shelf and grounding line. Local grounding occurs over submarine banks (shown with lower case letters for Crary, Mawson, Pennell, and Ross Banks), and adjacent to the Victoria Land Coast. Ice shelf is shaded. Inset depicts part of the model from Hughes *et al.* (1981, Fig. 6.10) with ice-sheet surface contours in 10² m. M, McMurdo Sound; LAV, Little America V; RI, Roosevelt Island. Ice streams in Marie Byrd Land are lettered A–E.

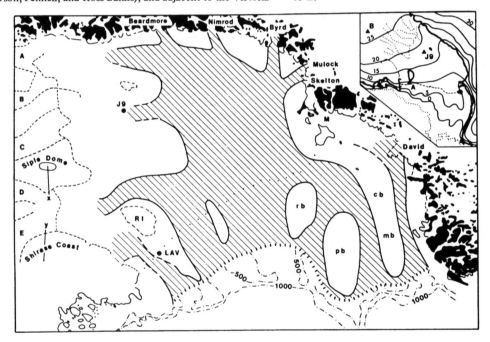

Figure 2.16. Flowline profiles: Byrd Land.
Curves on right side.
(1) Present day.
(2) 18 000 a BP from Drewry's (1979) reconstruction and today's slopes.
(3) 18 000 a BP for expansion to edge of continental shelf and today's slopes.
(4) 18 000 a BP profile from Hughes *et al.* (1981).
Curves on left side.
(1) Present day.
(2) and (3) Suggested profiles at 18 000 a BP.

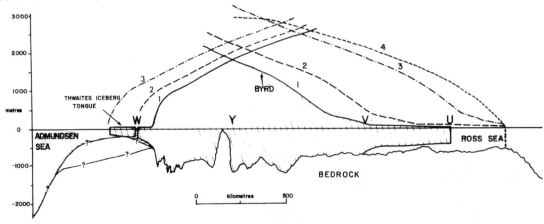

towards the Ross Sea, not only over the ice stream near V and the slope leading to the ice stream but also closer to the ice divide (Y). This may result from an accumulation rate that is up to three times greater along profile YW than over VY in Figure 2.16 as mentioned in section 2.3.

On Figure 2.16 we also show on the Ross Sea side of the ice divide our estimated profiles resulting from a sea-level lowering of 120 m according to Drewry's model (curve 2) and that of Denton & Hughes (1981), curve 4. On the other side of the ice divide some charts indicate a deeper continental shelf in the Amundsen Sea than the Ross Sea, as postulated in Figure 2.16. We suggest that the grounded ice from the Thwaites Glacier Tongue may have advanced into the Amundsen Sea by about 100 km (curve 2), or possibly up to as much as 250 km (curve 3). Then, assuming that no great changes in the general pattern of ice flow took place, we show the profiles over YW resulting from the different hypotheses in Figure 2.16. We have assumed that the surface profiles remained similar in form to the present, that is that lower slopes of an ice stream existed over the outer 250 km towards the Ross Sea, and generally lower slopes were maintained on that side. The position of the ice divide for different profiles on either side are indicated by pairs of figures referring to the left and right profiles in Figure 2.16 respectively. Thus 1.1 refers to the position of the present ice divide. For Drewry's model, the divide 3.2 would lie in approximately its present position, while either position 2.2 or 1.2 would lie upstream of Byrd Station on the present flowline. However, an advance of grounded ice to the edge of the continental shelf in the Ross Sea would carry the ice divide across Byrd Station to positions 3.4, 3.3, 2.3 or 1.3 and increase surface elevations to between 2500–2750 m at the ice divide. The general conclusion in chapter 5, based on the isotopic and total gas content in the ice core from Byrd Station, is that the coldest ice originated from an elevation perhaps 600 m above the present station elevation, of which about one third was due to downslope motion and the remainder due to a change in ice thickness. This conclusion would fit reasonably with any of the three ridge positions based on Drewry's model, but not with any of the positions due to grounding over the whole of the continental shelf of the Ross Sea. The minimum relict surface elevation present in the ice core on the latter model would be 1000 m above the present station elevation.

Glacial geology also provides evidence in the form of lateral moraines that is relevant to the surface form shown in Figure 2.16, in the same way as Mathews (1974) reconstructed profiles of the Laurentide Ice Sheet over Canada (see section 2.2). Thus reports by Denton & Borns (1974), Denton et al. (1975), Mercer (1968) and Mayewski (1975), of extensive 'fresh' lateral moraines at a few hundred metres above sea level along the outer Transantarctic Mountains facing the Ross Ice Shelf and Ross Sea, fit the concept of enlarged ice piedmonts as shown in Figure 2.16, but not the concept of a fully grounded ice sheet. In the McMurdo area, however, moraines and other evidence indicates the

presence locally of fully grounded ice between 75° S and 79° S as indicated in Figure 2.15.

Continental shelf sediments in the area covered by expanded ice shelf and ice rises in Figure 2.15 possess a complex pattern. Distribution and dating of the three principal lithological units that have been recognised is not at all clear and open to considerable debate, as given in Fillon (1975), Brady & Martin (1979), Drewry (1979), Kellogg, Truesdale & Osterman (1979), Kellogg & Kellogg (1981) and Yiou & Raisbeck (1981). The lack of consensus and the continued investigation of these sediments in the Ross Sea (Truswell & Drewry, in prep.) precludes their use in elucidating the recent glacial history of West Antarctic ice in our discussion.

Although gravitational anomalies may be related to the extent of rebound from former ice sheets, and hence to their size, measured anomalies from the recent Ross Ice Shelf Glaciological and Geophysical Survey do not clarify the position (Bentley, Robertson & Greischar, 1982), due to the influence of independent tectonic factors. Greischar & Bentley (1980) do, however, favour ice loading as an important and realistic explanation.

Interpretation of the Byrd Station Strain Network results, when combined with layering observed in radio-echo flights along the profile, led to the conclusion of Whillans (1976) that no great changes of the pattern of ice flow have taken place in this region over the past 20 000–30 000 a. However, limited changes of thickness would not be shown by his analysis.

The suggestion by Mercer (1968), that the apparent rise of sea level of around 6–9 m above present level at about 120 000 a BP, deduced from sea-level changes of coral terraces at Barbados, was due to a collapse of the West Antarctic ice sheet, should also be considered. This is the only mass of ice or water likely to produce a world-wide rise of sea level of this magnitude. There is some evidence of a similar rise around 5000 and 6000 a BP (Fairbridge, 1961; Schofield & Thompson, 1964), but it is clear from the Byrd ice core (Figure 1.2) that no such massive collapse of the West Antarctic ice sheet took place at that time. Furthermore, Clark & Lingle (1977) have shown that an instantaneous thinning of the West Antarctic ice sheet would cause submergence exceeding the world average at locations remote from Antarctica, when account is taken of immediate elastic uplift and reduced gravitational attraction on ocean water by the ice mass. Gradual reemergence will follow as viscous flow in the asthenosphere compensates for ice and water loads. At Hawaii, initial submergence of 125 per cent of the average world value would be reduced to 116 per cent after 1000 a and 109 per cent after 10 000 a. We believe that a similar response to the rapid melting of Pleistocene ice sheets may explain the apparently high sea levels of 120 000 a BP and 6000 a BP. Farrell & Clark (1976) discuss the general principles of such calculations.

We conclude that our reconstruction of conditions at 18 000 a BP over the western part of West Antarctica is supported by several lines of evidence involving appropriate time scales. Elsewhere around West Antarctica we lack dated evidence to confirm our model. However, thickening ranging from 600 m near the coast to 200 m at Whitmore Mountains previously noted seems appropriate for this period.

At other times during the last 10^6 a when sea levels fell comparably we would expect similar expansion of the ice sheet to that discussed above. If the sea-level fall was greater than 120–130 m and sustained for 5000 a or more, greater thickening will have taken place. The extreme limit to such thickening is shown in Figure 2.16, curve 3, when full grounding to the continental break of slope is likely to have taken place. Advances that are likely to have exceeded that of Figure 2.15 are indicated by the extreme lowering of sea level shown in Figure 3.3 at stages 6, 12, 16 to 18, at times around 170 000, 450 000 and 640 000 a BP.

Although we propose to model the state of the Antarctic ice sheet at any time in the past million years on the basis of present conditions for interglacial periods and our 18 000 a BP reconstruction for those periods known to be cold from the marine record, some caution is necessary. Firstly, the profiles discussed in section 2.2 relate mainly to stable ice sheet conditions. Wilson (1964) proposed massive surges of the Antarctic ice sheet of East Antarctica as a cause of ice ages. Budd & McInnes (1978) have developed computer studies of surging glaciers to illustrate the proposed phenomena. When they apply similar parameters to the East Antarctic ice sheet, they find it should surge with a period around 30 000–40 000 a. Periodic surging of the West Antarctic ice sheet has been proposed by Hughes (1975) in a mechanism that involves long-term changes of sea level and oceanic and atmospheric circulations as well as ice flow. Field evidence, such as the approximate agreement of temperature profiles with steady-state calculations do not favour these hypotheses, but they must be kept in mind, as well as our preferred pattern of semi-stable profiles which respond primarily to sea-level changes.

Apart from the direct effects of changes of sea level, a major change in the slope of ice streams or outlet glaciers would produce considerable changes of form of the Antarctic ice sheet. Thus, if Lambert Glacier (Figure 2.13(*b*)), which appears to be almost afloat at present, had in the past been completely grounded due to a lower sea level, its present low surface gradient would have been drastically altered. This could have resulted in surface ice elevations of perhaps 1000–2000 m higher at the head of the glacier around 73° S and would explain the high level glacial till on Mount Menzies. At the same time, one would expect generally higher elevations further inland. This in turn would have caused the ice divide (ABC in Figure 2.9) to migrate towards the Lambert Glacier.

We conclude that much of the evidence favours a form of the Antarctic ice sheet during ice ages that is explicable in terms of lower sea levels. However, some evidence suggests that ice-surface levels 200–300 km inland, in both East and West Antarctica, have been some hundreds of metres higher than can be explained by lower sea levels in the past and present-day profiles. Further work, including dating of relevant events, especially on lithological units on the continental shelf, is needed to provide satisfactory evidence.

2.5

The Greenland ice sheet

G. de Q. ROBIN

Introduction

At the Cambridge workshop in 1973, it was appreciated that the low $\delta^{18}O$ values in the ice core from Camp Century dating from the Wisconsin period could be due partly to ice flowing from high inland elevations, in contrast to the present situation where the top 75 per cent of the core appears to be of relatively local origin. The hypothesis was advanced that a high ridge of ice ran from the ice divide of central Greenland, to the north of Camp Century, to the large ice sheet covering the Canadian Arctic Islands. Ice flowing southwards from this ridge, from an elevation of 2500–3000 m then produced the very low $\delta^{18}O$ values found in the ice core at Camp Century. A tentative correction for such topographic control was shown in Dansgaard, Johnsen, Clausen & Gundestrup (1973, Fig. 10).

Following the workshop, Paterson (1977) reviewed geological and geophysical evidence in favour of the development of such a ridge, which it was suggested might have reached an elevation of 2500–3000 m over Kennedy Channel. Blake (1977) interprets glacial erratics and evidence of erosion up to 550 m a.s.l. on Pim Island and Cape Hershel as so young that they indicate a massive ice sheet filling Kane Basin at a late stage in the Wisconsin. However, Bradley & England (1977) consider that the most recent evidence from moraines in northern Ellesmere Island, near the Robeson Channel, does not support the existence of such a ridge in the late Wisconsin period.

As an alternative to the development of the high ridge Paterson (1977) and Robin (1977) have both suggested that a more limited seaward expansion of the ice sheet would change the ice flow pattern sufficiently to cause substantial volumes of ice from higher elevations in central Greenland to move over the site of Camp Century. For this to occur, the inland ice would have to expand southwards into Melville Bugt to the present 500 m bathymetric contour, and some westward advance into the much deeper Inglefield Fjord is also needed.

Whichever hypothesis is correct, it is clear from studies of the total gas content of the Camp Century ice core completed since the workshop (see sections 3.9 and 5.3), that the low $\delta^{18}O$ values in the ice core are partly due to a high elevation of origin rather than to cold climatic conditions alone. Even our knowledge of the present day flow of ice upstream of Camp Century (Figure 4.5) is not sufficient to be sure that present flowlines show a flux of inland ice passing through the site of Camp Century, or whether the present flow effectively starts from the ice divide in the Camp Century region.

In this section we discuss available knowledge and ideas on the past and present form and flow of the Greenland ice sheet in order to find the most likely answers. We are concerned also with interpretation of isotopic data obtained along traverses across central Greenland and at various sites in this area.

Present-day form and flow

Figures 4.1 and 4.2 show the general surface contours of Greenland as well as ice temperatures and accumulation rates. The main feature of the topography is the roughly north–south ridge that runs down the centre of Greenland, rising from around 2000 m at 80° N to 3200 m at 72° N, and in the south, where the island is narrower, a surface dome of 2800 m is present at 64° N due to higher subglacial topography than in the north. Apart from very thorough profiling across Central Greenland by the European glaciological expeditions (Figure 2.17(a)), the surface of the inland ice has not been well mapped until recently. Maps of the inland ice from a joint radio-echo survey by the Technical University of Denmark and the US National Science Foundation are not yet generally available. The same equipment and aircraft were used as in the NSF–SPRI–TUD programme. In addition, a start has been made in the use of altimetry from the GEOS III satellite for mapping the ice sheet south of 62° S. Brooks et al. (1978) report that a repeatability of the order of 2 m can be attained in this work. Data so far produced indicate a similar situation to that in Antarctica, with a tendency for relatively sharp 'ice divides' to form along the highest ridges.

The most detailed study of glaciological parameters in Greenland is that of Expéditions Glaciologique International au Groenland (EGIG). A triangulation network was installed from the west coast across the ice divide to Station Jarl Joset and was measured in 1964, 1968 and 1974. The ice surface profile, and the speed of ice movement resulting from these studies are shown in Figure 2.17, from Hofmann (1974). It should be noted that ice movement does not fall to zero on the crest of the ridge, but is about $4 \, m \, a^{-1}$ to the north according to velocity vectors in Hofmann (1974). This probably indicates the error of measurement, since on the crest of the ridge at Crête we would expect the velocity to fall to zero, or show some very slight movement to the south, the direction of slope of the ridge. Since the east–west component of velocity falls to zero at Crête, errors in this direction appear small. Irregularities of surface slopes and sharp changes of strain rate increase towards the west

coast, but these have only a second-order effect on the speed of ice movement. The maximum ice movement shown in Figure 2.17(b) is $109 \, m \, a^{-1}$, while its steady increase from the ice divide to the coast is in line with that expected from an approximately steady state, as expressed in equations (1.5) and (1.11). The steady-state hypothesis was also used in Robin (1955) to analyse the temperature profile at 76 km (point T) on Figure 2.17, and this gave a calculated velocity of movement of $113 \, m \, a^{-1}$, close to the measured velocity shown in Figure 2.17(b). Velocities must increase from this vicinity towards the Equip Sermia Glacier discharging into Equip Sermia Fjord, while elsewhere velocities are expected to decrease to the order of $10 \, m \, a^{-1}$ at the ice margin where this is dammed by the coastal mountains, since if the steady-state conditions apply, the marginal velocity equals $N/\tan \alpha$, where N is the ablation rate and α the surface slope.

For many studies to the south of 75° N it appears sufficient to treat ice flow as a two-dimensional problem, while to the north of that latitude flow is broadly radial in pattern towards coasts to the west, north and east. Since one major concern is to interpret the ice core at Camp Century, we discuss the problem of divergence of flowlines in this region in some detail with station data presented in chapter 4. It is apparent that we cannot tell the extent to which present ice flow at Camp Century is local and the extent to which it is influenced by flow of ice from further inland. In chapter 5 analyses are given for both local and flowline models.

Past form and flow

A broad survey of geological, geophysical and glaciological evidence of quaternary glaciation(s) of Greenland is given in Weidick (1976). The time of formation of the ice sheet is not well known from terrestrial evidence, but the presence of glacial marine sediments in the North Atlantic marine cores first appeared around 3 Ma BP according to Berggren (1972). The oldest moraines in Iceland dated by K.Ar methods are approximately 2.6 Ma old. Studies of $\delta^{18}O$ ratios in marine benthic fauna by Shackleton & Opdyke (1977) indicate that the last lengthy period of stable sea level ended about 3.2 Ma BP and shortly afterwards a change of $+0.4^o/_{oo}$ in their $\delta^{18}O$ values suggests a sea-level change of -40 m which they attributed to ice sheets formed on land. This is about one quarter of that stored at the maximum of later glaciations, and about six times the volume of water now locked in the Greenland ice sheet. Since the Greenland ice sheet is likely to have formed either during or even before the onset of the earliest major Pleistocene glaciation around the North Atlantic, it was probably in existence around 3 Ma BP. Whether or not it disappeared during subsequent interglacial periods is not certain, but at lower levels in ice cores at Camp Century and Devon Island, ice is found of higher $\delta^{18}O$ values than that of the Wisconsin ice above, suggesting that both ice masses were present during the last interglacial period. Certainly, once formed, an ice sheet tends to create its own climate due to its high elevation and albedo which helps it to survive subsequent warmer periods. However, the Greenland ice sheet, which loses only about half its annual ice budget into the sea and half by surface melting on land, is not as stable climatically as the Antarctic ice sheet.

It appears from glacial sculpturing of the outer coastal regions around Greenland, that the island has been almost completely covered by ice during at least one glaciation. Shoals less than 100 m depth, believed to be moraine deposits, exist from 50–100 km off the west coast of central Greenland. Evidence of erosion and deposition by ice on the sides of coastal mountains from 300–900 m above present sea level, and to over 1000 m in some ice-free areas, support this model in many areas. Weidick's map (1976, Fig. 365) of the presumed extent of the inland ice at maximum glaciation shows the inland ice in most areas extending some 50–100 km offshore of the present coastline at around the present 100 m depth contour, while Nares Strait is completely filled with inland ice. Form lines in central Greenland show no postulated shift of the position of the north–south divide, although to the west of the site of Camp Century

the filling of Inglefield Fjord with inland ice produces some changes of surface contours.

As mentioned in the introduction to this section, the extent of the last glaciation remains uncertain. Some rise of the marine limit has taken place since 9000 BP in eastern, western and northern Greenland. Figure 2.18 shows data from west Greenland and eastern north Greenland. The uppermost marine levels for which no dates are shown on both Figures 2.18(*a*) and (*b*) are likely to have been deposited during earlier glaciations, so we can draw deductions about ice advance during the last glaciation only from the location of dated deposits (Weidick, personal communication).

Figure 2.18, based on Weidick (1976), shows such data from western and eastern north Greenland. It appears from this and other data that uplift in western Greenland ceased about 5000 a BP whereas slow uplift continued in north Greenland until 2000 a BP. This contrasts with Scandinavia and North America where the central glaciated areas are still being uplifted, and

Figure 2.17. (*a*) Surface and bedrock profiles across central Greenland measured along the main east–west profile of Expéditions Glaciological Internationale au Groenland.

(*b*) Surface movement along profile in (*a*). The point T shows the value deduced in Robin (1955) from the observed tempera-

ture gradient to 125 m depth by use of a steady-state model of the ice sheet and climate. The dashed line from 400–570 km gives our estimate of velocities. (Profiles from Hofmann (1974, Fig. 3); reproduced by permission of *Zeitschrift für Gletscherkunde und Glazialgeologie*.)

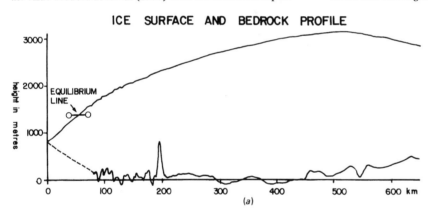

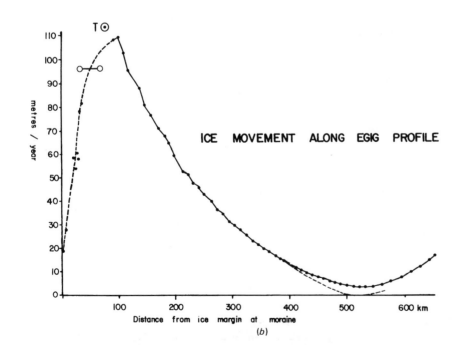

where a decrease in gravitational anomalies from the margin to the centre indicate that uplift is not yet complete. In west Greenland, the most negative gravity (Bouger) anomalies occur close to the highest marine limit. Weidick (1976) suggests that the evidence of uplift may indicate that the load of inland ice has been steady, or even slightly expanding, during the last 5000–6000 a. In section 5.3 we will see that analysis of gas content and $\delta^{18}O$ data in the ice core from Camp Century shows that the ice in north west Greenland has been thickening over the past 2000 a. However, in west Greenland we will see in section 4.2 that current wastage of ice in the ablation zone exceeds the supply of ice from inland, and although this reflects conditions in the twentieth century, the ice sheet may have been close to balance in the somewhat colder, preceding centuries.

The extent to which the ice sheet of northern Greenland expanded during the Wisconsin ice age is not certain, and was probably less than in earlier glaciations (Weidick, personal communication). However Figure 2.18(*b*) shows that a significant expansion took place in the Independence Fjord region, and a similar expansion in the Camp Century region would have been sufficient to produce the changes in the flow regime discussed earlier. To the west of Camp Century dated deposits suggest that deglaciation had taken place by 8500 a BP according to Davies, Krinsley & Nicol (1963), while Blake (1972) says this had taken place by 10 000 a BP. The former is more consistent with the well-dated ice core data from Camp Century (section 5.3) which indicates a major fall in the elevation of the ice surface from 10 000–8000 a BP.

On Ellesmere Island Figure 2.2 confirms that a considerable ice sheet must have covered much of the region during the Wisconsin period. England (1976) shows an uplift of 60 m or more over most of Ellesmere Island since 7500 a BP, with the greatest uplift seen towards the Kane Basin–Robeson Channel region. This agrees with Blake's (1977) interpretation, mentioned in the intro-

duction, that the Kane Basin was filled with a massive ice sheet late in the Wisconsin.

Overall, dated evidence of marine uplift and the greater ice elevations shown from ice core studies indicate that an enlarged ice sheet must have existed over northern Greenland during the late Wisconsin. It is useful to discuss factors limiting the expansion of ice sheets.

In reviewing factors controlling the formation, flow and decay of ice shelves around Antarctica, Robin (1979) has summarised factors that limit seaward expansion to three zones. The first zone is where broad, slowly moving glaciers, such as piedmont glaciers, terminate at cliffs resting on a beach between high and low water levels. At this stage, calving of ice limits further advance because of rapid melting of the base of the ice cliffs by sea water, especially during summer months. This first zone is found along much of the west coast of the Antarctic Peninsula to about 68° S. When the velocity of ice movement rises above some 10–50 m per year, ice in the Antarctic can advance beyond the low tide line, thus forming a second zone in which relatively little additional melting takes place until the ice can float. The third zone begins where the ice reaches a thickness of 200–400 m in water depths of about 160–340 m and ice shelves begin to form beyond the flotation (or grounding) lines. For ice shelves to continue as stable features, rather than as floating glacier tongues that break off periodically, suitable protection and anchoring points are needed from islands, shoals and ice rises. Ice shelves in Antarctica become less frequent north of around 70° S, although on the east side of the Antarctic Peninsula they extend to about 63° S. The most northerly ice shelves occur in areas where the mean annual temperature at sea level is no higher than −10° C.

In southern Greenland where the present equilibrium line approaches 2000 m above sea level, it seems unlikely that even at the ice age maximum ice would extend everywhere to sea level. In central west Greenland

Figure 2.18. Shoreline diagrams from (*a*) West Greenland (Holsteinsborg – Søndre Strømfjord) and (*b*) eastern North Greenland (Peary Land; Jørgen Brønlund Fjord – entrance of Independence Fjord). (After Weidick (1976, Fig. 374(*a*)). Reproduced by permission of the Geological Survey of Greenland.)

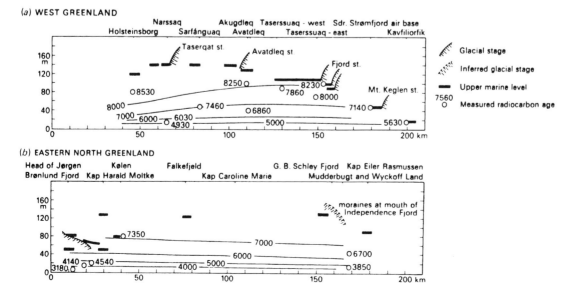

with the present equilibrium line around 1500 m, and the margin of the more slowly moving ice extending to an elevation about 600 m, it seems reasonable that during the glacial maximum the ice margin would expand to the shoreline of that time. This would be about 120 m below the present sea level. In north western Greenland the present equilibrium line is about 700 m, and it may have been near sea level at the glacial maximum. One would expect the faster moving glaciers to expand out to at least the flotation (grounding) line, and the filling of fjords and the strait between Greenland and Ellesmere Island with faster moving ice shelves or even grounded ice is probable. Whether or not the ice on the south of the Thule Peninsula would have advanced at that time beyond the tidal zone, to a depth of 200–400 m, would depend on the sea temperature and ice movement, but such an advance seems likely at least in areas where the present ice margin reaches the sea. When account is taken of the general lowering of sea level, we can expect ice to have extended to the present 500 m contour. Such an extension to the south of Camp Century appears necessary in order to account for the increased surface elevations recorded in the ice core as will be described in section 5.5.

Considerations similar to the above led Grosswald, Hughes & Denton (1978) to suggest that at 18 000 a BP, a continuous ice shelf covered the Arctic Ocean and North Atlantic as far south as the Iceland Faroes ridge and all of Baffin Bay as a minimum. Although our discussion suggests that conditions favourable to the formation of ice shelves may have existed in northern Baffin Bay at that time, the evidence from central west Greenland does not favour the existence of an ice shelf at this latitude.

GLACIOLOGICAL PARAMETERS, THEIR MEASUREMENT AND SIGNIFICANCE

3

3.1

Isotopic composition of the ocean surface as a source for the ice in Antarctica

N.J.SHACKLETON

Introduction

Although the most important processes through which water is isotopically fractionated between the ocean and the Antarctic ice sheet occur in the atmosphere, the ocean surface cannot be taken as uniform in space or time. Firstly, local variations in the evaporation–precipitation balance give rise to a pattern of sea surface isotopic composition resembling the better-known surface salinity variations which arise from the same mechanism. Secondly, the isotopic composition of the whole ocean reservoir changed significantly with each advance of northern-hemisphere ice sheets. Thirdly, sea surface temperature and other factors controlling isotopic fractionation into the marine atmosphere have changed during the Pleistocene glacials. This section gives a brief survey of knowledge of these three areas.

Regional variations in the isotopic composition of ocean water

The three principal molecular species with which the student of isotopic fractionation in natural water is concerned are:

$H_2{}^{16}O$

$HD{}^{16}O$

$H_2{}^{18}O$

In order to describe completely the processes involved, all three species must be considered; that is to say, both δD and $\delta^{18}O$ are measured. Both may be expressed as deviations per mil ($^0/_{00}$) from the SMOW (Standard Mean Ocean Water) standard. δD is defined as

$$1000 \left(\frac{HD{}^{16}O/H_2{}^{16}O_{sample}}{HD{}^{16}O/H_2{}^{16}O_{SMOW}} - 1 \right)$$

and $\delta^{18}O$ is defined as

$$1000 \left(\frac{H_2{}^{18}O/H_2{}^{16}O_{sample}}{H_2{}^{18}O/H_2{}^{16}O_{SMOW}} - 1 \right)$$

The pioneer mass spectrometric determinations of the distribution of these isotopes in the sea are by Epstein &

Mayeda (1953) for oxygen and Friedman (1953) for hydrogen. For the purposes of the present paper the available information can be described in terms of oxygen isotopes only, for which the second major source is Craig & Gordon (1965).

Over the surface of the ocean, salinity varies according to the local balance between evaporation and precipitation, and to the efficacy of vertical mixing. Because of the preferential evaporation of isotopically lighter molecules, areas of high surface salinity also show positive δ values (higher ^{18}O content). Thus the relationship between s (salinity) and δ is an indicator of the nature of the isotopic fractionation processes.

Roughly speaking, we may distinguish three regions: a subtropical region (trade winds) over which evaporation exceeds precipitation, separating the tropical region from the higher latitudes, in which precipitation exceeds evaporation. The processes by which the vapour transported poleward is progressively depleted in ^{18}O are discussed in Chapter 1; they result in the fact that marine precipitation is isotopically lightest at high latitudes. By contrast, the greater portion of the evaporated moisture that is re-precipitated at low latitudes has not been subjected to the same isotopic depletion and so is isotopically rather closer to ocean surface water. Hence the slope of a plot of δ versus salinity is small at low latitudes, but greater at high latitudes (Figure 3.1).

Epstein & Mayeda (1953) interpreted the higher-latitude slope in terms of the mixing of marine water and glacial meltwater, but Craig & Gordon (1965) show that the significant contribution is made by direct precipitation combined, in the case of the northern hemisphere, with continental runoff. They make the point clearly by estimating that direct marine precipitation on the Southern Ocean (south of 45° S) exceeds the meltwater contributions by a factor of 46.

Figure 3.2 shows a map of world surface salinity variation (from Sverdrup, Johnsen & Fleming, 1942) with some representative values for surface water. There are not sufficient data published to provide a contoured map of δ values.

Figure 3.1. ^{18}O content of ocean surface water plotted against salinity of same waters (from Craig & Gordon, 1965). Circles, high latitude; triangles, low latitude.

Variations in mean ocean $\delta^{18}O$ during the past million years

Like the Antarctic ice sheet today, the ice sheets which accumulated on Northern America and Scandinavia during the glacial episodes of the past million years were composed of isotopically light ice. The first attempt to estimate the magnitude of this effect was by Emiliani (1955). Using a value of about 35×10^6 km^3 for the volume of water extracted from the oceans (about 2.6 per cent of the ocean volume) and an isotopic composition of $-15^0/_{00}$ (an estimate of the isotopic composition of snow falling on the formerly glaciated areas) he arrived at a figure of $0.4^0/_{00}$ for the change in ocean isotopic composition. Olausson (1965), Shackleton (1967), Dansgaard & Tauber (1969) and others have improved on this estimate in the light of growing knowledge of isotopic processes in the atmosphere and of flow in ice sheets; all these workers arrive at a mean composition of major Pleistocene ice sheets isotopically lighter than $-30^0/_{00}$ and a resultant change in ocean isotopic composition exceeding $1^0/_{00}$. Considering an ice sheet in three dimensions (Shackleton & Kennett, 1975) shows that the mean isotopic composition of the ice sheet is determined mainly by the isotopic composition of the snow falling in the central part of the ice sheet. The mean value for Pleistocene ice sheets can be estimated from the height of the centre of former ice sheets determined as outlined in section 2.2 together with the atmospheric lapse rate. Thus the thickness of the ice sheets enters the calculation both when estimating the volume of water removed and when estimating its isotopic composition. Thus this thickness contributes the greatest uncertainty to the estimates.

The change in ocean isotopic composition can also be estimated by measuring the isotopic composition of calcium carbonate fossils deposited in isotopic equilibrium with sea water. The isotopic fractionation factor between water and carbonate is temperature dependent, but the temperature uncertainty may be reduced by analysing benthonic fossils from the deep ocean (Shackleton, 1967; Shackleton & Opdyke, 1973). At low temperatures the isotopic composition of carbonate varies by $0.27^0/_{00}$ per degree (Shackleton, 1974). Figure 3.3 shows the isotopic record of benthonic

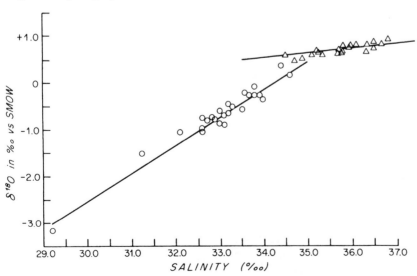

foraminifera from core V19-29, taken in the eastern Equatorial Pacific where bottom temperature is at present about 1.76 °C (from Ninkovich & Shackleton, 1975). Although this core yielded a particularly clear record, the overall isotopic range between the past glacial maximum and today, $1.65^0/_{00}$, has been well established in a number of cores from sites with bottom temperatures ranging from 4 °C to below 1 °C.

Bottom water is formed around Antarctica at a temperature of about -0.4 °C (Sverdrup *et al.*, 1942)

at a rate which is limited by the rate at which the wind blows ice northwards, permitting the formation of fresh sea ice and the release of cold, high salinity water necessary for Antarctic Bottom Water formation (Gordon, 1975). It is difficult at present to put an upper limit on the possible increase in the rate of bottom water formation during a glacial period, and its subsequent effect on the temperature of Equatorial Pacific deep water. However this is unlikely to produce a lowering of more than 1 °C, so that at least $1.4^0/_{00}$ of the change observed in

Figure 3.2. ^{18}O content of marine surface waters (from Epstein & Mayeda, 1953 and Craig & Gordon, 1965), $^0/_{00}$ to SMOW standard, superimposed on ocean salinity map (from Sverdrup *et al.*, 1942, Chart VI).

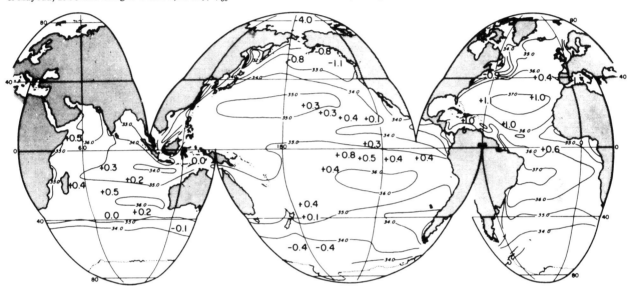

Figure 3.3. Upper: ^{18}O content of benthonic foraminifera (*Uvigerina* sp.) in East Equatorial Pacific core V19-29 (from Ninkovich & Shackleton, 1975). Lower: ^{18}O content of planktonic foraminifera (*Globigerinoides sacculifer*) for Western Equatorial Pacific core V28-238 (from Shackleton & Opdyke, 1973).

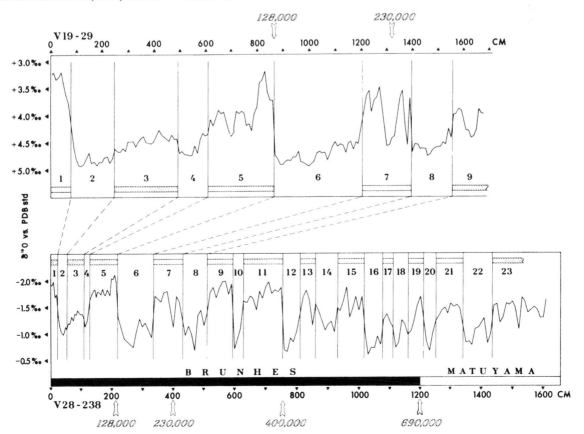

Figure 3.4. Estimated sea surface temperature 18 000 a BP (from
McIntyre *et al.*, 1976)

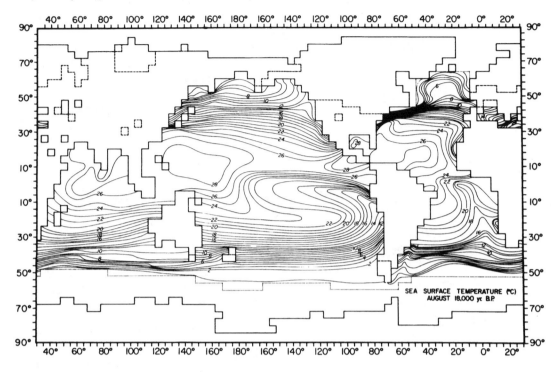

Figure 3.5. Comparison of oceanographic features in Atlantic–
Indian sector of the Southern Ocean today and 18 000 a BP
(from Hays *et al.*, 1976, Fig. 20).

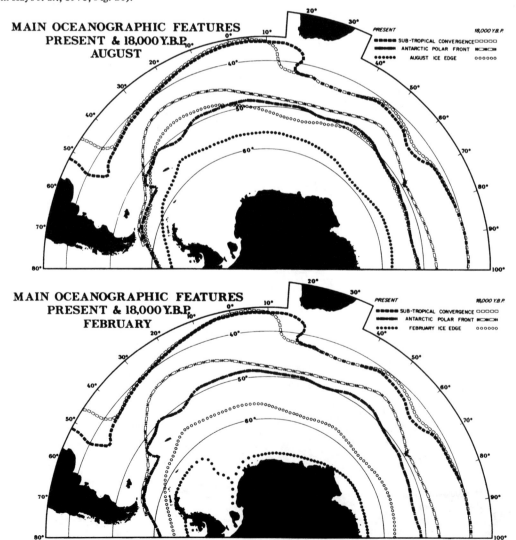

Figure 3.3(*a*) represents ocean isotopic change.

Figure 3.3(*b*) shows a longer record, covering almost a million years, derived from the analysis of planktonic foraminifera in the western Equatorial Pacific (Shackleton & Opdyke, 1973). It is shown by Shackleton & Opdyke (1973, 1976) that differences between the upper part of this record and the record shown in Figure 3.3(*a*) are due to the relatively poor stratigraphic resolution in the more slowly accumulating sediment. However, for periods exceeding some 30 000 a Figure 3.3(*b*) is a reliable depiction of changes in ocean isotopic composition. It will be appreciated that the chief interest in Figure 3.3 lies in its detailed record of Pleistocene glacial events. The basis for the time scales used is discussed in the sources for the figure.

Ocean surface temperature changes

Interest in climatic change has grown immensely in the past few years (National Academy of Sciences, 1975). When it is appreciated that 70 per cent of the lower boundary of the atmosphere is bounded by ocean, the value of reconstructions of past sea surface temperature as a means of understanding past climatic regimes is obvious (see Gates & Imbrie, 1975). Much of the effort of the CLIMAP project was directed towards surface temperature reconstruction (Moore, 1973; Cline & Hays, 1976). Figure 3.4 (McIntyre *et al.*, 1976) shows a surface temperature reconstruction for about 18 000 a BP, while Figure 3.5 (Hays, Lozano, Shackleton & Irving, 1976) shows oceanographic conditions around Antarctica at the same time. Although the pattern of climatic change during the past million years has been highly complex, and remains very largely unknown, Figures 3.4 and 3.5 may be used with the expectation that (a) they are a reasonable representation of conditions between 75 000 and 15 000 a ago, and (b) for only about 10% of the past million years would a map of present-day conditions be more appropriate than these glacial reconstructions.

It is widely agreed that between about 115 000 and 125 000 a BP world climate was broadly as it is today. At the time of writing (1976), considerable uncertainty surrounds the interval between 115 000 and 75 000 a BP, a point to be borne in mind while interpreting the earliest portions of the ice core isotopic records.

3.2

Atmospheric processes

G. de Q. ROBIN & S.J. JOHNSEN

The atmospheric processes that determine the variation of δ values with latitude, altitude and temperature were discussed in some detail in chapter 1. In this section we consider the physical processes governing the isotopic composition of individual snowflakes in more detail. These have been discussed in Dansgaard (1964), by Koerner & Russell (1979) and elsewhere.

When humid air masses are cooled, precipitation starts to form on condensation nuclei, either as small water droplets if the temperature is warmer than around $-15\,°C$ and as ice crystals at colder temperatures. If thermodynamic equilibrium exists between the vapour and the ice or water, so that water molecules are freely exchanged in both directions, the lighter molecules will be more abundant in the vapour owing to their higher mobility. The ratio is expressed by a fractionation factor (α) which is the ratio between the vapour pressure (P) of the lighter component and that of the heavier one. For deuterium

$$\alpha_D = P(H_2O)/P(HDO)$$

and for ^{18}O we put

$$\alpha_{18} = P(H_2O)/P(H_2^{18}O)$$

Table 3.1 shows values of α_D and α_{18} for both vapour–water and vapour–ice transitions based mainly on results of Majoube (1971*a*, *b*). These figures differ from earlier results given in Dansgaard (1964). Dansgaard shows that the δ value of vapour ($\delta_v\,°/_{00}$) in equilibrium with water or ice of δ value $= 0$ (Standard Mean Ocean Water) will be $\delta_v = 1000((1/\alpha)-1)$. When the vapour starts to condense the first condensate will have a δ value $= 0$. While Table 3.1 shows the magnitude of changes during evaporation and condensation, it does not indicate the relationship between δ values and temperature in the atmosphere or in the condensate. To do this we must take account of any changes in the source of moisture and of the history of the humid air masses. At the surface of the ocean, continuous evaporation will result in a slight enrichment of the surface layers in the heavier molecules HDO and $H_2^{18}O$, but this will be countered by diffusion within the water, either by molecular motion or by eddy diffusion. Stewart (1975) carried out extensive observations of isotopic changes accompanying the evaporation of liquid water drops, and showed that diffusion through the boundary layer in the gas phase around the water droplet appeared to be dominant. Quantitatively, the effect appears small compared with the values given in Table 3.1, since it produces an additional factor of 1.018 for $H_2^{18}O$ and 1.012 for HDO, that is a further change of only one to two per cent, and the change is expected to be similar over both water droplets and ice. In addition, a major factor in relation to the atmosphere is that since the oceans contain around 97.5 per cent of the water on earth and the atmosphere slightly under 0.01 per cent,

provided the ocean waters remain reasonably well mixed they provide a source of water vapour of almost constant isotopic δ values. Departures from uniform composition of ocean surface waters were seen in the previous section to be relatively small compared with changes in δ value of water vapour in the atmosphere and in precipitation therefrom.

When considering condensation from the vapour in the atmosphere, the relatively small reservoir of vapour and the large changes in the amount of vapour in the atmosphere have a large effect on the isotopic composition of the remaining water vapour. Figure 3.6, from Robin (1977), shows the variation of water vapour pressure with temperature over ice, together with a plot of net annual accumulation on the Antarctic ice sheet

Table 3.1. *Fractionation factors for the equilibrium exchange between vapour and condensate of HDO and H_2O (α_D) and between $H_2^{18}O$ and $H_2^{16}O$ (α_{18})*

Temperature, θ °C	α_D		α_{18}	
	Vapour–water	Vapour–ice	Vapour–water	Vapour–ice
+30	1.074		1.0090	
+20	1.085 (1.079)		1.0098 (1.0092)	
+10	1.098		1.0107	
0	1.112 (1.106)	1.133	1.0117 (1.0112)	1.0152
−10	1.129 (1.124)	1.152	1.0128 (1.0123)	1.0169
−20	1.149 (1.147)	1.174	1.0141 (1.0135)	1.0187
−30		1.200		1.0207
−40		1.229		1.0228

Based on measurements and equations of Majoube (1971a, b) and of Merlivat & Nief (1967) for α_D: vapour–ice. Figures in brackets are those used in Dansgaard (1964).

Figure 3.6. Water vapour pressure over ice (full line) and annual net accumulation of ice on the Antarctic ice sheet, both plotted against temperature. Crosses show data plotted against mean annual temperature of surface layers of ice sheet. Circles show same data plotted against mean annual temperature of atmosphere above the surface inversion layer. (From Robin, 1977, Fig. 10.)

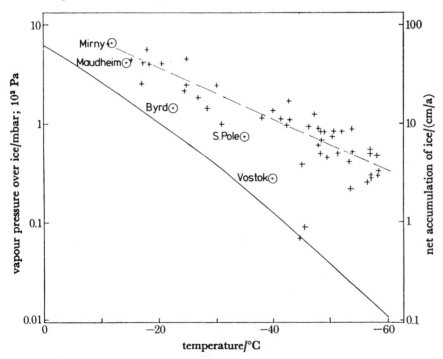

against temperature at the same location. The agreement between the vapour pressure curve and the variation of net accumulation with the mean free air temperature indicates that it is the amount of water vapour that can be carried in the free atmosphere at any temperature that exerts the dominating control over the amount of precipitation on the ice sheet. If a moist parcel of air moves inland from the coast at Mirny with a temperature of $-5\,°C$ it will precipitate around 97 per cent of its water vapour content before it reaches Vostok Station with an atmospheric temperature of $-40\,°C$. Similarly a saturated parcel of air approaching Antarctica from the north at a constant pressure will lose around two thirds of its water vapour content as its temperature falls from $+10\,°C$ to $-5\,°C$.

If thermodynamic equilibrium were maintained between water vapour in the atmosphere and in the condensate of water droplets or ice, small changes in the δ value of the condensate (δ_c) could be calculated from changes of vapour pressure ratios (fractionation factors) shown in Table 3.1. However, thermodynamic equilibrium is not maintained, and a solution that assumes that all the condensate is removed from the vapour as soon as it is formed gives a better approximation to the process over Antarctica. This is sometimes referred to as the Rayleigh condensation process (e.g. Dansgaard, 1964). It can be more clearly described as a 'moist adiabatic Rayleigh' or more simply pseudo-adiabatic, process in meteorology (Radok, personal communication).

Equilibrium process

The theory of condensation in a closed two-phase system is given in Dansgaard (1964) for both vapour–water and vapour–ice transitions. It considers both the equilibrium and Rayleigh problems mentioned above.

If equilibrium is maintained between the total liquid and vapour phases at any stage, δ_c is the first small amount of condensate assumed to be zero δ value, then δ_c^* for the mean value of the total condensate can be shown to change by

$$\delta_c^* = \frac{1}{\alpha_0}\frac{1}{\epsilon(F_v+1)} - 1 \qquad \epsilon = \frac{1}{\alpha} - 1 \qquad (3.1)$$

where α_0 is the value of α, the fractionation factor, at the beginning of the process, and F_v the remaining fraction of the vapour phase at the time for which δ_c^* is being determined.

Similarly for the vapour phase, its δ value (δ_v^*) is

$$\delta_v^* = \frac{1}{\alpha\alpha_0}\frac{1}{\epsilon F_v + 1} - 1 \qquad (3.2)$$

We also have for $F_v = 0$ that $\delta_c^* = (1/\alpha_0) - 1$ and that $\delta_v^* = (1/\alpha_0) - 1$ for $F_v = 1$.

Since condensation is caused by cooling, α increases as the process takes place as seen in Table 3.1.

The composition of liquid and vapour phases is shown by curves δ_c' and δ_v' for equilibrium processes in Figure 3.7. However, these curves are approximate since values of α are not well known at low temperatures.

Rayleigh process

If condensation of the vapour proceeds with immediate removal of the condensate from the vapour after its formation, the corresponding δ values for the condensate δ_c (water or ice) and vapour δ_v originally in equilibrium with Standard Mean Ocean Water (SMOW) will be found from the basic equations

$$\left.\begin{aligned}\frac{d(\delta_v+1)}{\delta_v+1} &= (\alpha-1)\frac{dF_v}{F_v}\\ \text{and}\\ (\delta_c+1) &= \alpha(\delta_v+1)\end{aligned}\right\} \qquad (3.3a)$$

We put $\alpha = \alpha_m$ and obtain

$$\left.\begin{aligned}\delta_v &= \frac{1}{\alpha_0}F_v^{\alpha_m-1} - 1\\ \delta_c &= \frac{\alpha}{\alpha_0}F_v^{\alpha_m-1} - 1\end{aligned}\right\} \qquad (3.3b)$$

where α is the fractionation factor at the condensation temperature θ for which δ_c and δ_v are calculated, α_0 is that at the initial temperature θ_0 at which condensation commences and $\alpha_m = \frac{1}{2}(\alpha_0 + \alpha)$ which is a close approximation. As condensation proceeds δ values become increasingly negative $(\delta_c$ and $\delta_v \to -\infty$ for $F_v \to 0)$ at a much greater rate than when equilibrium is maintained between vapour and condensate, as shown in Figure 3.7.

Differentiation of equations (3.3) with respect to temperature gives

$$\frac{d\delta_c}{d\theta} = \left(\frac{1}{\alpha}\frac{d\alpha}{d\theta} + \frac{(\alpha-1)}{F_v}\right)\left(\frac{dF_v}{d\theta}\right)(\delta_c+1) \qquad (3.4)$$

Figure 3.7. Isotopic fractionation of the remaining vapour and of the condensate as a function of the remaining fraction, F_v, of vapour.
$\delta_v^*(\delta_c^*)$ equilibrium between the total liquid and vapour phases during isothermal condensation.
$\delta_v'(\delta_c')$ same as above during condensation by cooling.
$\delta_v(\delta_s)$ sublimation by cooling. δ_s is the mean composition of the solid phase.
$\delta_v(\delta_c)$ sublimation or condensation by cooling under Rayleigh conditions. δ_c is the composition of newly formed condensate. (From Dansgaard, 1964, Fig. 1.)

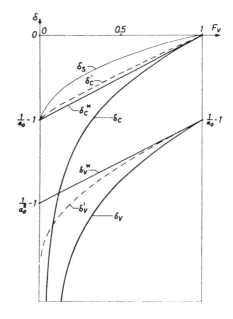

from which the gradient $d\delta/d\theta$ may be calculated. Some results for the vapour/ice transition from Dansgaard (1964) are shown in Table 3.2.

He also quotes a further calculation for isobaric cooling of vapour to water from $+20\,^{\circ}$C and $\delta = 0$ to $0\,^{\circ}$C followed by moist adiabatic Rayleigh cooling from $0\,^{\circ}$C to $-20\,^{\circ}$C. This produces a gradient of $0.67^{0}/_{00}/^{\circ}$C at $-20\,^{\circ}$C which is slightly less than the interpolated figure of $0.70^{0}/_{00}/^{\circ}$C for $-20\,^{\circ}$C from Table 3.2 because removal of condensate started at a higher temperature; this may correspond better to the average case of inland precipitation on the Antarctic or Greenland ice sheet. Dansgaard (1964) gives results of calculations based on isobaric cooling from $+20\,^{\circ}$C that match the slope of the $\delta-\theta$ curve in Figure 1.4 (Picciotto *et al.*, (1960) reasonably well, while curve 2 of Figure 1.5 shows how the gradient $d\delta^{18}O/d\theta$ changes with the temperature of precipitation due to moist adiabatic processes over the Antarctic ice sheet. Peel & Clausen (1982) reach a similar conclusion to explain isotopic gradients over the Antarctic Peninsula, namely an air mass with sub-tropical characteristics is cooled in part isobarically to $-4\,^{\circ}$C and subsequently adiabatically as snow is precipitated over the peninsula.

Our discussion so far considers δ values in relation to the temperature of condensation. The latter may vary considerably in an individual cloud system. Dansgaard (1953) measured rain from a warm front in which the air was cooled under moist adiabatic conditions from $+12\,^{\circ}$C to $-8\,^{\circ}$C. The corresponding change in measured $\delta^{18}O$ value was $8^{0}/_{00}$ but when corrected for evaporation and exchange the figure rose to $11^{0}/_{00}$, in agreement with calculations similar to those of Table 3.2 for the water phase. A somewhat wider range of δ values is to be expected in snow falling from colder frontal cloud systems over polar ice sheets. Furthermore, when dealing with ice crystals rather than water droplets, appreciable mixing by diffusion within ice crystals is not expected. Initial formation of ice crystals will generally take place in higher, colder air than that through which they later fall to the surface. As they fall, δ values of the succeeding layers of ice on the snowflake will become higher as vapour is condensed on to the crystal from moist air at increasing temperatures. However, if the cloud base does not reach ground level, evaporation of outer layers of ice crystals may take place between the cloud base and

ground, and this will preferentially remove ice with higher δ values, corresponding to the lower (warmer) part of the cloud layer. Such a process could explain the small inverse isotopic altitude effect reported by Lorius, Merlivat & Hagemann (1969) for the coast of Terre Adélie where the δ value tends to rise slightly with altitude to an elevation of 1000 m (see section 3.3).

Another feature in Table 3.2 is that values of $d\delta D/d\theta$ remain about 8.0 times that of corresponding values of $d\delta^{18}O/d\theta$ at all levels, as distinct from the ratios between fractionation factors shown in Table 3.1. This gives confidence in the model and the calculations involved since this ratio shows little variation in precipitation at all latitudes.

On the Antarctic coast of Syowa Station, Kato, Watanabe & Satow (1978) found that daily variations of $\delta^{18}O$ values were caused mainly by the supply of $\delta^{18}O$-rich water vapour during the approach of cyclones. The seasonal variations of new snow at Syowa Station, and of drifting snow at Mizuho Camp on the plateau, are stated to be controlled not only by the seasonal variations of atmospheric temperature, but also by changes in distance from the coast to the open sea. This is apparent from a phase lag between mean monthly temperatures and mean monthly δ values at Mizuho Camp. Figure 3.8 shows that individual isotherms of sea surface temperatures over the Southern Ocean vary in latitude seasonally by a similar amount to the changes in mean latitude of the northern edge of the pack ice (Budd, personal communication). Also shown are mean monthly $\delta^{18}O$ values from precipitation collected at the South Pole by Aldaz & Deutsch (1967) and the results of Kato, Watanabe & Satow (1978). In spite of considerable fluctuations, the seasonal amplitude of δ values is around the same magnitude at both coastal and plateau stations and their seasonal trend is similar to that of sea surface isotherms. A source of atmospheric water vapour of constant mean temperature that changed latitude seasonally with the isotherms would produce the observed effects. The main seasonal variations of the δ values thus appear to be due to seasonal changes of the radiative cooling (isobaric) of the atmosphere as it moves from the source of the vapour to the coast of Antarctica. Further changes of δ values as the atmosphere is forced up over the ice sheet are much the same in summer and winter. The same argument applies

Table 3.2. *Variation in $^{0}/_{00}/^{\circ}C$ of isotopic composition of newly formed ice condensed from water vapour with immediate removal of condensate. Initial condition $\delta = 0^{0}/_{00}$ at $0\,^{\circ}C$*

(1)	(2)	(3)	(4)	(5)	(6)	(7)
	Isobaric cooling			Moist adiabatic cooling		
Temperature range	$d\delta D/d\theta$	$d\delta^{18}O/d\theta$	Ratio (3)/(2)	$d\delta D/d\theta$	$d\delta^{18}O/d\theta$	Ratio (6)/(5)
$0\,^{\circ}$C to $-20\,^{\circ}$C	7.7	0.97	7.9	6.2	0.73	8.5
$-20\,^{\circ}$C to $-40\,^{\circ}$C	7.0	0.88	8.0	5.3	0.67	7.9

Figures from Dansgaard (1964).

Figure 3.8. Upper curves: Seasonal variations of mean latitude of the sea ice edge and of mean sea surface isotherms around Antarctica (from Budd, personal communication). Lower three curves: Mean monthly isotopic δ ¹⁸O values of precipitation at Syowa Station and Mizuho Camp (Kato *et al.*, 1978) and at the South Pole (from data in Aldaz & Deutsch, 1967).

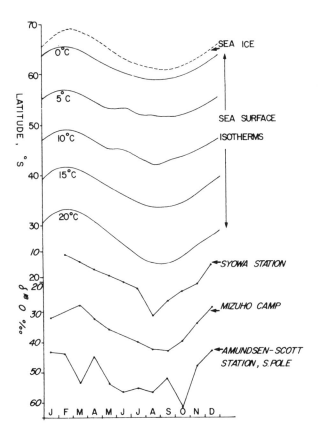

Figure 3.9. Mean δ ¹⁸O of precipitation around Greenland plotted against mean air temperature at ground or ice surface level. Open circles are for stations above 1000 m altitude, closed circles are for low-altitude stations. (From Dansgaard *et al.*, 1973, Fig. 2.)

to the Greenland ice sheet, although the more complex geography of the northern hemisphere makes it difficult to present the relevant data.

While the conclusions above relate to seasonal variations averaged over oceans around Antarctica, they indicate that variations of mean annual δ values and temperatures at individual Antarctic stations are also likely to be dependent upon the mean distance to the source of water vapour.

Surface δ values in the Greenland region

Measurements of the geographical distribution of δ values are often related to corresponding measurements of mean annual temperature and elevation to derive relationships between these factors. Our preceding discussion indicates that this is only an empirical relationship that will be present if there are reasonably close links between the δ values of snow when it reaches the surface, the mean condensation temperatures at which snow was formed and the mean annual surface temperature at the site of deposition. Over large polar ice sheets this appears to be the case except for coastal regions.

Figure 3.9, from Dansgaard, Johnsen, Clausen & Gundestrup (1973), shows the relationship between mean annual temperatures and mean δ values of precipitation. At coastal stations precipitation has been collected in precipitation gauges throughout the year, while on inland stations δ values are mean values from samples collected from ice cores or pits that often cover a number of years of net accumulation. Slopes of 1⁰/₀₀/°C for the two curves for coastal stations are said to agree with thermodynamic calculations of the isobaric cooling of precipitating air masses travelling over the oceans from the source area of the vapour. The authors consider that curves for south and west Greenland indicate higher condensation temperatures than for the far north and north east of Greenland, and that the latter

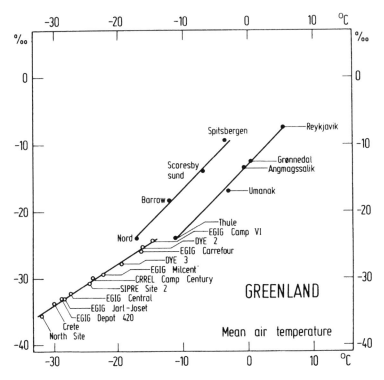

may take up at least part of their vapour from the Arctic Ocean. On the inland ice, the lower slope of $0.62^0/_{00}/^{\circ}C$ corresponds roughly with the fractionation during moist adiabatic cooling of an air mass as shown in Table 3.2. There appears to be no latitude effect for these inland results, as the points from the EGIG profile at roughly the same latitude lie on the same line as points from southern and northern locations on the ice sheet.

Studies of Koerner & Russell (1979) on the Devon Island ice cap, and by Koerner (1979) over the Queen Elizabeth Islands covering a wider area of the Canadian Arctic, examined mean δ values of snow deposited during one accumulation year (August/May–June), so that results were obtained to levels well below the equilibrium line on glaciers and ice caps. On slopes facing northern Baffin Bay they found δ value–elevation gradients that were readily explained by moist adiabatic cooling (0.59 and $0.64^0/_{00}/100\,m$), but elsewhere this relationship did not apply and there was an inverse relationship on many traverse lines. They point to a similarity between their results and those of Lorius *et al.* (1969) in Terre Adélie which were mentioned earlier.

If precipitation is coming from weather systems that have passed over high ground, the lower layers of the atmosphere are likely to experience some föhn effect that will dry descending air in the lee of ridges. This will increase evaporation of the outer layers of ice crystals that have been formed by frontal or other systems above the level of the mountains. This effect could readily explain the decrease of δ values with decreasing elevation on many of the profiles, especially if moisture is carried primarily by south or south east winds over this region. Koerner (1979) also concludes that some moisture is also supplied to the icecaps of the Arctic Islands from the Arctic Ocean. He obtained a linear regression fit of the mean δ value along each glacier traverse against distance (x km) from Baffin Bay (south east) which gave

$$\delta^{18}O = -26.00 - 0.0195x \qquad (3.5)$$

His correlation coefficient $r = 0.78$ is significant at <1% level. Although some aspects of his interpretation of the results may be queried, it must be pointed out that his figure of a decrease of δ value of around $1^0/_{00}/$ 50 km from source also fits the seasonal change of δ values around the Antarctic coastline shown in Figure 3.8. There is a seasonal north–south movement of sea surface isotherms (0–15 °C) averaging some 6.7° of latitude. This corresponds to a seasonal fluctuation of around $14^0/_{00}$ in $\delta^{18}O$ values in coastal snowfall. Combining these figures gives a $1^0/_{00}$ change of $\delta^{18}O$ value per 50 km increase in latitudinal distance to the source of water vapour.

3.3

Antarctica: survey of near-surface mean isotopic values

C. LORIUS

Measurements

Oxygen-18 and deuterium concentrations in natural waters are determined by mass spectrometry techniques (Dansgaard, 1969; Nief, 1969). Various standard samples are available for calibration purposes and under good conditions the accuracy of measurements is of the order of $\pm 0.2^0/_{00}$ ($\delta^{18}O$) or $\pm 1^0/_{00}$ (δD).

Comparative measurements on Antarctic samples (Epstein & Sharp, 1967; Lorius & Merlivat, 1977) show that δD and $\delta^{18}O$ are linearly related, the relationship being very similar to the one given from more general surveys (Craig, 1961; Dansgaard, 1964):

$$\delta D^0/_{00} = 8\delta^{18}O^0/_{00} + 10$$

This equation will be used for comparative purposes in the present paper.

Results
Geographical distribution

A collection of mean δ value measurements of surface snow layers over Antarctica from Morgan (thesis, University of Melbourne 1980) are shown in Figure 3.10, with place names added to aid interpretation of Figure 3.11. Most measurements were made on samples collected along traverses over the inland ice and from the vicinity of permanent stations, most of which are located in coastal areas.

Mean surface values from snow samples covering several years accumulation in coastal areas are typically of the order of $-20^0/_{00}$ for $\delta^{18}O$ ($\delta D = -150^0/_{00}$); they decrease inland, with the lowest reported value being $-58^0/_{00}$ at Plateau Station (not shown on Figure 3.10).

Mean δ values and mean surface temperature; inland areas

The geographical distribution of δ values in falling snow is influenced by many atmospheric parameters related to the origin and thermodynamic history of air masses. Particularly important is the difference between the condensation temperature at a given time and location and the temperature at which condensation started to form in the same air mass (Dansgaard, 1964; Dansgaard, Johnsen, Clausen & Gundestrup, 1973; and section 3.2). This temperature effect is reflected in the general decrease of Antarctic mean δ values observed with increasing surface elevation (Gonfiantini et al., 1963; Lorius, 1963; Picciotto, 1967; Lorius, Merlivat & Hagemann, 1969; Dansgaard et al., 1973; Lambert et al., 1977; Lorius & Merlivat, 1977), although systematic differences are found between various geographic areas.

The temperature of formation of any specific snow sample is generally not known accurately, so to obtain a practical empirical relationship mean firn temperatures at 10 m depth are used instead. However, the mean temperature near the surface of the ice sheet is poorly correlated with the condensation temperature at which precipitation is formed in clouds, since the strength of the temperature inversion over Antarctica varies with seasons, location and climatic areas (Dalrymple, 1966; Lorius et al., 1979 and chapters 1 and 6). However, the empirical relationship between the δ value and surface temperature mean values is approximately linear. Lorius & Merlivat (1977), using data from −20 to −54 °C, find

$$\delta D^0/_{00} = 6.04\theta(^\circ C) - 51$$

converting this to $\delta^{18}O$ gives:

$$\delta^{18}O^0/_{00} = 0.755\theta(^\circ C) - 7.6$$

Figure 3.10. Isotopic $\delta^{18}O$ values measured in surface layers of the Antarctic ice sheet, after Morgan (1980). Place names added to indicate location of measurements plotted in Figure 3.11.

This relationship is represented by line 1 on Figure 3.11, where selected Antarctic data using continuous samples of several years accumulation have been plotted for both coastal and inland sites. (For references, see caption of Figure 3.11.)

At elevations greater than 1000 m and distance to the coast greater than 100 km, Figure 3.11 shows that many published values are in reasonable agreement with line 1, in particular those obtained in Terre Adélie and in Wilkes Land (Lorius et al., 1969). However Marie Byrd Land stations (area 3 on Figure 3.11) appear to have comparatively lower δ values, which is also true for Queen Maud Land stations SA9 and HB 3, Pole of Relative Inaccessibility, Plateau Station and South Pole; this is confirmed by numerous results published for Queen Maud Land (Picciotto, 1967) in a preliminary form which prevents them from being used with precision, but which show a change of $\delta^{18}O$ between −45 and −58$^0/_{00}$ in an area where the temperature range is −46 to −58 °C. Factors which could influence the distribution of δ values plotted against temperatures include:

(1) variations in depth and intensity of the temperature inversion layer, as already mentioned;

(2) different origins and thermodynamic histories of precipitating air masses (Alt, Astapenko & Ropar, 1959, report tracks of depressions which are more complicated for Marie Byrd Land than for Wilkes Land where the trajectories originate more directly from the sea);

(3)   special processes controlling formation of precipitation in cold central areas with a downward transport of water vapour from higher atmospheric levels (Schwerdtfeger, 1969).

The linear δ–temperature relationship obtained in Terre Adélie, which also fits Wilkes Land results, suggests that the whole area is subject to relatively uniform meteorological conditions. The $\delta^{18}O$–temperature gradient of $0.75^0/_{00}/^\circ C$ is somewhat smaller than that shown on Dansgaard *et al.* (1973) for the Antarctic ($1^0/_{00}/^\circ C$). This may be due to the fact that the later figure was obtained by including some of the stations located in the δ-depleted areas. Gordiyenko & Barkov (1973) give a slightly different equation based on Russian measurements ($\delta^{18}O^0/_{00} = 0.84\theta(^\circ C) - 7.5$) without details about the stations selected.

Other factors which may influence the mean δ value of near-surface deposits in Antarctica should be mentioned; for instance, a spatial change in the rate of accumulation has been found anti-correlated with δ value (Lorius, 1963) possibly due to transport of snow by drift (Dansgaard *et al.*, 1973). This is discussed later.

Despite all the restrictive factors already mentioned, it must be pointed out that the observed slope ($0.75^0/_{00}/^\circ C$ for $\delta^{18}O$) lies between the calculated $\delta^0/_{00}$ changes for a 1 °C cooling of the temperature of condensation under moist adiabatic conditions and the slope under isobaric conditions using simplified atmospheric models (see section 2.2 and Dansgaard, 1964). This value is lower than those obtained by observations of cloud

temperatures and δ values of precipitation at Roi Baudouin Station on the coast and at the South Pole Station which are respectively 0.9 and $1.4^0/_{00}/^\circ C$ for $\delta^{18}O$ (lines A and B, Figure 1.5).

Coastal regions

We shall discuss observations on the large Ross and Filchner–Ronne Ice Shelves separately. We see on Figure 3.10 that mean $\delta^{18}O$ values at most coastal stations lie between −17 and $−21^0/_{00}$. These figures apply also to the coastal slopes of the ice sheet up to elevations around 1000 m. This was pointed out for Terre Adélie in Lorius *et al.* (1969) and by Lorius & Merlivat (1977) whose figures for samples covering 10 years or more show a mean value of δD of $−152.7 \pm 2.5^0/_{00}$ for three stations from 220–270 m elevation compared with $−146.4 \pm 3.7^0/_{00}$ for three stations between 680–790 m. Kato, Watanabe & Satow (1978) also state that $\delta^{18}O$ values of drifting snow inland from Syowa station at elevations below 1000 m are almost independent of elevation. In Figure 3.12(*a*) we see that $\delta^{18}O$ values are slightly higher near the summit of Roosevelt Island (around 800 m) and on the Siple Dome (about 81.5° S 151° W) than on the lower ice shelf nearby. Dansgaard *et al.* (1973) suggested that redistribution of deposits by drift might cause these relatively constant values below 1000 m. However, this is not supported by evidence at Dumont D'Urville (40 m elevation) where the mean δD value of precipitation, excluding drift over two years, was $−153.3^0/_{00}$. In Terre Adélie, the elevation of around 1000 m is about the level of low precipitating clouds as they move from the

Figure 3.11. Mean near-surface δ values plotted against mean air temperature at ground level. Key to symbols shown on diagram. δ values are from Lorius *et al.* (1969, Table 1) and from the following: Dansgaard (unpublished?) for D45, D59, D80; Gonfiantini (1965) for Roi Baudouin; Epstein, Sharp & Goddard (1963) for Little America, Wilkes S2; Dansgaard *et al.* (1973) for Horlick Mountains, Halley Bay and Byrd; Vilenskiy *et al.* (1970) for Mirny; Vilenskiy *et al.* (1974) for Molodezhnaya and Vostok δ ^{18}O values; Epstein & Sharp (1967) for Eights; Aldaz & Deutsch

(1967) for South Pole; Picciotto (1967) for Plateau Station; Picciotto *et al.* (1968) for Pole of Relative Inaccessibility; measurements by Lorius (unpublished) on samples from Wolmarans from SA9 and Peel from HB3. Line 1 for inland stations from Lorius & Merlivat (1977), line 2 for coastal stations from data in Budd & Morgan (1973). Points in Area 3 are from Marie Byrd Land and Area 4 from coastal stations. Some temperature values are from Bentley *et al.* (1964).

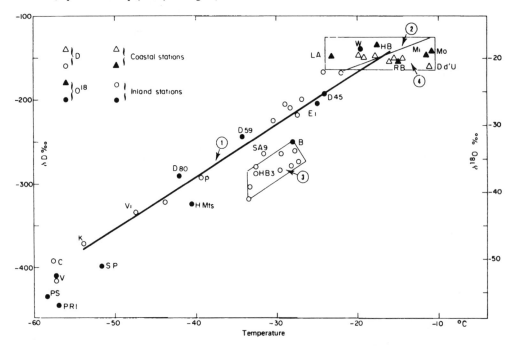

Figure 3.12. (a) Areal variation of the mean δ ^{18}O values of the upper firn over the Ross Ice Shelf, from Clausen *et al.* (1979).

(b) Ten-metre temperatures on the Ross Ice Shelf. Measured temperatures have been corrected for seasonal variations to give equivalent average values. From Thomas *et al.* (in press).

(a)

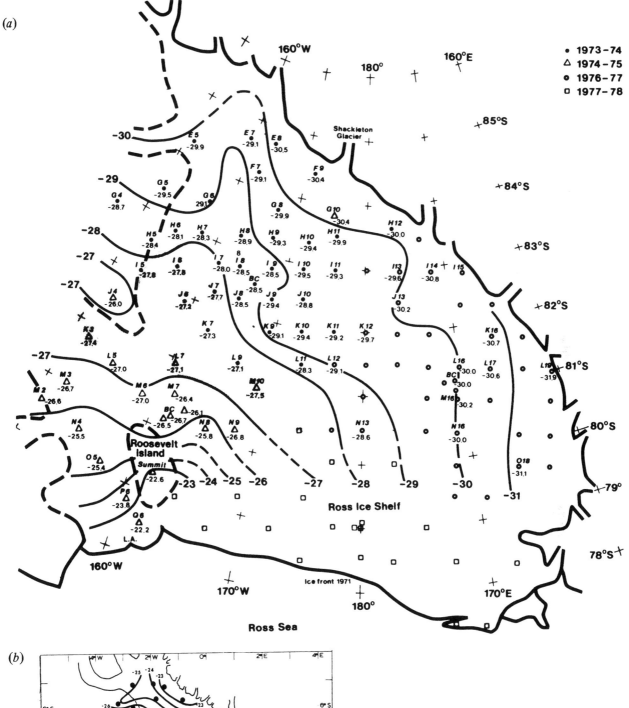

(b)

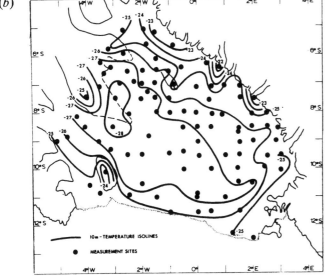

ocean to the continent, where the ice sheet reaches 1000 m elevation around 40 km from the coast. Condensation conditions over this 40 km zone may therefore be reasonably constant in cloud above 1000 m and they will not be affected by the inland surface elevations below 1000 m or related to the decreasing temperature with elevation in this zone. The apparent slight fall in δ values with decreasing elevation below 800 m may be due to the evaporation of outer layers of snow crystals falling through drier air below the cloud level, as suggested in section 2.2 in relation to the observations of Koerner (1979) for the Canadian Arctic islands.

At Law Dome, the above effect does not appear as Budd & Morgan (1973, 1977) find that over an elevation range from 300–1400 m the mean gradient of $\delta^{18}O$ is $0.65^{0}/_{00}/^{\circ}C$ along two lines, a gradient that is slightly lower than that at higher altitudes on the main ice sheet of Terre Adélie and Wilkes Land inland of Law Dome.

Ice shelves

Figure 3.10 indicates that over ice shelves, δ values decrease with distance from the sea, but not as rapidly as they decrease over rising inland ice. Figure 3.12(*a*) shows a more detailed δ value distribution over the Ross Ice Shelf from Clausen, Dansgaard, Nielsen & Clough (1979). This indicates that the main supply of moisture apparently moves over the ice shelf from the north east, as δ values fall both to the south and west of Roosevelt Island. The isotopic data can be compared with isotherms on Figure 3.12(*b*) from Thomas, MacAyeal, Eilers & Gaylord (in press). Although there is some tendency for both δ values and temperatures to fall together in the north east corner of the Ross Ice Shelf, further to the south and west temperatures increase towards the Transantarctic Mountains although δ values continue to fall. This leads to conclusions similar to Koerner (1979) regarding δ value distribution over the Canadian Arctic Islands (see section 2.2). It appears that mean δ values decrease with distance from the source of moisture as a result of isobaric precipitation processes over level ice shelves, in contrast to the dominance of orographic processes over the rising ice sheets. The temperature rise towards the Transantarctic Mountains is governed by different atmospheric processes that may not be associated with occurrence of precipitation. The isotherms of Figure 3.12(*b*) suggest that katabatic winds down major glaciers exert a warming effect in the region where these discharge on to the level ice shelf. The overall temperature distribution suggests that the cold ground inversion is broken down during winter as one approaches the Transantarctic Mountains by strong south west 'barrier' winds similar to those described by Schwerdtfeger (1975) along the Antarctic Peninsula. The mean decrease of δ values with distance appears to be around $1^{0}/_{00}$ per 65 km, which is slightly less than the $1^{0}/_{00}$ per 50 km found by Koerner (1979) over the rougher Canadian Arctic Islands. Although very little data is shown on Figure 3.10, over the Filchner–Ronne Ice Shelves, a similar gradient with distance appears probable.

Representative nature of mean near-surface δ values

In cold polar glaciers the δ values of accumulated snow are not disturbed by melting and percolation processes and the observed changes at shallow depths at a given location are due mainly to seasonal fluctuations in the isotopic composition of precipitation which are smoothed during firnification (Johnsen, 1977*a*, and section 3.4). However, mean annual values may differ from year to year depending on various factors, such as climatic conditions involving, for example, unusually high summer δ values (Picciotto, Deutsch & Aldaz, 1966), rate of snow accumulation (Vilenskiy, Teys & Kochetkova, 1974) and snow drift. At Molodezhnaya, Vilenskiy *et al.* reported differences between annual mean δ values of up to $1.3 \, \delta^{18}O^{0}/_{00}$ over a five-year time period; mean values over a three-year period obtained at two nearby sites differed by $1.1 \, \delta^{18}O^{0}/_{00}$. Results from Terre Adélie (Lorius & Merlivat, 1977) also give an idea of the areal scattering of δ values. For instance, sampling at stations which are a few hundred metres distant from each other give values which agree to within $\pm 5^{0}/_{00}$ for δD ($0.6^{0}/_{00}$ for $\delta^{18}O$). At a scale of 10 km we have found differences of up to $20^{0}/_{00}$ ($2.5^{0}/_{00}$ for $\delta^{18}O$) which are not connected with temperature changes.

Standard deviations listed in section 6.1 indicate that a ten-year sample from coastal sites in Antarctica should indicate the long-term mean δ values to an accuracy of $\pm 0.5^{0}/_{00}$. This does not apply at high elevations on the inland plateau as shown on Figure 6.2 which indicate a much larger variability of mean δ values where accumulation rates are very low. Results from Dome C indicate an accuracy of the long-term mean δ values of a ten-year sample to be only just below $\pm 1.0^{0}/_{00}$.

Geographically, we have seen that in Antarctica the local, regional and large-scale δ-value distributions vary in accordance with a number of parameters. The mean isotopic concentrations in a given area cannot therefore be derived from theoretical calculations or experimental relationships obtained in other areas and must be measured in each zone investigated. Furthermore, due to the variety of the atmospheric processes involved, it must be pointed out that the geographic distribution which is actually observed may have been different under other climatic conditions.

3.4

Diffusion of stable isotopes

S.J.JOHNSEN & G. de Q.ROBIN

Molecular diffusion along isotopic gradients in polar ice sheets

When surface melting is not significant, we have seen that polar ice sheets preserve the detailed stratigraphy of isotopic layering for many thousands of years. Under these conditions, disappearance of isotopic layering is due to two types of diffusion (Johnsen, 1977a). Mass exchange by diffusion of water vapour and sintering in the upper layers of porous snow smooth out and tend to obliterate closely spaced isotopic layering over a period ranging from months to decades. Many thousands of years later, very slow diffusion of water molecules in solid ice will eliminate any annual layering that survived the first process. The same process will also tend to smooth out any discontinuities formed subsequently by irregular deformation in the ice mass. Effects of the two processes were seen in the Camp Century ice core in Figure 1.7.

In sub-polar glaciers, with a firn layer that remains porous, surface meltwater will percolate irregularly into the top few decimetres or metres during summer then refreeze, thus obliterating the annual isotopic layering more rapidly than in polar ice sheets. Nevertheless, the mean isotopic composition over a few metres depth will remain unchanged and still give a long-term record of climatic changes as will be seen in the results from Devon Island in chapter 4. However, if the surface of a sub-polar ice cap is non-porous, surface run-off of meltwater in summer will remove seasonal accumulation selectively. Since this can have a considerable effect on the mean isotopic composition which we cannot readily estimate, such ice masses are unsatisfactory for isotopic studies of past climate. These factors were shown in studies of the Meighen Ice Cap in the Canadian Arctic by Koerner, Paterson & Krouse (1973).

Vapour diffusion in firn

The primary process of densification in the top few metres of firn is the mass diffusion of water vapour that enables large ice crystals to grow as the smaller crystals

evaporate. Johnsen (1977a) considers that this process is no longer important when the density exceeds $550 \, \text{kg} \, \text{m}^{-3}$. His calculations of the changing amplitude of $\delta^{18}O$ cycles in firn match observed changes reasonably when a mean total diffusion length of water molecules in firn (L_0) of 80 mm (ice equivalent) is used, irrespective of the temperature or accumulation rate at the site. It indicates that after firnification, water molecules are, on average, displaced vertically by some 80 mm (ice equivalent) from their initial positions in the ice core, disregarding sinking of the firn and thinning due to vertical strain. The major parameters in this process are clearly the total change in density, and especially the final density at which the process stops.

The amplitude $A_{\delta s}$ of a $\delta^{18}O$ cycle in a surface layer of equivalent ice thickness λ changes to an amplitude A_δ at depths below that of the critical density given by

$$A_\delta = A_{\delta s} \exp\left(-2\pi^2 L_0^2/\lambda^2\right) \qquad (3.6)$$

In layers of 0.6, 0.3 and 0.2 m of equivalent ice thickness, the amplitudes of $\delta^{18}O$ cycles will be reduced by 30, 75 and 96 per cent respectively by the time vapour diffusion ceases to be effective due to increasing ice density with depth. Such layer thicknesses are typical of annual layer thicknesses over much of the Greenland ice sheet. They are preserved in isotopic stratigraphy to depths exceeding 1000 m, whereas shorter period layering is not. In Antarctica the mean annual layer thickness is about 0.16 m so that, apart from higher accumulation zones which are mainly in coastal regions, annual layering is not preserved. This applies even at Vostok Station at a mean temperature of $-57 \, °C$ where Dansgaard, Barkov & Splettstoesser (1977) have shown that isotopic annual layers averaging around $0.025 \, \text{m} \, \text{a}^{-1}$ of ice disappears in the top metre of firn in about 10 a.

Diffusion in solid ice

The process of diffusion in solid ice at great depths is several orders of magnitude slower than in porous firn. Figure 3.13, from Hobbs (1974, p. 382), shows experimental values of the volume diffusion

Figure 3.13. Volume diffusion coefficients (D_v) for ice as a function of temperature in K, from Hobbs (1974, Fig. 5.13). (a) from ^{18}O measurements of Delibaltas *et al.* (1966) (full line and filled circles); (b) from tritium measurements of Itagaki (1964) (dot/dash line). (c) from tritium measurements of Blicks, Dengel & Riehl (1966) (dashed line).

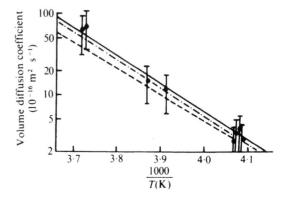

coefficient (D_v) as a function of temperature. These were obtained by measuring the rate at which water molecules labelled with atoms of hydrogen (^{3}H) and oxygen (^{18}O) diffused into polycrystalline ice. Possible reasons why results from labelling with oxygen and hydrogen were so close are discussed in Hobbs (1974) who favours a vacancy diffusion mechanism involving complete water molecules for self-diffusion in pure ice. However, results may be dominated by diffusion along grain boundaries rather than within single ice crystals. In any case, layering that has survived vapour diffusion in the surface layers is not affected by the much slower diffusion in solid ice until vertical compression of annual layering has greatly increased the δ^{18}O gradients in the ice mass.

According to Hammer *et al.* (1978), in polar ice the total mean diffusion length L_t at time t may be expressed in terms of the thinning factor $F_f = \lambda_t/\lambda_0$, the ratio of layer thickness at time λ_t to its thickness at the time of deposit λ_0, by

$$L_t = F_t \left\{ L_0^2 + 2 \int_0^t D_v(t') \cdot F_t^{-2} \, dt' \right\}^{1/2} \qquad (3.7)$$

where L_0 is the mean total diffusion length in firn (80 mm). The δ amplitudes A_δ at time T are then given as before by equation (3.6) except that the value L_t is substituted for L_0. In Figure 3.14 we show calculated effects for Camp Century, plotted against a time scale for the ice core derived by Hammer *et al.* (1978). As well as the mean total diffusion length (L_t), we show its two components, one due to diffusion in firn (L_0) modified by subsequent layer compression to L_f, the other being the mean total diffusion length in ice L_i which increases with time as layers become thinner and the ice warmer. At 8700 a, L_t amounts to 40 per cent of λ, and from equation (3.6) A_δ falls to 0.2 per cent of $A_{\delta s}$, which is only three times the measuring accuracy and therefore at the limit of detection. This agrees with the disappearance of layering as seen in chapter 1.

Deconvolution of diffusion effects

When annual layers are partly obliterated as the result of diffusion, as in Figure 1.7, it is possible to estimate the original annual cycle, provided that a sufficient number of measurements of high accuracy are available. We have already seen that equation (3.6) predicts the disappearance of layering satisfactorily. This justifies the use of the equations (3.6) and (3.7) to reconstruct the original δ profiles at the time of deposition in as much detail as possible.

Johnsen (1977a) treats the diffusion process as mathematically equivalent to putting the observed δ^{18}O profile $\delta_0(z)$ through a symmetrical filter F_1 of the transfer function (frequency response)

$$\bar{F}_1(k, L) = \exp\left(\tfrac{1}{2}k^2 L^2\right) \qquad (3.8)$$

where k ($= 2\pi/A$) is the wave number, and $\bar{F}_1$ the Fourier transform of F_1. The actual profile $\delta_a(z)$ is related to the depositional profile by

$$\delta_a(z) = F_1^* \delta(z)$$

where * means convolution (filtering). To obtain the

original 'depositional' δ values, we have to carry out the converse process, deconvolution, on the observed data. In doing this our filter $\bar{F}$ has to be the inverse of $\bar{F}_1$, i.e. $\bar{F} = \bar{F}_1^{-1}$, except that at higher frequencies strong damping has to be introduced to avoid blowing up the noise level in the record. An optimum filter for extracting the signal in the presence of noise, making allowance for the finite length of samples, is then employed to deconvolute the $\delta_a(z)$ record from the ice core. We show results of deconvolution from the firnification process from Johnsen (1977a) in Figure 3.15 for an ice core from Crête obtained in 1974. In the observed data, the δ amplitude is reduced considerably by 4 m depth. Layering is not clear at some levels in the observed δ profile from 59–65 m depth (C), but it is recovered clearly in the deconvoluted profile (B), in which the annual δ amplitudes are similar to the value in the top 2 m.

Confirmatory evidence that we are dealing with annual layers also comes from detailed comparison with the dust particle content (Hammer *et al.*, 1978). In Figure 3.16 we show profiles of dust particle content against δ^{18}O profiles for another section of the core from Crête. Both observed and deconvoluted δ profiles are shown. The dust particles were studied by use of a light scattering technique, as distinct from earlier studies when

Figure 3.14. Total mean diffusion length L_t (in mm of ice equivalent) of water molecules in the Camp Century ice core as a function of time since deposition. This is calculated from equation (3.7) and from the ice flow model (equation 5.7). λ_t is the layer thickness at time t, L_f the component of diffusion length due to processes in the firn modified by layer compression and L_i the component due to diffusion in solid ice. After Hammer *et al.* (1978), Fig. 8).

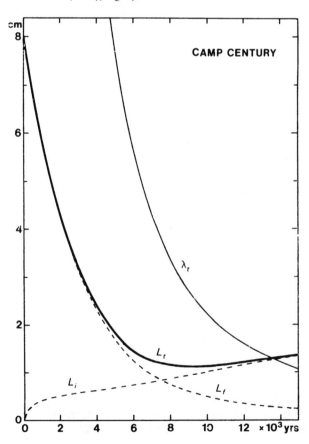

melted samples were passed through a Coulter counter, a method which requires larger sampling volumes. The dust particles studied in the Crête sample ranged mostly from 0.1–2.0 μm in size. They were apparently carried to the ice sheet from continental deserts by the tropospheric circulation. The dust content of the ice cores shows a seasonal maximum in spring months in most years.

Use of a diffusion length L_t for deconvolution of isotopic profiles affected by diffusion within solid ice has been attempted with some success (Johnsen, 1977a). However, results are less reliable than for upper layers, since unlike L_0 for the firnification process, L_t is a function of temperature and time (Figure 3.13), and it is more difficult to determine its precise value in a given case.

Diffusion across discontinuities

In addition to diffusion of annual layering discussed above, we consider the effects of diffusion across discontinuities in the ice core formed when ice of one δ value is brought into contact with ice of a different δ value. Paterson *et al.* (1977) have suggested that very high δ gradients of $9^0/_{00}$ in about 30 mm of ice core some 2.3 m above bedrock on Devon Island (Figure 4.20) were of tectonic origin, possibly due to shear within the ice mass. Robin (1977) suggested that the sharp negative spike in δ values some 26 m above bedrock in the Camp Century core may be due to injection of ice of different δ value into another ice mass due to

the irregular folding of basal ice as will be described in section 3.10.

In addition to mechanisms described by Boulton in section 3.10, Figure 3.17 illustrates on a simple two-dimensional model another way in which sharp isotopic 'spikes' could be produced by shear near the base of an ice sheet. Suppose we have a layer of ice 'C', which may have been deposited in an ice age, above and below which lie ice of different characteristics 'W'. To the left of the diagram we show ice motion parallel to sinusoidal bedrock undulations. As this ice passes the centre of the diagram, we postulate that ice motion near bedrock changes to horizontal shear over a discrete plane Z', so that the ice below this plane remains stationary. The pre-existing undulations are shifted by a distance OX in relation to the lower layers, resulting in an irregular distribution of layer C on the right of the diagram. Vertical profiles of cold and warm ice, plotted against height above bedrock, as they would be identified by boreholes at different points along OX, are plotted above these points. To the left of the origin, profiles (1) and (2) of height above bedrock, remains constant, but to the right, the thickness of the cold layer varies with position. We see on profile (6) that a layer of warm ice overlies a layer of cold ice, producing a cold

Figure 3.16. The mid-section is a δ ^{18}O profile representing the period AD 1765–1805 at Crête. Ambiguities (e.g. AD 1784, 1789 and 1802) in the interpretation of annual layers have been solved by cross checks with the microparticle profile to the left, showing one peak of fallout per year, and with the deconvoluted δ ^{18}O profile to the right that is first order corrected for diffusive smoothing in the firn.

Figure 3.15. (A) δ-profile of top 4 m at Crête showing reduction in δ amplitude with depth by diffusion in firn.
(C) δ-profile from 59–65 m after diffusion in firn is complete.
(B) Same profile after deconvolution using $L_0 = 80$ mm and filtering out higher frequency noise (after Johnsen, 1977a, Fig. 7.)

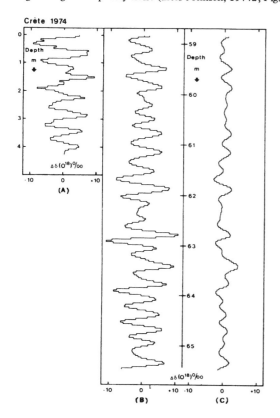

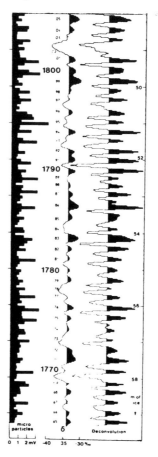

'spike' on our profile, while on other profiles (3) to (7), layer thicknesses are changed.

If discontinuous shear at a sharp interface takes place, then unless the shear planes are parallel to the depositional layering in ice at all levels, it is clear that an effect along the lines shown in Figure 3.17 must occur. We see that resultant discontinuities in the time scale down a vertical ice core can readily produce the cold 'spikes' observed beneath the Wisconsin ice of the Camp Century and Devon Island cores. As well as producing gaps in an isotopic profile, as recognised by Paterson *et al.* (1977), it can result in repeating different time sections of the column. The effect can of course occur at all levels, including the top of a 'cold' layer, where gaps are also reported on Devon Island. Any marked difference in hardness between different layers might be expected to affect the extent of shear surfaces. Two points may be noted from existing isotopic data. Several major 'spikes' beneath the Wisconsin ice at Camp Century and Devon Island are of the order of 10 cm thick, and although they may have been thicker when first formed, if our explanation is correct, it appears that discontinuous shear surfaces may not penetrate far into the Wisconsin ice. The same conclusion comes from the comparison of the two isotopic profiles in the bottom 6 m of the Devon Island ice core by Paterson *et al.* (1977). Of their Wisconsin ice layer, some 70 per cent of the ice column is repeated as a continuous layer in both cores, with sections 'missing' or 'added' both above and below this level.

We need a quantitative idea of the magnitude of diffusion processes from a sharp interface that could result from shear. We first consider the case of diffusion across a sharp boundary in static ice without vertical strain. The mathematics involve the same diffusion equations as for heat transfer (Carslaw & Jaeger, 1959).

For a single boundary at $z = 0$, dividing two ice masses of δ values δ_1 and δ_2 at time $t = 0$, the distribution of δ values (δ_t) at time t due to diffusion is given by

$$\delta_t = \tfrac{1}{2}(\delta_1 + \delta_2) + \tfrac{1}{2}(\delta_1 - \delta_2)\operatorname{erf}\frac{z}{2\sqrt{(D_v t)}} \quad (3.9)$$

Calculations were made taking $D_v = 10^{-8}\,\mathrm{m^2\,a^{-1}}$, its value at $-10\,^{\circ}\mathrm{C}$ (Figure 3.18). We note from Figure 3.13 that error bars on individual measurements of D_v show an uncertainty of around ± 50 per cent, while from the number of measurements and comparison with other studies tabulated in Hobbs (1974, p. 384), our value of D_v is seen to decrease by a factor around 10 from $-10\,^{\circ}\mathrm{C}$ to $-30\,^{\circ}\mathrm{C}$. We also note that the uncertain temperature and strain-rate history of deep ice limits the accuracy of interpretation.

In addition to diffusion in static ice, we consider the effect of the vertical strain rate $\dot{\epsilon}_z$ on diffusion across a horizontal surface, $z = 0$. The steady-state solution for an interface between ice masses of isotopic values δ_1 and δ_2, where outward diffusion from the interface is in balance with the inward strain rate at all points to produce a constant profile, is given by

$$\delta^{18}\mathrm{O} = \tfrac{1}{2}(\delta_1 + \delta_2) + \tfrac{1}{2}(\delta_1 - \delta_2)\left[\operatorname{erf}\sqrt{\left(\frac{|\dot{\epsilon}_z|}{2D_v}\right)}z\right]_{-\infty}^{+\infty}$$

(3.10)

The form of the curves is the same as the time solution of equation (3.9) so we present numerical solutions in Figure 3.18 for both cases. The relation between time and strain rate for the same curve is $t = 1/(2|\dot{\epsilon}_z|)$, thus the curve for 10 000 a of diffusion in static ice is the same as the curve for $0.5 \times 10^{-4}\,\mathrm{a^{-1}}$ strain rate. The diffusion length L_{ss} for steady-state

Figure 3.17. The left-hand lower section shows isotopically 'warm' (W) and 'cold' (C) layers of ice deforming by parallel flow over sinusoidal bedrock undulations. The profile measured by an ice core in relation to height above bedrock is shown in the upper section by curves (1) and (2). The right-hand section shows the effect of introducing horizontal shear at height Z' to ice previously being deformed as on the left-hand section, together with the resultant profiles that would be found in ice cores as different positions (3) to (7). Complications due to mass continuity requirements at the junction between sections are not considered.

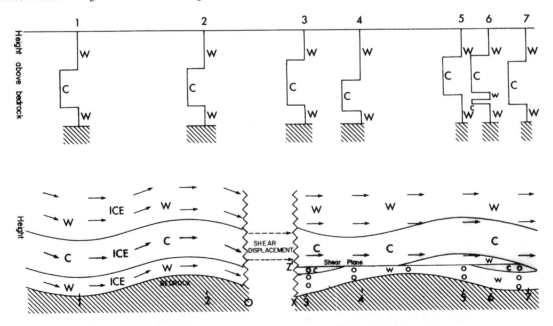

problems is given by the value of z for which $z\sqrt{(\dot{\epsilon}_z/(2D_v))}=1$. This is the point at which $\frac{1}{2}(\delta_1-\delta_2)=0.843$ of its final value on Figure 3.18. For a strain rate of $10^{-5}\,\mathrm{a}^{-1}$ at $-10\,^\circ\mathrm{C}$ the diffusion length $L_{ss}\approx 0.45$ cm. At $-30\,^\circ\mathrm{C}$ the figures fall by a factor of three. Thus in the upper levels at Camp Century, where $\dot{\epsilon}_z\approx 3\times 10^{-4}\,\mathrm{a}^{-1}$ and $\theta\approx -25\,^\circ\mathrm{C}$, we find $L_{ss}\approx 0.3$ cm; at 100 m above bedrock, where $\dot{\epsilon}_z\approx 0.8\times 10^{-4}\,\mathrm{a}^{-1}$ and $\theta\approx -16\,^\circ\mathrm{C}$, we get $L_{ss}\approx 2$ cm; 25 m above bedrock, if $\dot{\epsilon}_z\approx 0.2\times 10^{-4}\,\mathrm{a}^{-1}$ and $\theta\approx -14\,^\circ\mathrm{C}$, $L_{ss}\approx 5$ cm.

These figures provide a useful guide for assessing the role of diffusion in solid ice when interpreting isotopic profiles in ice for which vertical strain rates can be estimated. In general, the total diffusion is small except in the lowest layers.

We next consider diffusion from a thin slab of ice, of thickness $2a$ and isotopic value δ_1, within a large

Figure 3.18. Diffusion between ice masses of isotopic values δ_1 and δ_2 at $-10\,^\circ\mathrm{C}$ below and above the plane $z=0$ respectively. Each curve shows diffusion as a function of distance from this plane both in terms of years elapsed since diffusion commenced in a static mass (1) or alternatively as the steady-state distribution which applies for the stated strain rates (2). Curve (a) shows diffusion over 1000 a in static ice or steady-state diffusion for $\dot{\epsilon}_z=5\times 10^{-4}\,\mathrm{a}^{-1}$; curve (b) ditto for 10 000 a or $\dot{\epsilon}_z=5\times 10^{-5}\,\mathrm{a}^{-1}$; curve (c) ditto for 100 000 a or $\dot{\epsilon}_z=5\times 10^{-6}\,\mathrm{a}^{-1}$.

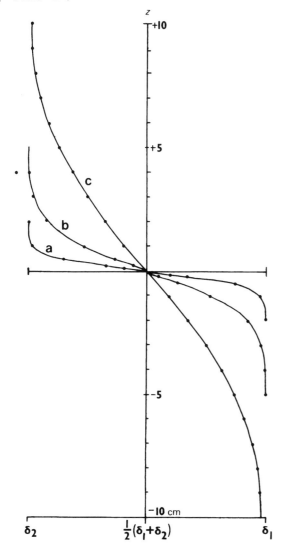

mass of static ice of value δ_2. Equation (3.9) is then modified to

$$\delta_t=\delta_1+\tfrac{1}{2}(\delta_1-\delta_2)\left[\operatorname{erf}\frac{a-z}{2\sqrt{D_v t}}+\operatorname{erf}\frac{a+z}{2\sqrt{D_v t}}\right]$$
(3.11)

We again take $D_v=10^{-8}\,\mathrm{m^2\,a^{-1}}$. Results for slabs 2 cm and 5 cm thick ($a=1$ cm and 2.5 cm) are shown in Figures 3.19 and 3.20. The former indicates that any large spikes in the δ profile of a width less than 3 cm at half amplitude will have been formed during the past 10^4 a or so. For a slab of width $2a=50$ cm, diffusion from either side can be treated separately, but for a slab with $2a=5$ cm, a figure typical of several observed spikes, some lowering of the peak of the spike due to diffusion will have taken place over times longer than 10^4 a. In Figure 3.20, by courtesy of Professor Dansgaard, we have superimposed on results for a slab 5 cm thick a plot of $\delta^{18}\mathrm{O}$ values in the cold spike, 25 m above bedrock in the Camp Century ice core, which has been interpreted as a possible sharp cold spell around 90 000 a BP by Dansgaard, Johnsen, Clausen & Langway (1972). The observed δ values which were sampled over 1 cm intervals appear to match our diffusion curves for static ice at $-10\,^\circ\mathrm{C}$ for a time from the slab formation of a little less than 10 000 a BP. At the temperature of $-13\,^\circ\mathrm{C}$ of the basal ice, this time should be approximately 15 000 a. Our diffusion length for ice in a steady state undergoing vertical strain at this depth was $L_{ss}\approx 5$ cm, which roughly matches the observations. However, this would involve diffusion over a much

Figure 3.19. Diffusion at $-10\,^\circ\mathrm{C}$ of a slab of ice 2 cm thick and isotopic value δ_2 placed within a static ice mass δ_1 at zero time. The curves show the δ distribution after the stated times have elapsed.

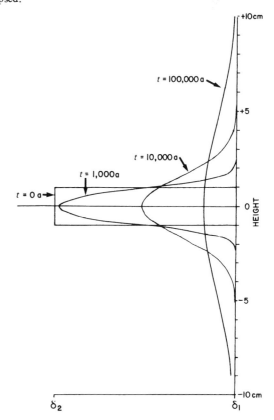

longer time, in which case the extreme negative δ value of $-39^0/_{00}$ would have disappeared. We conclude that the cold spike was formed in the isotopically warmer ice towards the end of the last ice age, say around 15 000 a BP, when the size of the ice sheet and pattern of ice flow were changing rapidly.

Results from other locations should be mentioned briefly. At Byrd Station the vertical strain rate is about $0.7 \times 10^{-4} \, a^{-1}$ and the temperature at upper levels about $-28 \, ^\circ C$, so the diffusion length over the upper half of the ice core should not exceed 1 cm. The effect of diffusion in solid ice on mean δ values over 0.5 m lengths of ice core will therefore be negligible for most purposes and can be neglected when considering most of the results in chapter 4. At Vostok and Dome C, where temperatures are below $-50 \, ^\circ C$ in the upper layers of ice and vertical strain rates are of the order of $10^{-5} \, a^{-1}$, steady-state diffusion lengths in solid ice will be of the order of 1 mm in the upper levels. However at all stations, these diffusion lengths increase towards bedrock as the ice temperature increases.

We turn again to results of Paterson *et al.* (1977), from Devon Island, to see if we can estimate strain rates or the time of formation of the sharp isotopic discontinuities in the lowest 5 m of the core. Figure 4.20 shows jumps of $\delta^{18}O$ value exceeding $5^0/_{00}$ in 50 mm or less on the 1972 core, at heights above bedrock around 2.0, 2.3, 2.65, 4.0, 4.1 and 4.3 m, and one on the 1973 core at 2.7 m. If we assume that whatever the cause of these sharp jumps a steady-state equilibrium has been reached

Figure 3.20. The continuous curves show the diffusion at $-10\,^\circ C$ of a slab of ice 5 cm thick and of isotopic δ value $-40^0/_{00}$ placed within a static ice mass of δ value $-29^0/_{00}$ after stated time intervals. The vertical bars show the measured $\delta^{18}O$ values over 1 cm intervals across an isotopic 'spike' 25 m above bedrock in the Camp Century ice core shown in Figure 4.10.

between isotopic diffusion and vertical strain rate, we can then make a rough estimate of the latter from Figure 3.18. In practice we estimate diffusion lengths from the isotopic profile, and by assuming these to be those for the steady state (L_{ss}), the strain rate is given by $\{\dot{\epsilon}_z\} = 2D_v/(L_{ss})^2$. Diffusion lengths, which can only be estimated roughly from the isotopic profiles, range from 1 mm on the lower side of the spike at 2.3 m on core 72, through 10 mm at 2.7 m (core 73) to 25 mm on the lower side of the spike at 2.0 m (core 72) and on the upper side of the spike at 4.3 m (core 72).

Strain rates derived from the above model are plotted on Figure 3.21 against height above bedrock. In view of the approximate estimates involved a considerable scatter is to be expected, but most results cluster around a mean value of $2 \times 10^{-5} \, a^{-1}$ at 4 m and $10^{-5} \, a^{-1}$ at 2 m, which suggest that our model is appropriate and these are approximate strain rates.

We note in Figure 3.21 that two points (2.0 and 2.3 m) lie outside the main group. It appears improbable that a strain rate of $220 \times 10^{-5} \, a^{-1}$ on the lower side of the spike at 2.3 m could be sandwiched between points with strain rates that are two orders of magnitude less. We therefore interpret this point in terms of time of diffusion in static ice as given in equation (3.10), which suggests a time of diffusion of around 500 a.

Our resolution of δ values with depth is so coarse that it is fortunate that the cut in the ice core that separated adjacent samples at this point must have been very close to an interface. While we conclude that the interface between the two isotopic masses has formed within the last thousand years, it is also possible that active shear is still present at this interface. A similar conclusion but with poorer time resolution applies to the sharp changes of δ value on the upper side of the spike at 2.0 m, where a somewhat higher estimate of

Figure 3.21. Strain rates necessary to maintain the observed isotopic gradients in the lowest 5 m of the ice cores from Devon Island.

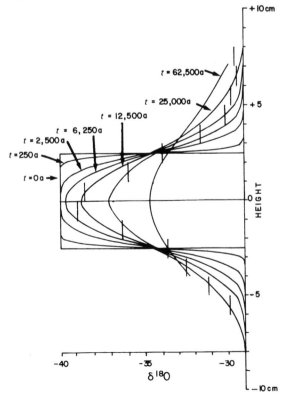

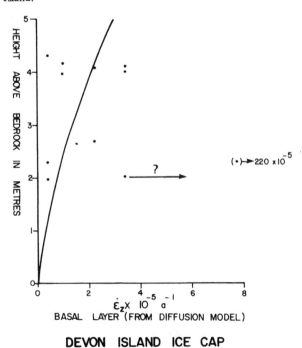

DEVON ISLAND ICE CAP

diffusion length could be due to sampling across an equally sharp interface.

Another possibility is that the two sharp interfaces were formed at the same time by injection of some 0.2 m of isotopically warmer ice into the centre of one broader spike of colder ice at some time during the past 1000 a, to form the two spikes at 2.0 and 2.3 m. Our suggested age is also consistent with the extreme negative $\delta^{18}O$ value on the 2.0 m spike. This is almost the same value as that of the lower side of the 2.3 m spike. Had the event occurred much earlier, we should have expected diffusion to have decreased the $\delta^{18}O$ shift on such a narrow spike according to Figure 3.19.

Paterson (1976) has measured both vertical strain rates $\dot{\epsilon}_z$ and shear strain rates τ_{xz} in the 1972 borehole. His mean vertical strain rate over the lowest 50 m is around $50 \times 10^{-5} a^{-1}$, and it varies relatively slowly above this level to a maximum value around $130 \times 10^{-5} a^{-1}$ around 100 m depth (200 m above bedrock). The horizontal τ_{xy} is also low over most of the borehole, reaching a maximum of $3.4 \times 10^{-3} a^{-1}$. However, this increased greatly to around $10^{-2} a^{-1}$ over the bottom 5 m.

The large shear strain rate that was observed can only be explained satisfactorily by the presence of active shear surfaces in the lowest layers of ice of the type that we have deduced from the isotopic evidence. However, it is difficult to estimate the vertical spacing and frequency of such shear surfaces since they are only obvious when they result in bringing together ice masses with different $\delta^{18}O$ composition. We suggest that a mean vertical spacing between 0.1–1.0 m at any time would appear consistent with the isotopic and strain rate data. We also note that only one or two out of ten isotopic jumps suggest possible present activity on shear planes, which indicates that active shearing along any given surface lasts only for a limited time.

3.5

Techniques for measuring temperatures in glaciers and ice sheets

W.S.B.PATERSON

Instruments
Thermistors

Most temperature measurements in ice are made with thermistors. They are cheap and robust and, with care, can measure to a precision of about 0.01 °C. A thermistor is a semi-conductor with a high negative temperature coefficient of resistance, typically about 5 per cent per °C. Temperature is determined by measuring the resistance; the thermistor is calibrated before, and if possible after, the field season. Thermistors with a resistance of at least several thousand ohms should be used so that small fluctuations in the resistance of the cable have a negligible effect on the measured resistance. For field use, the thermistors should be sealed in pressure-tight containers which should be as small as possible to minimize their thermal time constant.

Resistance can be measured with a standard Wheatstone Bridge; five decades are needed for precise measurements. The measuring current produces some heating in the thermistor and so must be kept very small. Thus the voltage on the bridge should not exceed 0.5 V or even less, according to thermistor resistance and desired accuracy. A very sensitive null detector is therefore needed; battery-operated solid-state models, robust enough for field use, are now readily available.

Quartz thermometers

A quartz thermometer consists of an electronic oscillator controlled by a quartz crystal whose resonant frequency varies with temperature. Some instruments have a digital read-out. Quartz thermometers have precision comparable to thermistors (0.01 °C) and are more convenient; on the other hand, they cost more.

Methods of measurement

Various types of drill are used in glaciers: drills that take core and those that do not, mechanical drills and electrically heated thermal drills, thermal drills that remove the meltwater from the hole and those that do not. Different types of borehole require different methods of temperature measurement.

Dry hole

In this type of hole, as made for example by the CRREL (US Army Cold Regions Research and Engineering Laboratory) thermal coring drill (Ueda & Garfield, 1969), the meltwater is removed from the hole as it is produced. Temperature can be measured by a single thermistor, lowered to different depths. A two-wire cable is sufficient; however, four wires are preferable, with two joined together at the bottom so that changes in lead resistance can be monitored. Because, at any depth, the air in the hole can be assumed to have the same temperature as the ice, to have the thermistor in contact with the ice is not essential. If it is, however, less time is required for each measurement because the thermistor approaches equilibrium temperature more rapidly. Bird (1976) has described a device, consisting of three lightweight springs, to keep the temperature sensor in contact with the ice.

Fluid-filled hole

This may be a dry hole subsequently filled with a fluid such as diesel oil to equalize the hydrostatic pressure in the hole and thus prevent closure by ice flow. Alternatively, it may be a hole drilled mechanically, in which case drilling fluid is needed to remove the ice chips. Temperatures in such a hole can be measured with a single thermistor, as in a dry hole, except that the thermistor must be in a pressure-tight case and that any joins in the cable must be leak-proof.

Permanent cable installation

Some thermal drills take neither core nor meltwater from the hole. Such drills were originally designed for use in ice at the melting point but have been used successfully in ice at temperatures down to about $-8\,^{\circ}$C (Classen & Clarke, 1972). The meltwater refreezes above the drill, but sufficient heat is dissipated in the power cable to prevent its freezing in. To make temperature measurements in such a hole, a multicore cable, with thermistors spaced at intervals along it, is attached to the power cable and goes into the ice with it. On completion of drilling, the drill, power cable, and thermistor cable are allowed to freeze in. Because, with this system the thermistors cannot be recalibrated after the measurements, it may be desirable to have two thermistors at each depth as a check. A single wire can serve as return lead for all the thermistors. However, if it is broken by shear in the ice, as sometimes happens, no measurements are obtained. Thus it is preferable to use separate wires at each depth. Classen & Clarke (1972) have described how to prepare thermistor cables.

An alternative technique employs a hot-water drill; water is heated to $15\,^{\circ}$C at the surface and fed down a hose (Gillet, 1975). On completion of drilling, the drill and hose are withdrawn and a thermistor cable installed immediately afterwards and allowed to freeze in.

The Philberth probe

In this device (Philberth, 1962; Aamot, 1968) a thermal drill and thermistor are combined in a single unit. During drilling, the wires freeze in above the drill but additional wire uncoils from inside the unit as it descends. To make a temperature measurement, drilling is stopped for a few days and readings are made periodically. The equilibrium temperature is found by extrapolation of a graph of temperature against time. During the period of temperature measurement the probe freezes in. However, once the power is switched on again, sufficient heat is generated along the sides of the unit to melt it free. Temperatures down to a depth of 1000 m in the Greenland ice sheet have been measured in this way (Philberth, 1970).

Measurement of core temperature

It may be possible to determine borehole temperature by measuring the temperature of a core immediately after it is brought to the surface. Paterson (unpublished) tried this on the Devon Island ice cap where the ice temperature is about $-20\,^{\circ}$C. The core was obtained with a thermal drill; a thermistor was inserted in a small hole drilled into the centre of the core. Measured temperatures were several degrees above the borehole temperature. The method might be worth trying, however, in ice where the temperature is near the melting point, or in cores taken with a mechanical drill.

Sources of error

Sources of error can be grouped into two categories: first, the thermistor reading may not be the temperature of the ice on the wall of the borehole and, second, borehole temperature may not be the steady-state temperature of the surrounding ice. There are more than one source of error in each category.

Thermistor drift

'Drift' of a thermistor is change of calibration with time. It occurs in only a small percentage of thermistors. The chance of it is reduced if the thermistor is first cycled several times through a wide range of temperature. To check for drift, thermistors should be recalibrated after use whenever possible.

Self-heating of thermistor

The heating produced by passing different amounts of current through a thermistor can be measured in the laboratory. Field measurements can then be corrected. Laboratory and field conditions may differ, however, and so it is preferable to make the field measurements at a current low enough for self-heating to be negligible.

Thermistor not at ice temperature

After a thermistor is lowered to a new position in a borehole, it takes time to attain the new temperature. The time varies with the temperature gradient and the depth interval between measurements. The best method is to make a measurement as soon as the thermistor has been lowered to the new position and to repeat the measurements every few minutes thereafter. The equilibrium temperature can be obtained by extrapolation of a graph of temperature against time. Bird (1976), using

a quartz thermometer sensor in contact with the borehole wall, found that measurements over a total period
of 15 min were adequate. Paterson (unpublished), using
a thermistor not necessarily touching the ice, makes
measurements every 5 or 10 min for about an hour.

Perturbation by drilling

The ice is warmed by drilling, particularly by
a thermal drill, and takes some time to return to its
equilibrium temperature. Paterson (1968) measured
temperatures 10 days, one year and two years after
drilling a borehole in ice at about $-16\,^{\circ}\text{C}$ by thermal
drill with removal of meltwater. The one-year and two-
year measurements did not differ significantly from each
other and were presumed to represent the equilibrium
value. The ten-day measurements were too high by
amounts between $0.03-0.12\,^{\circ}\text{C}$. Jarvis & Clarke (1974)
derived theoretical cooling curves for a borehole in
which the meltwater was allowed to refreeze. For ice at
about $-5\,^{\circ}\text{C}$, temperatures measured ten days after the
end of drilling were predicted to be within $0.2\,^{\circ}\text{C}$ of the
equilibrium value. These figures are merely examples;
the time needed to obtain equilibrium will vary according to the ice temperature and whether the meltwater is
allowed to refreeze in the hole. To make sure that equilibrium values are obtained, it seems best to make the
measurements one year after drilling.

Convection in borehole

Convection is expected in sections of a borehole
where the temperature gradient exceeds a certain critical
value which can be calculated by a formula given by
Krige (1939). Paterson (unpublished) has made calculations for air-filled and diesel-filled boreholes of diameter
160 mm, as made by the CRREL thermal coring drill. He
concluded that an air-filled hole would be stable but that
convection was likely in an oil-filled one. Paterson
observed a decrease in temperature gradient near the
bottom of a borehole in the Devon Island ice cap, one
year after it had been filled with diesel oil. This could
indicate convection. Temperatures measured in fluid-
filled holes should therefore be regarded with some
suspicion until this matter has been further investigated.
Because one term in the formula for the critical gradient
is inversely proportional to the fourth power of the hole
diameter, to reduce the diameter is an effective way of
reducing the chance of convection.

Air flow in firn

The upper 50–100 m of boreholes in polar ice
sheets are in firn which is permeable to air. An upward
current of air can sometimes be felt at the top of such
a borehole; the air is believed to enter through the firn,
driven by pressure differences at the surface. To see if
such air flow had any significant effect on borehole
temperatures, Paterson (unpublished) measured temperatures on different days at depths between 15 and 50 m
in a borehole in the Devon Island ice cap. Day-to-day
differences did not exceed $0.01\,^{\circ}\text{C}$ which was comparable with the error of measurement and therefore not
significant.

3.6

Accumulation rate measurements
on cold polar glaciers

C. LORIUS

Methods

Determination of the depth of particular layers whose age is known

Stake measurements. The oldest and simplest
method of measuring snow accumulation is to read the
emergence of pole markers at different dates. Measurements are very accurate, especially when it is possible to
take into account the settling of the reference snow
layer – although this is often negligible ($5\,\text{mm}\,\text{m}^{-1}\,\text{a}^{-1}$
at the South Pole station according to Giovinetto &
Schwerdtfeger, 1966). Measurements are usually based
on networks of stakes (up to 100) which are distributed
over areas of the order of a few square kilometres in
order to reduce errors caused by the areal variability of
snow accumulation; errors are also reduced by increasing
the length of the observation period. However, this is
only practicable at occupied stations or at sites where
repeated visits are made. Available data from such
studies do not go back beyond the last 20 years. (For
further references, see Bull, 1971.)

Radioactivity measurements. The fallout of
artificial radionuclides released by nuclear bombs formed
very well-marked layers in early 1953 (Greenland) and
early 1955 (Antarctica). References, and a detailed
description of the method, are given by Picciotto,
Crozaz & De Breuck (1971). The β activity due to
natural radionuclides present in the snow layers is much
smaller than the β activity of artificial fission products,
thus enabling us to identify these marked horizons by
measuring gross β activity in a continuous series of firn
samples of a few hundred grams taken down the ice core
to below the depth of the radioactive layer. The method
has been used in recent years at a large number of
stations and has led to very reliable measurements in all
areas of cold polar glaciers; accuracy is good since it
depends mainly on the increment chosen for cutting
core samples. Additional reference layers, such as those
of 1964–5 and 1970–71 have been used in Antarctica
(Crozaz, 1969; Clausen & Dansgaard, 1977; Lambert
et al., 1977). Determinations of selected radionuclides,

such as strontium 90 and tritium, although more complicated to perform, may be useful for other applications. Artificial radioactive fallout, when measured in detail at favourable sites, shows a clear seasonal pattern connected with stratosphere–troposphere exchanges; this has been observed particularly at a number of Greenland stations (Merlivat, Ravoire, Vergnaud & Lorius, 1973).

Volcanic layers. In Greenland, reference layers with higher electrical conductivity due to fallout of soluble volcanic debris have been identified. This method could be applied in the vicinity of active volcanoes when the chronology of the main eruptions is known. In Greenland all great northern hemisphere volcanic eruptions, at least since 1783, resulted in layers deposited shortly after the eruptions showing elevated specific electrical conductivity (Hammer, 1977; Hammer, 1980; Hammer, Clausen & Dansgaard, 1980). This is mainly due to high concentrations of H_2SO_4, the sulphate originating from gaseous sulphur components released from the volcano and oxidized in the atmosphere prior to wash-out. These layers can also be detected from chemical analysis. It has been shown that the eruption in 1963 of Mount Agung in Bali ($8°$ S $115°$ E) was recorded in the Antarctic snow layers in the form of higher sulphate concentrations (Delmas & Boutron, 1978).

Radio-echo layering. In section 3.11 conductivity changes in ice layers caused by volcanic fallout are considered to be the most likely cause of the deep internal reflections shown in radio-echo profiles of polar ice sheets. If certain layers on such records can be tied to specific dates, or a general chronology of layers established, this would be of great value in assessing the variation of long-term accumulation rates over the ice sheet.

Observation of periodical phenomena
Stratigraphy. Seasonal changes in meteorological conditions allow us to determine the annual snow deposit by observing changes in grain size, density, hardness, or the occurrence of particular stratigraphic layers such as radiation crust, melt features, or depth hoar horizons (for details of techniques see Kotlyakov, 1961; Benson, 1962; Langway, 1970). This method gives good results in areas of moderate snow balance and climatic conditions. However, interpretation of stratigraphic data is more difficult in the extreme conditions which prevail over large areas of the Antarctic plateau, where summer temperatures are too low to form clearly differentiated seasonal layers, and where a low rate of snow accumulation, together at times with erosion by the wind, may cause a gap in the time sequence of layers. This blurring or elimination of seasonal boundaries may lead us to overestimate the rate of snow accumulation by an unknown factor (Koerner, 1971). Interpretation of stratigraphy is more difficult when one has to use cores instead of pits, and the difficulty also increases with depth because features induced by seasonal changes are smoothed out to a varying

extent during diagenesis. In good conditions, however, it is possible to get continuous annual stratigraphy down to ice (Gow, 1968; Langway, 1970).

Stable isotope variations. In high accumulation areas, deuterium and oxygen-18 concentrations of the firn layers show marked seasonal oscillations which are caused by changing conditions in the precipitating air masses (chapter 1). As shown in papers by Dansgaard *et al.* (1973) and Hammer *et al.* (1978), this method has been applied with success at a number of Greenland stations. Examples of results from Greenland are shown in Figure 1.7 and Figure 4.12. A sampling frequency of around eight samples per year is needed for satisfactory interpretation of ice cores. However, results at some Antarctic stations have been unreliable, and the method is unsatisfactory for detecting annual layering in low accumulation areas such as Vostok Station (Dansgaard, Barkov & Splettstoesser, 1977). Smoothing of isotopic layering in polar firn by diffusion of water molecules explains both the disappearance of annual layering in low accumulation areas and the reduced seasonal amplitude of δ values at depth compared with the top layers (see section 3.4). These processes linked with density cease by a depth of 10–30 m and layering is then preserved until diffusion in solid ice becomes significant at great depths.

Due to the above factors, there was a tendency in earlier studies to overestimate accumulation rates in parts of Antarctica (Epstein, Sharp & Goddard, 1963; Picciotto *et al.*, 1968; Lorius *et al.*, 1970). However, in favourable conditions the method is reliable and provides outstanding records of past precipitation with so few ambiguities that dating of a continuous record back to 548 AD has an accuracy of ±3 years (Figure 4.12, and Hammer *et al.*, 1978). This accuracy can only be obtained on a continuous record, but when the ice core is incomplete due to losses in drilling, sample values along the core provide an estimated age accuracy of ±3 per cent down to 1000 m at Camp Century (Hammer *et al.*, 1978, and Figure 1.9).

To convert these layer thicknesses to an accumulation rate record involves correction for (1) density variations, (2) accumulation rate deviations upstream and (3) total vertical strain since the time of deposition (Reeh *et al.*, 1978). The density corrections are only significant at shallow depths, while (2) is not significant near an ice divide or ice summit. Total vertical strain (3) can be taken from measurement of present-day surface strains on the assumption that they have been constant, which appears reasonable near an ice crest, or corrections can be calculated on the assumption of uniform vertical strain and the fractional depth of the present layer in relation to the ice thickness at its point of deposition along lines indicated by equations (1.7) and (1.15). Although at great depths the errors introduced by such assumptions increase rapidly, at moderate depths errors are small. Reeh *et al.* (1978) obtained secular trends of accumulation rate at three sites in central Greenland (Figure 3.22). The long-term linear trends are -4 ± 2 per cent per millenium for Crête

Station close to the ice divide using surface strain measurements, 0.0 ± 3.5 per cent per millenium for Milcent on the western slope of the ice sheet using vertical strain rates calculated along similar lines to those discussed above, and +2 ± 6 per cent per millenium for Dye 3 to the east of the main ice divide where still less basic data is available to estimate the three corrections. Shorter period trends are shown by putting the basic data through 30-year and 120-year filters. There is considerable agreement between trends at Crête and Milcent over much of the later record, but less with Dye 3. This is attributed to changes in the circulation pattern, since precipitation reaches Milcent from the south west but at Dye 3 it comes from the south east, whereas Crête probably receives precipitation from both directions.

Microparticle studies. Periodicity in the micro-particle content of polar firn and ice layers has been observed by Thompson (1973) and others. While studies on the Byrd ice core suggested layer thicknesses at depths that are greater than those expected from glaciological modelling (Thompson, Hamilton & Bull, 1975), at Camp Century the observations reported in Hammer *et al.* (1978) agree with and confirm isotopic studies of annual layering. The earlier work done with Coulter counters required larger sampling volumes to provide

Figure 3.22. Accumulation rate records from three Greenland ice-sheet stations, determined mainly from seasonal δ¹⁸O cycles. The annual accumulation data series have been smoothed by digital low-pass filters with cut-off periods of 120 a (heavy curves) and 30 a (thin curves). (Reeh *et al.*, 1978, Fig. 1. Reproduced from the *Journal of Glaciology* by permission of the International Glaciological Society.)

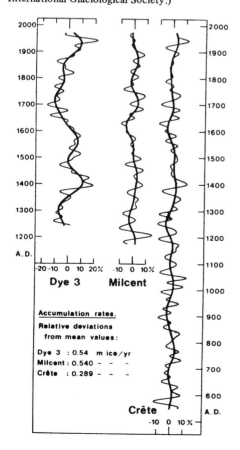

meltwater, as opposed to the light scattering techniques used by Hammer *et al.* (1978). In Figure 1.9 we have seen that dust layering has been used to extend the layer thickness curve towards bedrock at levels where diffusion in solid ice has obliterated the isotopic layering. The method appears promising for low accumulation areas in polar regions, since dust particles will not be subject to diffusion processes in firn or ice to an extent comparable with that of water molecules. However, until now no convincing results have been obtained from low accumulation areas.

Other impurities. Chemical analysis can provide information on annual layering similar to that of dust particles. This has been shown from measurements both in Greenland (Langway, Klouda, Herron & Cragin, 1977) for such elements as sodium, magnesium, calcium, potassium and aluminium, and in Antarctica on shallow samples (Boutron & Lorius, 1979) for marine origin impurities (sodium and magnesium) and also at greater depths by Herron & Langway (1979b). However, due to the very low concentrations, measurements are generally difficult to perform and are time consuming. It is hoped that recent technical improvements will overcome this problem.

Techniques for measuring the decrease in concentration of natural radioactive isotopes
The decrease with time of the activity of various radionuclides has been used for dating ice at depth. The long-lived isotopes ³²Si (Clausen, 1973) and ¹⁴C (Oeschger *et al.*, 1966) offer very immediate and promising prospects. Since they have not been applied to the measuring of accumulation rates, these methods, which require large quantities of ice, will not be discussed here, although the development of accelerator ion counting appears very promising and of potential use for other radioisotopes.

Tritium: Natural ³H, which has a 12.3-year period and is produced in the atmosphere by cosmic radiation, has been used in a limited number of cases (Aegerter, Oeschger, Renaud & Schumacher, 1969). However, owing to nuclear tests which have caused tritium content to increase by several orders of magnitude, this method requires very careful procedures for sampling and measurement.

Lead 210: ²¹⁰Pb (half-life = 20 years) is produced in the atmosphere by the decay of ²²²Rn and is present in minute amounts in atmospheric precipitation; it has been used for dating firn layers back approximately 100 years (Picciotto *et al.*, 1971). Both the techniques in use, β activity of ²¹⁰Bi and α activity of ²¹⁰Po, require low-level counting equipment and elaborate radio-chemical separations on samples generally weighing a few kilograms. Extreme precautions must be taken to avoid possible contamination; sampling is preferably made on a continuous section, each sample covering several years to average out variations in ²¹⁰Pb fallout. This method also relies upon the assumption that the specific activity of ²¹⁰Pb and the rate of accumulation

remain constant; these conditions seem to be fulfilled
in both polar ice sheets since, with proper sampling
procedures, a linear decrease of the activity logarithm
was observed at a number of stations when plotted
against depth (in water equivalent). The accuracy is of
the order of ±10 per cent. However, some recent results
of Sanak & Lambert (1977) raise some questions about
the above assumptions.

Indirect methods

The rate of snow accumulation can also be derived
by calculation from various indirect measurements. For
example, vertical strain rate measurements have allowed
Cameron (1971) to get reliable accumulation data at
Byrd Station; Kohnen (1971) and Bentley (1975) have
proposed methods to determine the long-term accumulation rate from the temperature and the depths of the
seismic horizons.

In studies reported in chapter 5, observed ice
temperature profiles have been matched with profiles
computed for different accumulation and surface
warming rates. The accumulation values giving the best
fit are in good agreement with currently observed
accumulation rates for Camp Century, Byrd and Vostok
Stations, although the calculated best-fit values refer to
a time span of at least 10^3 years.

Accuracy and representative nature of the measurements

Comparative accuracy of the methods

If we are to make comparative studies of
accumulation data, these data must first be converted
by using depth–density profiles. The error depends
mainly upon density values which are generally measured
with an accuracy of a few per cent.

The above discussions show that there are essentially two methods which are objective and enable us to
make precise determinations of the rate of snow accumulation over both the polar ice sheets: available stake
measurements cover the last 25 years at a few sites,
while the same period is covered by determination of
artificial radioactive fallout layers, a method which can
be applied all over cold polar glaciers. These measurements can be made with an accuracy of the order of
a few per cent.

The ^{210}Pb method allows determinations over
a 100-year range but requires some additional precautions. Stratigraphic and stable isotope measurements
involve a more subjective interpretation and do not give
reliable results in extreme conditions, particularly for
large parts of Antarctica where the accumulation is very
low. However, under favourable conditions, or when
properly checked against results from other methods,
they may enable us to considerably extend the time
range investigated.

These considerations are illustrated by results
obtained at two stations characterized by different
accumulation rates (Table 3.3).

Representative nature of measurements

Temporal and areal variations both contribute to
the variability of snow accumulation in a given area.

Areal variability. Large-scale features of the
accumulation distribution are controlled by meteorological parameters and orographic factors, which explain
for instance the general decrease of accumulation from
the coast (up to $900\,\mathrm{kg\,m^{-2}\,a^{-1}}$) to the high Antarctic
plateau (down to less than $30\,\mathrm{kg\,m^{-2}\,a^{-1}}$). The local
areal variability was estimated at $30\,\mathrm{kg\,m^{-2}\,a^{-1}}$ from
data on stake networks covering a few square kilometres
and stratigraphic observations at 24 Antarctic stations,
the mean accumulation for the entire continent (excluding the Antarctic Peninsula) being $150\,\mathrm{kg\,m^{-2}\,a^{-1}}$
(Giovinetto, 1964).

From numerous data obtained by various authors
using the same observation techniques, Bull (1971)
concluded that over distances of up to 5 km the areal
variability could easily be as much as 20 per cent.
Variations of accumulation have in particular been
observed due to the combined action of wind and surface
undulations; mass balance is generally greater when the
slope is less. Whillans (1975) measured such variations
upstream of Byrd Station; over part of his network
(100 km) the accumulation rate fluctuated by about
10 per cent around the mean value ($160\,\mathrm{kg\,m^{-2}\,a^{-1}}$),
apparently caused by slope changes as little as one part
in a thousand over a distance of a few kilometres. At
Byrd Station, surface depressions accumulate 30–50 per
cent more snow than the exposed crests (Gow, de
Blander, Crozaz & Picciotto, 1972). Approaching the ice
divide, the accumulation increases continuously over the
last 40 km, up to $240\,\mathrm{kg\,m^{-2}\,a^{-1}}$, while the slope
decreases moderately. This latter effect may be due in
part to atmospheric phenomena other than the interaction of slope and deposition of drift snow. This shows
possible limitations in areal extrapolation of accumulation data.

In east Antarctica extensive stake records of
accumulation from the coast to Plateau Station via
Mizuho Plateau were reported by Yamada, Okukira,
Yohoyama & Watanabe (1978) for the years between
1968 and 1974. Above the ablation zone, conditions
from around 550–1700 m show similar variability to
that of the preceding paragraph. From 1700–3500 m
variability of annual accumulation is much greater, with
many of the stakes (2 km apart) recording net ablation
rather than net accumulation in individual years. From
3500 m to Plateau Station (3673 m) variability was
smaller and no ablation areas were present even though
the accumulation was very small. Radok & Lile (1977),
in analysing measurements of R. Dingle in 1967–8,
found the annual accumulation at Plateau Station to be
$0.028\,\mathrm{m\,a^{-1}}$ of ice, a similar rate to that at Vostok
Station. Of the precipitation at Plateau Station, different analyses show that from 70–87 per cent fell as
'diamond dust' ice crystals under clear sky conditions,
and only 13–30 per cent came from clouds.

Temporal variability. On the basis of Antarctic
stake measurements, Bull (1971) concluded that over
five-year periods temporal variability could greatly
exceed 20 per cent; for short periods at the South Pole,
Giovinetto & Schwerdtfeger (1966) give a value of 15 per
cent, which compares favourably with the 25 per cent

estimated as a maximum for Antarctic stations (Giovinetto, 1964).

Accumulation measurements on a stratigraphic section or a core include both the effect of local temporal and local areal variability. Topographic features, such as sastrugi or undulations several kilometres across, may vary in space with time. Consideration of ^{210}Pb results over the past 100 years, using average values over 10-year intervals, suggests that accumulation over this period has remained constant and close to values obtained from gross β measurements. Bull (1971) summarized most of the long series of accumulation data from stratigraphic studies in Antarctica: no systematic trend has been observed except at the South Pole Station (Giovinetto & Schwerdtfeger, 1966) where accumulation appears to have increased over three successive 66-year periods starting from 1760 (respectively 54, 68, 75 kg m^{-2} a^{-1}). Results from Greenland (Figure 3.22) show somewhat smaller variations, of 11, 5 and 4 per cent from the long-term trends at Dye 3, Milcent

and Crête respectively, after periods shorter than 120 years have been filtered out. However, at Dye 3, and possibly at Milcent, Reeh (personal communication) considers it possible that some of the variability may be due to areal variability related to surface undulations and to temporal effects of different strain histories at different levels resulting from flow over bottom undulations.

Layer thicknesses shown in Figure 1.9 fit the basic model of strain rate used in section 5.2 and indicate that, at Camp Century, mean accumulation rates have not varied to any great extent (say ±10 per cent) over the past eight thousand years. Prior to that, layer thicknesses at Camp Century are deduced from dust content and the results are more scattered than from isotopic layering immediately above. Since the ice in these layers has probably come from further inland on a much thicker ice sheet (see section 5.3), vertical strain corrections are difficult to apply.

Robin (1977) has shown that, in Antarctica, the

Table 3.3. *Comparison of accumulation measurements by different methods*

(a) Pole of Relative Inaccessibility (Antarctica)

Time interval	Method	Accumulation rate m a^{-1} water	m a^{-1} ice	Reference
1959–64	Stake and reference	<0.036	<0.039	Picciotto *et al.* (1968)
1965	Mark measurements	0.031 ± 0.005	0.034 ± 0.005	Picciotto *et al.* (1968)
c. 1945–65	Pit stratigraphy	0.12	<0.13	Picciotto *et al.* (1968)
–1965		0.029	0.032	Koerner (1971)
c. 1945–65	δ^{18}O ratios	≈0.09	≈0.10	Picciotto *et al.* (1968)
1955–65	Radioactive fallout horizon	0.030 ± 0.010	0.033 ± 0.0011	Picciotto *et al.* (1968)
c. 1886–65	^{210}Pb radioactive decay	0.031 ± 0.003	0.034 ± 0.003	Picciotto *et al.* (1968)
c. 1886–65	^{210}Pb ^{210}Bi	0.035 ± 0.005	0.038 ± 0.005	Picciotto *et al.* (1968)
c. 1886–65	^{210}Pb ^{210}Po	0.029 ± 0.003	0.032 ± 0.003	Picciotto *et al.* (1968)

(b) Camp Century accumulation

Time interval	Method	Accumulation rate m a^{-1} water	m a^{-1} ice	Reference
*1964–47	Firn stratigraphy	0.312	0.342	Crozaz, Langway & Picciotto (1966)
*1964–47	Fission products	0.312	0.342	Crozaz *et al.* (1966)
*1962–55	Firn stratigraphy	0.318 ± 0.057	0.349 ± 0.062	Mock (1965)
*1964–1800	^{210}Pb	0.32 ± 0.03	0.35 ± 0.03	Crozaz & Langway (1966)
*1965–1876	Firn stratigraphy	0.367	0.403	Mock (1968)
*1966–3	Trend surface	0.35	0.38	Mock (1968)
12 years prior to 1966	δ^{18}O isotope annual variations	0.32	0.35	Johnsen, Dansgaard, Clausen & Langway (1972)
6 years around 1850 adjusted for strain	δ^{18}O isotope annual variations	0.35	0.39	Johnsen *et al.* (1972)
0–10 000 a BP	Best fit to annual layering 0–1100 m	0.35	0.382	Hammer *et al.* (1978)

* After Mock (1968).

main factor governing accumulation rate variations over the ice sheet appears to be the temperature of the free atmosphere above the ground inversion layers (Figure 3.6). He suggests that during the last ice age this effect would cause accumulation rates to fall to two-thirds of present rates under otherwise similar conditions (see also Lorius, Merlivat, Jouzel & Pourchet, 1979). This effect has not been included in temperature profile calculations in chapter 5, but it may help to explain the remaining discrepancies between the observed and calculated temperature profile at Byrd Station (Figure 5.56). However, more observations and calculations are needed to settle this point.

Our broad conclusion is that the actual snow accumulation measurements made over a few decades can be considered valid for periods of 10^3–10^4 years over local areas of a polar ice sheet. Spatial extrapolation or interpolation must, however, be carried out with much care as the accumulation rate may vary in a somewhat irregular manner when larger areas are taken into consideration.

Figure 4.4 shows the general distribution of net accumulation over the Antarctic continent and indicates the density of observation in the different regions.

3.7

Ice movement

I.M.WHILLANS

Ice flow is important to most of the problems discussed in this monograph. Its effects are critical to the age–depth relationship for ice cores, and horizontal flow means that deeper ice originated at progressively greater distances from the core hole site. Because of this, the best interpretations of core hole results require an understanding of the ice flow leading to the core hole site and of the strain that the ice has undergone.

This section describes how ice motion is measured and, where measurements are not available, how estimates of this motion can be made. Then, using data from central west Antarctica, measured rates of motion are compared to the rates that would be inferred without those data, thus providing an indication of the accuracy of the movement estimates.

Then, to show how flowline data may be used, the data from central west Antarctica are used to test some concepts of ice sheet flow and to calculate the thickness balance and the age of ice at each depth. On the whole, the ice flow is consistent with our understanding and this part of the ice sheet is thinning slowly, at least in part, because of the recent climatic warming.

Measurements of absolute movement

Surface velocity is measured by the repeated determination of the position of a marker, such as a pole, in the glacier surface. For small valley glaciers this motion can be determined to a high accuracy using conventional survey techniques by reference to stationary points around the glacier. Motion determination is much more difficult for large ice sheets because often there is a lack of convenient stationary points, and most early measurements are in coastal or mountainous areas close to exposed bedrock (for example, Swithinbank, 1963; Budd, Landon-Smith & Wishart, 1967; Bauer, 1968; McLaren, 1968). A few survey networks have been carried long distances from exposed rock (Mock, 1963; Dorrer, 1970; Hofmann, 1974; Budd, 1977; Seckel, 1977; Naruse, 1978), but this is an expensive technique. Astronomical and aerial triangulation techniques have been tried (Chapman & Jones, 1970; Morgan, 1970), but slow glacier movement, inadequate precision, and technical problems with the aerial photography, lead to large uncertainties in the resulting velocities.

Sometimes satisfactory velocities can be obtained using the substrate under the glacier as a reference for velocity measurements. Where an open hole to the base is available, measurements of the progressive tilting of the hole provide the velocity at each depth (for example, the Camp Century, Greenland, velocity profile, Figure 4.7). On the Brunt Ice Shelf, Thomas (1973) was able to measure movement with respect to a distinctive magnetic anomaly associated with the rock beneath the sea. The same principle (Walford, 1972; Doake, 1975) has been used with radar reflections from the bottom: the pattern in radio-echo strength is believed to be fixed with respect to the base of the ice sheet if the bedrock interface is frozen, as in Doake (1975), and so the movement of the ice surface through this pattern is a measure of the motion.

It is usually more difficult to measure the vertical component of velocity by the techniques described above. Paterson (1976) was able to obtain a vertical velocity on Devon Island ice cap by measuring the displacement of notches in the wall of a borehole with respect to bedrock at the base of the hole. Walford, Holdorf & Oakberg (1977) are also developing a method using vertical changes of radio-echo strength as a reference for determining the vertical component of motion.

Starting about 1972, there have been many velocity measurements based on the use of 'Doppler' satellites. The United States Navy have placed six such Navy Navigation Satellite System satellites into orbit. These satelites broadcast time and identification information. A receiving station on the earth's surface can detect the Doppler frequency shift in transmitted signals as the satellites broadcast time and identification information. recordings of several satellite passes, the position of the receiving station is precisely determined in three dimensions. Repetition at a later date provides the movement of the station. Since the satellites are in polar orbits, the technique is particularly valuable in the high latitudes of the large ice masses where passes are frequent and in many directions. Position precision is about one metre in horizontal and vertical directions when about 30 suitable passes and just one receiver are used. For Byrd Station the satellite-determined movement is $12.7 \, \mathrm{m \, a^{-1}}$ (Chapman, personal communication), which compares very well with the $12.9 \, \mathrm{m \, a^{-1}}$ result that was measured with respect to the ice crest (Figures 3.23 and 3.24).

Measurements of relative movement

Absolute movement measurements on ice sheets and ice shelves are generally accurate to no better than one metre per year. Strain-rate calculations require much more precise relative movement measurements, and these are obtained by repeated surface surveys between poles set in the ice-sheet surface. Standard techniques are used for measuring angles by theodolite and distances by instruments that utilize radio-frequency

modulated radar wavelength, visible light, or near-infra-red rays.

One of the difficulties encountered in surface surveys is due to the surface relief of ice sheets. The horizon is usually only 1–10 km distant, and this distance must be the maximum length of the sides of the surveyed figures, although some limited advantage is obtained by surveying from elevated platforms. Especially troublesome is a near-surface air-temperature inversion that refracts light rays in a vertical sense. The errors due to the temperature inversion may be additive as the survey changes elevation.

The surveys of the Byrd Station Network (Figure 3.23) were made using theodolites and radar-wavelength distance-measuring equipment set 3 m above the snow surface. This technique provides relative horizontal positions accurate to about 0.05 m. The poles are 3 km apart and relative movements between neighbouring poles were about 1 m during the 3 years between surveys. Surveying precision was not sufficient to detect vertical motion.

In East Antarctica a comparable programme on the Mizuho Plateau measured horizontal strain rates and obtained elevation changes accurate to less than 0.9 m (standard deviation), which was sufficient to detect a major secular change in the ice sheet (Mae & Naruse, 1978; Naruse, 1978).

Inferring flowlines from elevation contours

It is impractical to measure very many velocities, and flowlines are often deduced from elevation contours. Because glaciers move due to gravity, their movement is in the direction of the mean maximum surface slope, or perpendicular to smoothed elevation contours. Locally, the slopes deviate because of ice flow over irregularities in the substrate, but the mean movement should follow the direction of the surface slope averaged over about twenty times the ice thickness (Budd, 1968, 1970*a*). For example, in Figure 3.24, the variations caused by slope changes over distances of about 3 km (equal to the ice thickness) affect the surface velocities by only $0.05 \, \mathrm{m \, a^{-1}}$, or by about 1%, and the thickness change at 70 km has no large effect on the surface velocity. Naruse (1978) obtained similar results for the Mizuho Plateau, East Antarctica.

Figure 3.23 shows flowline direction and elevation along the Byrd Station Strain Network as well as the elevation contours drawn by Behrendt, Wold & Dowling (1962) using data that were available before the construction of the Network. In this region, the data of Behrendt *et al.* were principally obtained by aircraft, from the barometric elevation of the aircraft minus the radar-determined height above the snow surface. Their estimated accuracy is ±100 m. Surface slopes along the Byrd Station Strain Network were measured directly by theodolite, and the reliability of the elevation of the ice crest, at 0 km, with respect to Byrd Station, is of the order of 10 m.

The flowline determined from the strain network data approximates the flowline that can be drawn from the contours of Behrendt *et al.*, but the 1900 m contour does not exist at that location, and instead surface elevations increase northwestward (downward in Figure 3.23) from the station at the ice crest. The contour lines are misplaced by as much as 100 m vertically or 50 km horizontally, and applying this uncertainty elsewhere suggests that flowlines estimated from barometric elevation data may be 10° or more in error.

The principle of aerial mapping has been extended to cover most of the ice sheets using aircraft, balloons and a satellite. The air-borne radio-echo sounding programmes in Antarctica and Greenland have provided very valuable maps of ice sheet surface elevation (Drewry, 1975*b*; Overgaard & Gudmandsen, 1978; Rose, 1979). The elevations in Antarctica are internally consistent to 50 m or less. Radio altimetry from drifting meteorological (TWERLE) balloons has provided elevations to a precision of about 100 m (Levanon & Bentley, 1979). The precision of these techniques is mainly limited by secular air-pressure variations and uncertainties in the geographic position of the sounder. Profiling by satellite in southern Greenland shows repeatability to 2 m (Brooks *et al.* 1978). A repeatability of better than ±1 m was obtained from radio altimeter profiling by satellite over Antarctica according to Brooks (1982).

These mapping programmes are very important, and we are learning a great deal about ice flow patterns. The limit to helpful precision is about 2 m, because that is the amplitude of many of the small-scale perturbations to surface slope. Very small systematic errors arise from the references used for elevations. In Antarctica, for example, air pressures are low in the centre (Schwerdtfeger, 1970) so that uncorrected barometric elevations are about 100 m too large. Similarly, the satellite profiling refers the elevations to a standard describing the shape of the earth; these elevations may not be the same as those that would be determined by more traditional techniques that refer elevations to the surface gravity field. These problems lead to very small errors in elevation, and slopes and flowlines deduced from the data are little affected.

Continuity

As explained above, flow directions can be obtained from measurements of surface slope. The speed of ice flow can be obtained using these flowlines, measurements of accumulation rate, and an assumption about the thickness balance of the glacier. These quantities are related by the equation of continuity (Whillans, 1977),

$$B = A - M - \nabla(\bar{\rho} Z U) \qquad (3.12)$$

The thickness balance, B, is a measure of the net mass gain or loss of a stationary vertical column through the ice mass. Often called simply the mass balance, it is equal to the accumulation rate at the upper surface, A, less the bottom melting rate, M (or plus the basal freezing rate), less the horizontal gradient in mass transport due to flow. Z is ice thickness, $\bar{\rho}$ is the mean ice density, U is the density-weighted mean ice velocity, and ∇ is the operator

Figure 3.23. Elevation and ice velocity along the Byrd Station Strain Network, and 100 m elevation contours from Behrendt, Wold & Dowling (1962) and 5 m for the short contours (BSSN).

Distance scale is in kilometres from the pole with the highest elevation. For clarity, velocities are plotted beside the BSSN, instead of in it.

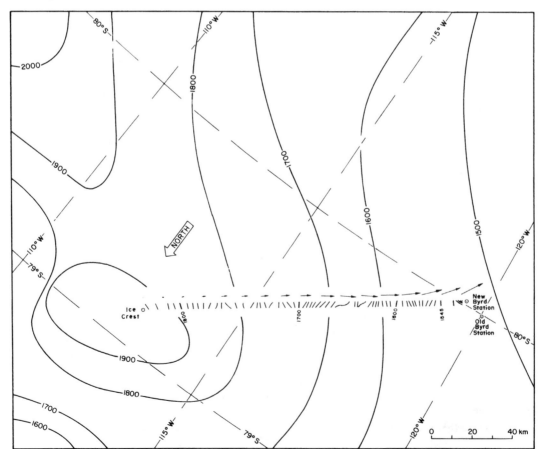

for the two horizontal (x, y) directions

$$\nabla = 1_x \frac{\partial}{\partial x} + 1_y \frac{\partial}{\partial y}$$

1_x and 1_y being unit vectors.

In the absence of any other indication, the thickness balance, B, is often assumed to be zero, and equation (3.12) is used to calculate the balance velocity (the velocity for supposed zero balance). Multiplying equation (3.12) by the distances, Y, between two arbitrary flowlines gives

$$0 = (A - M)Y - Y \frac{\partial}{\partial x}(\bar{\rho}ZU_x) - Y \frac{\partial}{\partial y}(\bar{\rho}ZU_y)$$

Placing Y inside the first differentiation, and constraining the x-axis to follow the flowline $(U_y = 0)$ gives

$$0 = (A - M)Y - \frac{\partial}{\partial x} Y\bar{\rho}ZU_x + \bar{\rho}ZU_x \frac{\partial Y}{\partial x}$$
$$- \rho ZY \frac{\partial U_y}{\partial y}$$

Because the velocity vectors are aligned along flowlines, $U_x \partial Y/\partial x =$ the rate of spreading $= Y\partial U_y/\partial y$, and the last two terms cancel. The equation is now integrated from a place where the velocity is known, such as at an ice divide, where $U_x = 0$:

$$U_x = \frac{1}{Y\bar{\rho}Z} \int_{\text{divide}}^{x} (A - M)Y \, dx$$

($B = 0$, x-axis along flowline, U_x is the balance velocity).

As is to be expected, the mass flux, $U_x Y\bar{\rho}Z$, through a vertical section (of area YZ) is equal to the

snow accumulation (less basal melting) integrated upglacier.

Values for bottom melting (M) or freezing $(-M)$ rates can be calculated using heat balance considerations, and they are usually negligibly small. Flowlines, and their spacing are obtained from elevation contours, as discussed earlier, and so the balance velocity can be calculated for any site where the thickness, Z, and upglacier accumulation rates, A, are known.

The balance velocity is compared, in Figure 3.24(a), to measured velocities for the flowline leading to Byrd Station. The balance velocity is an estimate of the average velocity through the ice thickness, which is expected to be smaller than the surface velocity. According to the measured borehole tilting at Byrd Station, and its extrapolation to the bed (thin, dashed line in Figure 3.25) the difference between mean and surface velocities is about 12%, or 1.6 m a^{-1} at Byrd Station. The measured surface velocities are larger than the balance velocities by more than this amount.

The first remarkable feature in Figure 3.24(a) is, however, that the balance and measured velocities vary smoothly and together and that they are similar in nature. This supports some implicit assumptions in our calculation of balance velocity: at least almost all of the ice is participating in flow and there is no major cross-flow at depth. This is important because some cross-flow was measured at Byrd Station (thin curve at left in Figure 3.25) and there had been concern that ice near the bed may be nearly stagnant. It seems that these potential difficulties are not critical, at least near Byrd Station.

Figure 3.24. Quantities plotted against distance from the ice crest along the BSSN. Byrd Station is at 162 km. (a) Measured surface velocity and balance velocity. Dashed lines indicate that the BSSN axis and the flowline deviate and so quantities cannot be compared with confidence. The balance velocity is defined somewhat differently than in Whillans (1977). (b) Maximum surface slope. (c) Ice thickness in metres as measured by radar (courtesy of the Scott Polar Research Institute).

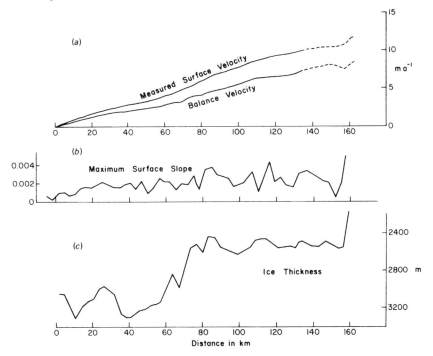

The second remarkable feature in the balance velocity–measured velocity comparison (Figure 3.24(*a*)) is that the balance velocities are only 15–20% smaller than the measured surface velocities corrected for internal shear. As is discussed below, the discrepancy is because the ice sheet is thinning slowly ($B < 0$).

For most places on the ice sheets, however, uncertainties caused by the low density and questionable accuracy of accumulation rate measurements (Bull, 1971; section 3.6) lead to much larger uncertainties about the reliability of calculated balance velocities. At the South Pole, Bentley's (1971) balance velocity of 6 m a^{-1} compares well with the measured value of about 8.5 m a^{-1} (USGS, 1976).

The situation on ice shelves where flowlines are not defined by surface contours is more difficult. On the Ross Ice Shelf some of Robin's (1975*a*) early estimates of movement are as much as a factor of three smaller than subsequently measured values (Thomas & Bentley, 1978*a*). The largest proportional errors were due to incorrect prediction of flowlines in slowly moving areas, but the predicted velocity of the largest mass influx from Byrd Land from ice stream 'B' (where flow direction was inferred from thickness isopachs) of around 500 m a^{-1} compared well with the measured speed of $\approx$500 m a^{-1} and with the measured direction. In general, on ice shelves a lack of knowledge of the thickness balance, primarily because of lack of knowledge of basal melting or freezing, makes estimates of the speed of movement inaccurate, although in many cases the direction of movement can be predicted from the geometry of the confining boundaries. Relict features on radio-echo sounding records on large ice shelves, such as old bottom crevasses and non-reflecting zones that appear on successive radio-echo profiles can also be used to define the direction of flowlines (Neal, 1979) as can some large-scale linear features that show up clearly on some satellite imagery (Crabtree & Doake, 1980).

Analysis of the ice flow near Byrd Station

The data collected along the Byrd Station Strain Network (BSSN) and the Byrd Station borehole tilting measurements provide the only detailed information we have on ice movement in the vicinity of a deep core hole in an ice sheet. The data have also been used in the analyses of section 5.5. They were used above to test some ideas on flow at depth, and have also been used to calculate the thickness balance of this part of the ice sheet, and to address the cause and significance of the internal radio-reflecting layers.

The thickness balance

The thickness balance has been calculated using the equation of continuity (equation 3.12) and this shows that the ice sheet is thinning slowly (Whillans, 1977). The magnitude of the calculated thinning rate is affected by the way in which measured surface velocities are corrected to represent mean velocities. Perhaps the simplest interpretation of velocity at depth is represented by the thin, dashed curve of Figure 3.25, which indicates a thinning rate of 37 mm a^{-1}. If the accumulation rate were 17% larger, the ice sheet would be in balance. A more extreme interpretation of velocity change at depth is represented by the dotted line of Figure 3.25. That interpretation leads to a thinning rate of 30 mm a^{-1}. The glacier all along the flowline leading to Byrd Station seems to be thinning at about the same rate.

These calculations indicate that although the ice sheet has a significantly negative thickness balance, it is not dramatically growing or collapsing. Confirmation of this, and insight into the cause of the internal radio-reflecting layers is afforded by a comparison of calculated isochrones with the shapes of the radar layering (Whillans, 1976).

Isochrones are buried surfaces of the ice sheet, and their depths can be calculated using two of the following sets of data: accumulation rate (burial rate),

Figure 3.25. Horizontal velocity against depth at Byrd Station from the measured surface velocity and borehole tilting data (Garfield & Ueda, 1976). Heavy line shows maximum measured shearing. Thin lines are velocity components in the directions 220° and 310° azimuth. Data stop at 1482 m. Two possible extrapolations of the profile to deeper depths are shown (see text).

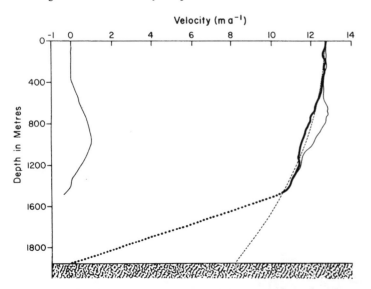

the velocity field in the glacier, and ice thicknesses. Figure 3.26 shows isochrones calculated using a steady-state ice-flow model (zero thickness balance) and the first two data sets but not ice thickness. The close similarity between the shapes of the isochrones and the radar layering supports the suppositions that the thickness balance has not been very different from zero for some tens of thousands of years and that the radar layering represents isochrones.

Limits to the use of the radio-layering to test the balance of the ice sheet arise because a uniform thickening or thinning produces layering that looks steady-state. Figure 3.26 shows isochrones calculated using measured strain rates and also a strain rate that is 15% slower for the 18 000 a isochrone. Both sets of isochrones follow the layering. In order that this test be helpful, there must be a horizontal gradient in accumulation rate, or ice thickness, or both.

In the case of the Byrd Station flowline, accumulation rates are substantially larger near the ice crest ($x = -10$ to $+40$ km in the figures) than elsewhere. Isochrones are buried more rapidly there and the relative depths of isochrones are greater, as are the radar layers. If there had been a major change in ice-sheet configuration there would have been a change in the pattern of accumulation rate associated with a change in the position of the ice divide, and a consequent anomaly in the radar layering.

Thus, this part of the ice sheet is thinning slowly and uniformly. Four possible causes for this come to mind: sea-level change, a change in the accumulation rate, an internal instability, and a temperature change at the surface.

The effect of a change in surface temperature is most easily addressed because there is information on the magnitude and timing of climatic temperature changes and it is possible to calculate the first effects on the ice sheet. There was a large increase in the isotopic ratio of deposited snow about 14 000 a ago, implying a surface temperature warming of about 8 °C (Figure 5.55). As this Holocene warmth penetrated into the ice sheet, the ice sheared more readily and flowed faster. The glacier is therefore flowing more rapidly than it did and ice flux by flow is now larger than replenishment by snowfall and the ice sheet is thinning (Whillans, 1978, 1981). The calculated rate of thinning by this model is about one-third that measured.

Other causes can contribute to thinning. For these, however, there is either no good independent evidence for the event, or a lack of adequate theory to describe how the ice sheet responds. Internal instability in the ice sheet has long been suggested, but there is no record of such an instability having occurred. Similarly, there is no good evidence of a change in accumulation rate, and, in fact, accumulation rates have been nearly constant for the last few centuries (Gow, 1968). Robin (1977) has inferred that accumulation rates may have increased at the end of the Wisconsinan glaciation, but that would have caused ice sheet thickening that is now complete (Raynaud & Whillans, 1982). Sea level certainly did change, and this could lead to thinning in the interior (Thomas, 1979), but it is not known how long it takes for such a change to affect ice flow in the interior.

Thus, the interpretation for the thinning is that part of the thinning is due to the Holocene warmth, and the cause of the rest of the thinning is not known.

Figure 3.26. Vertical section along the flowline leading to Byrd Station (162 km). Heavy lines indicate the location of four radar-reflecting horizons in the ice sheet. Thin lines are isochrones calculated from surface strain rate and accumulation rate data, and assumed zero-valued thickness balance. Ages in years are marked. The dashed line is the 18 000 a isochrone for strain rates 15% slower. Near Byrd Station the BSSN axis and the flowline deviate and the calculations are not reliable. These results differ from those in Whillans (1976) because here allowance has been made for a depth-varying strain rate.

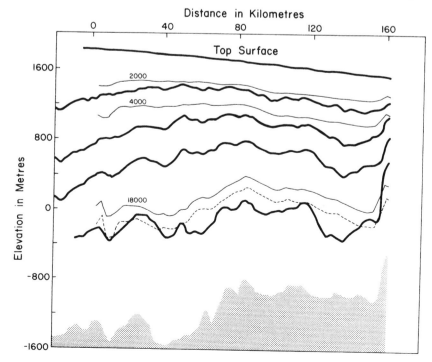

Ice flow model

Once the thickness balance is known, and a model developed for the cause of the imbalance, it is possible to make a reasonably accurate numerical model for the ice flow. For the BSSN, measured strain rates and velocities along the surface are used, and these are adjusted to describe the variation with depth and time. Because the flow at depth is in nearly the same direction as that at the surface (Figure 3.25), theory indicates that the strain rates must vary with depth in the same way as does the horizontal velocity. Before the start of the Holocene warming, the ice was colder at depth, and so strain rates were smaller in magnitude than present-day values. Horizontal velocities at depth are available only

at Byrd Station and a simple model is used that describes the effects of ice crystal orientation fabric and temperature as they change along the flowline. The details of the model are described in Whillans (1979).

Figure 3.27 shows the results of the flowline calculation. The calculation moves with a vertical column of ice as it travels along the flowline in time, so the left–right axis is both horizontal distance (bottom scale) and time (upper scale). The particle path originating at the ice divide (0 km) represents the calculated position of the substrate, and it has been made to conform, approximately, with the position of the substrate measured by radar. Starting at about 2000 a ago, the ice sheet has thinned by about 200 m.

Figure 3.27. Flowline calculation for the BSSN. Dashed lines are particle paths. Calculations follow a column of ice as it moves down the flowline in time (upper scale) and distance (lower scale). The surface elevation is adjusted so that the calculated position of the bed (deepest particle path) agrees with that determined by radar. The ice sheet has thinned by about 200 m.

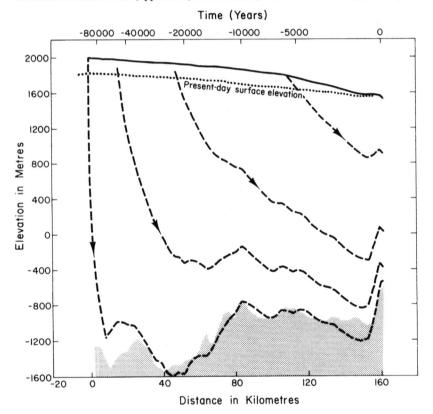

Figure 3.28. Depth–age scale for the Byrd Station deep ice core from the model of Figure 3.27. The scale becomes progressively more speculative, especially for ice deeper than 1400 m. (Figure 5.50 gives similar results from different computation.)

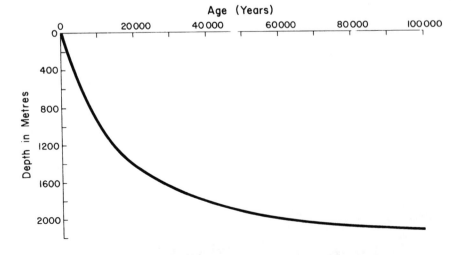

No mass balance changes for, during or preceding the Wisconsinan Stage are included in the model. This is probably unrealistic, but there are no good data at present on the nature of any such changes. The calculated time scale (Figure 3.28) for older than about 20 000 a is therefore very speculative, although ages may be expected to increase very rapidly with depth. The time scale for shallower and younger depths is dependent on the correctness of the concepts used in modelling the flow and on the correctness of the model describing the thickness imbalance, or thinning.

This model was published in 1979 and the Holocene warming effect is no longer considered sufficient to explain all the observed thinning (Whillans, 1981), and there is some evidence for a small thickening between 14 000 and about 5000 a ago (Raynaud & Whillans, 1982). The main features of the model of Figures 3.27 and 3.28 are, however, not affected.

The analysis of the BSSN provides an example of what can be learned using data from the length of a flowline leading to a deep core hole. The data have been used to test whether the ice flow is indeed simple, as is usually supposed, to calculate the thickness balance, and to assess the cause of the imbalance. The understanding, so gained, has then been applied to model the flow pattern with depth and time, and to calculate the time scale for the Byrd Station ice core. Work is continuing on these problems, but the potential for interpretation of the flow along the BSSN is limited because the Byrd Station ice core has not been reliably dated by other methods. If the core were independently dated, much better tests of the validity of the flow concepts could have been made.

3.8

Ice depths, including depths along flowlines

G. de Q. ROBIN

Measurements of depth of ice cores and boreholes

These are straightforward linear measurements, usually by some form of steel tape or its equivalent. Methods are so simple and direct that little has been written (Ueda & Garfield, 1969).

When using a rotary drilling rig, depth is measured by the number of drilling pipes. With coring devices lowered by wire, the normal method is to mark a fixed interval on the drilling tower of from 2–10 m, depending on the height of the tower, then to bind and mark a piece of tape on the wire for each standard length as the drilling head is lowered. This gives an accuracy of from 0.5–0.1 per cent, and has the advantage that the wire is under tension when distance marks are applied.

Another method, used sometimes for independent wires such as those used for temperature measurements in the top few hundred metres, is to mark distances on the wire laid out on the surface. Unless the wire is under tension when measured, this may introduce a subsequent error approaching 1.0 per cent in magnitude.

In general, direct measurements of depth have errors that are small compared to errors in related parameters.

However, unless inclinometer measurements of tilt of the boreholes have been made, there may be a significant difference between the vertical depth below the surface and the depth along the borehole. In the Camp Century borehole the difference between vertical and slant depth at bedrock due to borehole tilt was only 0.02 per cent (0.3 m) five years after the hole was drilled. In the Byrd borehole the difference was 1.1 per cent after drilling, the slant depth being 23 m greater than vertical depth.

Incremental depth measurements on borehole samples

In selecting samples along an ice core, the core is normally laid against a distance scale and samples are cut to an accuracy of better than 1 cm for positioning relative to the scale. For detailed relative measurements of stratigraphy, this is satisfactory. The major uncertainty that arises is due to incompleteness of the core recovery. Material is most likely to be lost at the top and bottom of each core sample, due to factors depending on the drilling technique, such as shattering of material with a mechanical drill, or melting off the top of a core when using thermal coring. The figure for completeness of core recovery, and the length of individual cores, give a general guide to possible gaps in the core. Thus a core recovery of 95 per cent, with individual cores of about 3 m long, means that on average 15 cm out of each 3 m core is missing. Unless the core is badly shattered, the major gaps will be at the end of each sample. Within one individual core, errors may be of the order of 1 cm, or 1 per cent, due to uncertainty introduced by cracking and other effects caused by removal of stress from the core when it is brought up from great depths.

Geophysical measurements of ice depth

For modelling studies such as those of chapter 5, we need to know the variation of ice depth along a flowline as well as the depth at any drilling site. The bulk of this information has been provided by radio-echo sounding, seismic reflection shooting and gravity measurements. The latter two provide spot measurements while radio-echo sounding normally provides continuous profiles. By running a network of soundings over a region, such as the 50 km network over western Marie Byrd Land (Rose, 1979), one can estimate possible depth variations along any probable flowlines upstream of a borehole. In the case of the Byrd Station Strain Network, surface markers made it possible to fly along the strain network which lies approximately along the flowline (Figure 3.23).

Geophysical techniques of ice depth sounding are well known and a brief summary only is presented here.

Seismic shooting

We need only mention reflection shooting methods for ice depth surveys. The method is described in all standard geophysical text-books (e.g. Dobrin, 1976). A general review of seismic and gravity soundings as used during the International Geophysical Year and the decade following is given by Bentley (1964).

Depth measurements by seismic reflection shooting are obtained by determining the time taken for compressional elastic waves from a small explosion to travel to bedrock and back to the surface where they are detected by seismometers. If clear echoes are present, the main limit to accuracy comes from errors in estimating the velocity of compressional (P) waves through the ice mass. These velocities vary with the ice density, temperature and fabric. Uncertainty over variations of crystal orientation with depth at different locations causes most uncertainty in depth determinations (Bentley, 1972). Gross errors due to misidentification of echoes on noisy records obtained in central Antarctica have also been made, but these errors have now been generally identified (Drewry, 1975c). Apart from gross errors, seismic depth measurements should be accurate to within 3 per cent at most locations.

Gravity measurements

Variation of gravity with location along an oversnow traverse will be dominated by the effects of changing ice depths, but some variations due to changing rock type will remain. Furthermore, a reading by gravimeter indicates the mean ice depth over a zone of radius comparable to the ice depth, which may be up to 3 or 4 km in central parts of the polar ice sheets. A gravity variation of one part in a million (1 milligal) will be produced by a changing ice thickness of 14 m underlain by rock of density around 2.8 g cm^{-3}, provided we are dealing with horizontal layers of ice and rock of large extent compared with the ice depth. In practice bedrock is rough, and comparisons with seismic measurements in west Antarctica are scattered about a mean variation of 20 m per milligal change of gravity according to Bentley (1964). A part of this scatter is due to comparison of an areal average thickness obtained by gravimeter with that along a seismometer line a few hundred metres long at most, so the comparison is not accurate. In practice, gravity values have been controlled by seismic depth measurements at intervals of not more than 50 km along a traverse.

The advantage of gravity measurements is that they are made rapidly and hence in greater number than seismic shots. They serve to provide detail along a traverse and show if the seismic depth at a particular location is typical for that area. Ice thicknesses derived from such measurements appear accurate to within 5 per cent if controlled by seismic soundings or 8 per cent if not controlled according to the analysis of Antarctic results by Drewry (1975c).

Radio-echo determinations of depth

This system is being used increasingly for measuring thickness of cold polar glaciers and ice sheets. A continuous profile of bedrock can be obtained from a moving vehicle or aircraft (see Figures 3.34–3.36), in much the same way as marine echo sounding is carried out from a ship. A general description of the equipment is given by Evans & Smith (1969), while application of the technique to the Antarctic ice sheet is reviewed by Robin (1972, 1975b). A more detailed account of the principles involved was presented by Robin, Evans & Bailey (1969).

Provided a reasonably continuous bedrock echo is obtained, the reliability is high. Proportional variations of the velocity of radio waves in ice with density, temperature and perhaps crystal fabric appear around half those for the seismic method. An accuracy of 2 per cent for depth measurements by radio-echo sounding is claimed. Careful checks between seismic measurements, borehole data and radio-echo soundings showed differences of <1.5 per cent at the Camp Century and Byrd Station boreholes (Drewry 1975c).

Comparison of different methods

Some comparisons between depths measured by seismic shooting, gravity, radio-echo sounding and borehole drilling have been mentioned above. In comparing published figures, navigational errors may also contribute to the difference between measurements of different groups. Robin *et al.* (1969) compared radio-echo depths with seismic soundings made within 200 m in north west Greenland during the same traverse. These show a maximum difference of 3 per cent between seismic shooting, radio, and borehole depths. When comparing airborne radio-echo soundings with seismic and gravity from surface traverses in east Antarctica, Drewry (1975c) found agreement between radio-echo and values from seismic-gravity profiles to be within ± 3 per cent for one third of his 58 comparisons, another third agreed within ±10 per cent, while the remaining values probably involved gross reading errors of seismic echoes. Drewry's comparisons involve navigational errors that may reach a maximum of a few kilometres. Schaefer (1972) found the average difference of seven radio-echo surface soundings from seismic soundings of Robin (1958) in Dronning Maud Land made 20 years earlier to be about 3 per cent, the greatest difference being 5 per cent with some navigational error likely. Unpublished reports of occasional differences of up to 10 per cent between radio-echo and seismic data have been made, and, if true, it is possible that in certain situations reflections from different horizons in or below the basal ice are being detected by the two methods.

Representative nature of ice depths used in calculations

The only situation where the ice depth at a given point can be considered as typical of the surrounding area is on a flat ice shelf with no undulations or crevasses in the vicinity. Figure 3.24 shows depth variations upstream of the borehole at Byrd Station. This is typical of a moderately rough bedrock.

However, if one is estimating depths along a flowline, unless one has a great deal of information on flow vectors, the major errors in estimating ice depths upstream of the borehole are most likely to be due to inability to fix the flowline, rather than to errors of depth measurement. Comparison of the earlier flowline studies upstream of Byrd Station presented by Budd, Jenssen & Radok (1971a) with the more detailed observations of Whillans reported in section 3.7 illustrate this point. In the earlier case, ice thicknesses used in calculation ranged up to 4200 m, compared with a maximum of 3200 m along Whillans' flowline. The problems of identifying a flowline are discussed in section 3.7.

The preceding paragraph applies to conditions in the ice sheet at the present day. However, in studying ice cores from Camp Century we need to know the condition of the ice sheet over the past 100 000 a. It will be seen from discussions in section 2.5 that there are major uncertainties to be resolved over past variations of size of the ice sheet in north west Greenland, and similar problems exist elsewhere. In section 2.4 (Table 2.2) we see that such errors are greatest near the margins and least in the interiors of ice sheets.

3.9

Total gas content

D. RAYNAUD

The idea that the total gas content of polar ice, V, could be an indicator of the surface elevation at the time of ice formation (Lorius, Raynaud & Dollé, 1968) has been applied to several polar cores (Budd & Morgan, 1977; Raynaud & Lorius, 1977; Raynaud, Lorius, Budd & Young, 1979; Raynaud & Whillans, 1982). Recently there have been important improvements to this method. The aim of this contribution is to discuss the measurement of V and to describe to what extent V is an indicator of the elevation of the ice formation site.

Total gas content measurements
Preparation of samples

To be representative of the gas volume trapped during ice formation the sample must neither have lost its original gas nor trapped extra gas during or after the recovery of the ice. Consequently the preparation of samples requires several precautions.

Boring can cause superficial melting (in the case of thermal drilling) and fracturing of the core. Contamination can also occur by the trapping of gas at the core surface after the recovery of the ice. Consequently, the sample must be cut from the internal part of the core and carefully inspected to detect any possible fractures. The study of thin sections from ice in the immediate vicinity of the sample makes it possible to check for fracturing and to obtain complementary information such as the abundance, size, shape and distribution of inclusions. It is useful to extract the gas soon after the sample has been prepared so as to avoid trapping gas at the ice surface.

Another possible source of error is the gas loss from the dissected bubbles at the boundary of the sample. In the case of an ice cube with a homogeneous distribution of bubbles, a calculation shows that the percentage of the gas volume thus lost depends mainly on the mean diameter of bubbles (Table 3.4). This diameter is typically 0.1–0.9 mm for polar ice.

Total gas extraction and measurement
Gas extraction by melting in kerosene. The procedure has been described in detail by Langway (1958).

The gas is extracted by melting the sample under kerosene in a vessel topped by a gas burette. As the ice melts, some of the gas initially entrapped in the ice rises and displaces the kerosene in the burette, and some gas remains dissolved in the melt water. The total gas volume is given by the sum of the gas collected in the burette and the gas dissolved in the melt water.

In order to perform accurate measurements, certain precautions are needed.

(1) Before beginning, the kerosene must be saturated with water and gas. If possible, the same batch of kerosene should be used throughout the measurement series, in order to avoid problems due to inhomogeneities of this liquid.

(2) The sample must first be agitated in the kerosene to expel any adhering air pockets.

If these precautions have been taken, the accuracy of the method depends mainly on the reading error of the burette and the error in the amount of air dissolved in the melt water. Langway (1958) gives a reading error of 0.5% and a maximum error of 5% for the amount of air dissolved. To obtain the total gas content (in $cm^3 g^{-1}$ of ice) one must know the weight of the sample. Generally the weighing error is small compared to the other uncertainties.

Gas extraction under vacuum. By this method the sample is normally placed under vacuum and melted, and the gas phase is extracted while the water vapour is trapped (Scholander, Hemmingsen, Coachman & Nutt, 1961; Matsuo & Miyake, 1966; Raynaud, 1976; Raynaud & Delmas, 1977; Stauffer & Berner, 1978). Coachman, Hemmingsen & Scholander (1956) extracted the gas by shaving the ice under vacuum and at cold temperature, which is technically a more difficult procedure. To perform accurate measurements one must obtain complete extraction of the gas phase without losing or producing gas in trapping the water vapour. An alternative means of ensuring a complete extraction of the gas phase is to sublime the ice in vacuum (Raynaud, 1976). Now at the Laboratoire de Glaciologie in Grenoble we extract the gas by melting the ice and refreezing the melt water under vacuum. The procedure is designed to get an effective gas extraction and to reduce any modification

Table 3.4. *Percentage of gas lost from the dissected bubbles at the boundary of the sample*

Mean diameter of bubbles (mm)	Sample size (g)	Percentage of gas lost (%)
0.1	50	0.4
	20	0.6
0.5	50	2.0
	20	2.7
0.9	50	3.5
	20	4.7

of the initial gas content in trapping the water vapour (Lebel, 1978; Raynaud, Delmas, Ascensio & Legrand, 1982). This method enables us to analyse a great number of samples with good relative precision (Raynaud & Lebel, 1979).

In order to measure the total gas content, the most accurate way is to collect the gas extracted by a Toepler pump in a McLeod gauge or a gas burette.

In the case of complete extraction, where no modification in the dry gas volume occurs in trapping the water vapour, the main sources of error in the determination of the total gas content (reduced to 1 g of ice) are:

(1) the amount of ice sublimated when the sample is placed under vacuum;

(2) the reading error of the McLeod gauge or gas burette;

(3) the weighing error of the sample.

Thus for a typical sample of 20 g, using a burette with an accuracy of $0.01\ cm^3$ and a balance with an accuracy of 0.001 g, we obtain from experimental tests the following errors: for (1) ±0.5%, for (2) ±0.5%, and the weighing error is negligible.

Comparison of different methods. Gas extraction by melting under kerosene is easier to perform, particularly in the field. The equipment is, furthermore, inexpensive and a great number of measurements can be made on a low budget. On the other hand, gas extraction under vacuum is a more accurate method for individual measurements, and the gas can be used for composition analysis. Such analysis can provide, among other types of data, important information linked with the representative nature of the total gas content, such as the occurrence of melting or presence of a gas phase associated with the ice matrix (and not with the air bubbles) when polar ice forms.

Representative nature of total gas content as an indicator of ice surface altitude
Parameters relevant to the total gas content at the ice formation site

In the 'dry snow' zones of polar glaciers the gas volume, V, reduced to STP and 1 g of ice, which is trapped mechanically in ice as the firn pores close off from the atmosphere can theoretically be obtained by the application of the Boyle–Mariotte law to each pore. Then, if n pores are involved

$$V = \frac{\theta_0}{P_0} \sum_1^n v_c \frac{P_c}{\theta_c} \qquad (3.13)$$

where θ_0 and P_0 are the standard temperature (273.16 K) and pressure, v_c the pore volume at close-off, and P_c and θ_c the pressure and the temperature (K) of the gas at close-off. The composition of the total gas enclosed in the polar ice indicates that part of the CO_2 measured is due to the decomposition of contaminating carbonate dust or linked to the solid matrix (Raynaud & Delmas, 1977; Stauffer & Berner, 1978; Raynaud *et al.*, 1982). This CO_2 excess could lead to an overestimation by a few tenths of 1% of the amount of gas mechanically trapped at the pore close-off, which is too small to affect total gas content measurements. As indicated by equation (3.13), V depends theoretically on the abundance of bubbles formed during the transformation of firn to ice and on v_c, P_c and θ_c. The latter two parameters depend on the elevation of ice formation site and on climatic conditions.

Temperature and pressure of the gas occlusion at the pore close-off. θ_c and P_c are mainly influenced by the temperature and atmospheric pressure at the snow surface but they depend also on the depth at which the pores close off. Consequently, the variations in these two parameters from one pore to another of the same ice sample are linked to the depth and time ranges which are involved in the close-off process. Very few data on the distribution of the closed pores in firn are available. Nevertheless, near the surface the densification of snow is mainly the result of grain packing until the critical porosity is reached (Anderson & Benson, 1963) and in these upper layers the concentration of closed pores can be assumed to be very small. Consequently, the close-off process would occur between the critical level (generally found in the 10–30 m depth range) and the firn–ice transition, which varies in polar ice sheets from about 50 m near the coast to 100–150 m in the central parts. Furthermore, if we take into account the temperature dependence of the time needed to reach the firn-ice transition (Gow, 1975), the time interval required to reach the close-off level of all the pores varies with the firn temperature (and consequently with the site) from less than 100 years at $-20\,^\circ C$ to several thousands of years at $-60\,^\circ C$.

θ_c is the firn temperature at the close-off depth. The firn temperature observed below the critical level is generally close to the mean annual temperature at the surface and shows small temperature–depth gradients which vary, in East Antarctica, from $+0.01\,^\circ C\,m^{-1}$ in the central parts to $-0.02\,^\circ C\,m^{-1}$ near the coast (Budd, Jenssen & Radok, 1971a). Thus, at one given site, if the mean annual temperature is constant, the maximum variation of θ_c from one pore to another would be $1.5\,^\circ C$.

P_c is the pressure of the air filling the firn pore as the latter close off from the atmosphere. Consequently, P_c depends on the atmospheric pressure at the surface corrected for the close-off depth. At present, the relative variation of the mean monthly atmospheric pressure measured in several representative stations of Antarctica (Schwerdtfeger, 1970) increases with the altitude and is less than $\pm 2\%$. Although the long-term variations are unknown, charts derived from a model for the development of atmospheric circulation (Lamb, 1972) suggest a maximum variation for the mean sea-level atmospheric pressure of about 15 mb over northern Greenland between 20 000 BC and the present day. These variations in the atmospheric pressure at the surface are more or less smoothed in the firn according to whether they are short or long term. Furthermore, at a given formation site, the relative variation of P_c from one pore to another, linked

to the depth range in which the close-off process occurs, increases theoretically with the altitude at the surface to reach a maximum value of about $\pm 1\%$.

In conclusion, for a given ice sample, if we assume that the mean annual temperature at the ice formation site is constant and that the time needed to reach the close-off of all the bubbles is more than one year, then the maximum relative standard deviation of the ratio P_c/θ_c, calculated from the range of the P_c and θ_c variations from one pore to another, is less than 1.5% of the mean P_c/θ_c value (Raynaud, 1976), and equation (3.13) can be approximated by

$$V = \frac{\theta_0}{P_0}\frac{P_c}{\theta_c}V_c \qquad (3.14)$$

in which $V_c = \Sigma_1^n v_c$. The relative error introduced, by assuming that P_c/θ_c is constant, is less than 1.5%.

Volume of pores at close-off. Equation (3.14) indicates that V depends on V_c, the total volume of the pores as they close off. The question is to know whether and to what extent V_c depends on the firnification parameters such as accumulation rate and temperature. Recently Raynaud & Lebel (1979) have measured the total gas content in ice formed under a wide range of temperature and elevation (see the following section). Using these measurements and estimates of P_c and θ_c, they calculated the corresponding mean pore volumes at close-off, $(\bar V_c)$ for each formation site by use of equation (3.14). Results indicate no significant correlation between $\bar V_c$ and accumulation rate but a consistent linear decrease of V_c with decreasing θ_c (see Figure 3.29(b)) at a rate of $7.4 \times 10^{-4}\,cm^3\,g^{-1}$ of ice per $^\circ C$ in the range -12 to $-53\,^\circ C$ (Raynaud & Lebel, 1979).

Geographical variations of total gas content

Three sets of representative measurements are available to determine to what extent V depends on the present-day conditions prevailing at the ice formation site. None of the results below have been corrected for gas loss by dissected bubbles.

Figure 3.29. (a) Mean total gas content ($\bar V$) of Antarctic and Greenland ice versus the elevation of the formation site. Standard deviations of the measured mean values of V are shown.
(b) Mean derived pore volumes at close off ($\bar V_c$) versus temperature at the formation site.
(From Raynaud & Lebel, 1979, Fig. 1.)

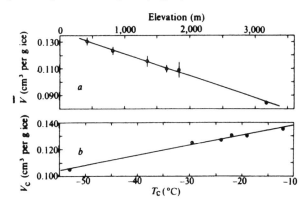

(a) The results presented by Budd & Morgan (1977) can be considered as experimental evidence on the present-day dependence of V on elevation of the ice formation site in the area of the Law Dome (Wilkes Land, Antarctica). The mean total gas content measured on ice samples taken near the firn–ice transition in several locations decreases clearly as the elevation increases. The rate of this decrease, given by the authors, is about 1.5×10^{-3} cm^3 g^{-1} of ice per 100 m.

(b) From measurements of V performed on samples from stations Milcent and Crête (Greenland), Miller (1978) reports a reduction in mean total gas content of 0.014 cm^3 g^{-1} of ice corresponding to a 727 m increase in the elevation of the ice formation site. This change corresponds to a rate of 1.9×10^{-3} cm^3 g^{-1} of ice per 100 m.

(c) 142 measurements of V have been recently carried out in the Laboratoire de Glaciologie (Grenoble) on ice sampled at six different drilling sites in Antarctica and Greenland. Estimated elevations and temperatures at the ice formation sites range respectively from 400–3200 m and from -12 to $-53\,^{\circ}$C. The results indicate (Raynaud & Lebel, 1979):

(1) the dispersion between individual V determinations for a given formation site is generally larger than experimental errors because of natural fluctuations of V_c, P_c and θ_c;

(2) the mean V values obtained for each formation site are proportional to the elevation of the site (Figure 3.29(a)); the mean total gas content decreases at a rate of about 1.7×10^{-3} cm^3 g^{-1} of ice per 100 m as the elevation increases, which is close to the gradients reported over smaller elevation ranges by Budd & Morgan and Miller;

(3) taking into account the variation of $\bar{V}_c$ with θ_c deduced from these results (see previous section) and equation (3.14), a general climatic cooling of $1\,^{\circ}$C at the ice formation site (P_c remaining constant) would imply a decrease in the mean total gas content of about 2×10^{-4} cm^3 g^{-1} of ice.

Conclusion

Our knowledge of the present-day geographical distribution of V indicates clearly that the total gas content of dry polar ice depends primarily on the elevation of the formation site and secondarily on the temperature. Thus an increase of only 100 m in the elevation would have the same effect on V as a general climatic cooling of 8.5 $^{\circ}$C. The total gas content of ice is, consequently, a very sensitive indicator of the elevation.

Since V depends on atmospheric pressure, the interpretation of total gas content in terms of past elevations of the ice formation site requires a knowledge of the sea level, the corresponding air pressure, and the air pressure–elevation gradient, during the past. Consequently, the interpretation becomes more difficult the older the ice, especially for periods beyond the Holocene.

Because the stable isotope composition of polar ice is sensitive both to elevation and climate changes, the comparison of isotopic and gas profiles measured along polar cores can be used, taking into account certain hypotheses, to obtain elevation variations of the formation site and climatic variations (section 5.3; Raynaud, Duval, Lebel & Lorius, 1979).

3.10

Debris and isotopic sequences in basal layers of polar ice sheets

G.S.BOULTON

Introduction

There are three processes which occur at the base of ice sheets and ice caps which may complicate the interpretation of isotopic profiles from these zones and which are also important in determining the evolution of subglacially derived sediments. They are:

(a) net mass loss by melting at the base of an ice sheet;

(b) net mass gain at the base of an ice sheet by net freezing to the glacier sole of subglacial water;

(c) complex modes of deformation in the basal ice in response to irregularities on the subglacial bed or inhomogeneities within the ice.

Their implications for the interpretation of isotopic profiles in ice sheets are:

Process (a)

melting of basal ice destroys the opportunity to sample what we expect to be the oldest ice;

Process (b)

the basal ice in a borehole may not have formed on the glacier surface thus preserving a record of surface conditions, but may have formed by basal freezing;

Process (c)

complex deformation patterns may locally invert or complicate the original depositional stratigraphy, thus producing an isotopic profile which does not faithfully reflect a sequence of sub-aerial events.

Prediction of the patterns of melting and refreezing at the base of an ice sheet and of zones where complex patterns of deformation might occur is an important prerequisite for any deep drilling. After cores have been obtained, it is important to have criteria which might help in identifying the results of these processes.

Patterns of melting and re-freezing at the bases of ice sheets

Several theories have been developed which are reasonably successful in predicting the internal tempera-

ture distribution in ice sheets at sites where they have been cored (Robin, 1955, 1976; Budd, Young & Austin, 1976; Jones, 1978). These allow us to calculate the distribution of internal and basal temperatures (e.g. Figure 3.30(*a*)). If the glacier sole is at the melting point, the rate of melting can be calculated. In the absence of melting/freezing, the basal temperature gradient $d\theta/dz$ will be

$$\frac{d\theta}{dz} = \gamma_G + \tau_b u/K \qquad (3.15)$$

where γ_G is the temperature gradient in the basal ice, τ_b the basal shear stress, u the basal ice velocity and K the thermal conductivity of ice. If a proportion c of the heat input is used for melting the ice, then the melting rate M is given by

$$M = -\frac{cK}{L\rho_i}\left(\frac{d\theta}{dz}\right)_b \qquad (3.16)$$

where L is the latent heat of ice, ρ_i is the density of ice and $(d\theta/dz)_b$ is the basal temperature gradient calculated from equation (3.15) but neglecting whether or not any melting/freezing is taking place. In a zone of melting beneath a two-dimensional steady-state ice sheet, the total rate of melting up-glacier of a point must be balanced by a subglacial flow of water past that point.

The production of water in a zone of melting will be

$$\int_{x_1}^{x_2} M\, dx \qquad (3.17)$$

where x_1 and x_2 are the up-glacier and down-glacier extremities respectively of the zone of melting.

Figure 3.30. (*a*) A prediction of the surface snow/ice facies, internal temperature distribution and basal thermal conditions in a transect through the western flank of the Greenland ice sheet at 70° N. The inputs to the model are the surface and bed profiles, the mean annual temperature and annual temperature amplitude at the snout, and the pattern of winter precipitation.

(*b*) A prediction of the areas and magnitude of basal melting and refreezing, the basal flowlines with respect to the bed, and the amount of ice lost and gained by melting and refreezing respectively. Regelation ice is expected to contain basally derived debris.

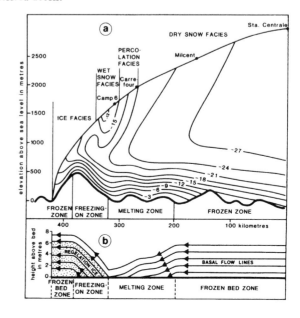

The basal meltwater discharge W, will be

$$W = \int_0^x (M - F)\, dx \qquad (3.18)$$

where F is the rate of freezing.

In many areas, as in Figure 3.30(a), the zone of melting does not extend to the edge of the ice sheet. It is followed first by a zone in which basal water, which is driven by the hydraulic gradient in the direction of ice flow, is refrozen to the base of the ice sheet releasing latent heat in the process. After sufficient distance, if the basal water supply has all been refrozen, the base of the glacier will again be frozen to the substratum. We thus have, in succession, a basal zone of melting, of net freezing-on, and of frozen substratum (Figure 3.30).

In the freezing zone, the rate of freezing (F) will be given by the equivalent of equation (3.16) for freezing. The thickness of the frozen-on layer (λ) will be governed by the rate of freezing and the ice velocity (u):

$$\lambda(x) = \int_{x_1}^{x_2} \frac{F}{u}\, dx \qquad (3.19)$$

Freezing-on of water at the base of the ice sheet does, however, change the temperature solution. In order to accommodate this added heat source, the following procedure is adopted (Williams & Boulton, in prep.). We consider individual segments of the glacier sole of length Δx, with W_i as the water input to each segment and W_0 the water discharge downstream from the same segment. There are five possible thermal conditions.

(1) If $W_i = 0$ and $\theta_b < \theta_m$
(where θ_b is the basal temperature, and θ_m the melting temperature)
then $M = 0$, and $F = 0$
and $-d\theta/dz = \gamma_G + \tau_b u/K$.

(2) If $W_i = 0$ and $\theta_b = \theta_m$

$$M = \frac{\gamma_G + \tau_b u + K(d\theta/dz)c}{L}$$

and $W_0 = M\Delta x$

(3) If $W_i > 0$, compute a trial θ_b^* with $F^* = 0$
and $-d\theta^*/dz = \gamma_G + \tau_b u/K$.
If $\theta_b^* > \theta_m$, $\theta_b = \theta_m$
and M is as in (2)
$W_0 = W_i + M\Delta x$

(4) If in (3) $\theta_b^* < \theta_m$, compute a second trial in which all of W freezes, so

$$\gamma_G + \tau_b u + \frac{LW}{\Delta x} = -K \frac{d\theta^{**}}{dz}$$

If the determined value of $d\theta^{**}/dz$ gives $\theta_b^{**} > \theta_m$, an amount $F\Delta x$ of the meltwater has frozen to bring θ_b to θ_m. Set $\theta_b = \theta_m$ and obtain $(d\theta/dz)_b c$.
Then $F = [-K(d\theta/dz)_b c - \gamma_G - \tau_b u]/L$
and $W_0 = W_i - F\Delta x$

(5) If in (4) $\theta_b^{**} < \theta_m$, then $\theta_b = \theta_b^{**}$,
$F = W_i$ and $W_0 = 0$.

This is incorporated in an iterative scheme to find a stable distribution of internal temperature and basal zones of melting and refreezing. Figure 3.30(a) is one such solution.

Figure 3.30(a) also includes another innovation. The sole inputs to this model are the surface and bedrock form of the ice sheet, mean annual temperature and temperature amplitude at the snout, the vertical atmospheric temperature gradient, and the pattern of winter precipitation over the ice sheet. From this data, an energy balance model is used to predict mass balance, snow facies, and 10 m temperatures in the ice sheet (Williams & Boulton, in prep.), the necessary conditions for calculation of thermal regime at the upper boundary of the ice sheet. A different modelling technique was used by Budd, Jenssen & Radok (1970) to predict a thickness of regulation ice of 7 m at Byrd Station in Antarctica compared with 5 m of debris-rich ice found in a borehole core from that site and presumed to be regulation ice.

Thus, from very limited information, it is possible to predict the areas in which basal ice will be lost by melting, the amount lost and the areas and thickness of basal regulation ice which will not yield an isotopic record of past atmospheric changes. These are necessary prerequisites for the choice of a suitable drilling site.

Origin of subglacially derived debris in polar ice sheets

Relatively thick, subglacially derived debris sequences (Table 3.5) have been described from the margins of many polar glaciers (*Greenland:* Koch & Wegener, 1917; Bishop, 1957; Goldthwait, 1960, 1971; Swinzow, 1962; *Baffin Island:* Ward, 1952; Goldthwait, 1951; Holdsworth, 1973; *Ellesmere Island:* Souchez, 1971; *Spitsbergen:* Gripp, 1929; Boulton, 1970; *Antarctica:* Souchez, 1966; Lorius, 1968; Rundle, 1973; Robinson, 1980) whilst debris-rich ice has been described in the deep ice cores to bedrock from the interiors of the Greenland ice sheet (Herron & Langway, 1979a) and the Antarctic ice sheet (Gow, Epstein & Sheehy, 1979). In contrast, temperate glaciers show only a very thin layer of subglacially derived debris (Table 3.5).

This contrast appears to support the suggestion that glacier thermal regime is an important control on the inclusion of basally derived debris by glaciers (Weertman, 1961b; Boulton, 1970). It seems likely that net basal freezing-on of large quantities of ice occurs in polar and subpolar glaciers in the way shown in Figure 3.30 and that debris at the glacier sole is incorporated in basal ice during freezing. However, in temperate glaciers the freezing-on mechanism operates on only a limited scale. Net basal melting occurs, but small thicknesses of several centimetres of regulation ice may survive for short periods, and it is in this thin regulation layer that subglacially derived debris is transported (Kamb & La Chapelle, 1964; Boulton, Morris, Armstrong & Thomas, 1979).

There is further strong evidence that the thick, basal debris sequences in polar glaciers are produced primarily by freezing-on.

(a) Assuming that basal regulation ice will contain debris, the predicted thickness of regulation ice at Byrd Station agrees well with the observed thickness of debris-bearing ice in the core from this site.

(b)   Thick basal debris sequences in polar glaciers only occur in those cases where a zone of freezing-on succeeds an up-glacier zone of melting. Where the glacier sole is frozen to its bed from a particular point to the upper extremity of the flowline, as in the snout region of the Meserve glacier in Antarctica (Holdsworth, 1974) and the sites of the Devon Island ice cap boreholes (Koerner & Fisher, 1979), no significant thickness of basal debris is observed.

(c)   There are two cases of glaciers where the freezing-on process is thought to occur and in which knowledge of the sequence of lithologies over which the glaciers have flowed enables us to show that this sequence is repeated in the thick basal debris sequences (Boulton, 1970; DiLabio & Shilts, 1979; see also Figure 3.31). It is difficult to imagine processes other than cumulative freezing-on which might achieve the same result. Unless subglacial geology is uniform over large areas, one would expect vertical lithological trends in a frozen-on debris sequence unless this has been disturbed by complex flow processes. Herron & Langway (1979a) have demonstrated that the basal debris sequence at Camp Century does show a systematic vertical change in lithology.

Although it has been suggested that debris incorporation can occur as a result of the development of thrust planes in basal ice (e.g. Harrison, 1957), there is no direct evidence of this, and the character of most basal debris sequences appears to deny it, although it may play an important role in changing the distribution of debris that is already in an englacial position (Boulton, 1970).

It should be emphasised that the level to which basally derived debris rises in an ice sheet is the lowest level to which isotopic profiles may be expected to reflect atmospheric conditions. It also seems that basally derived debris is a good indicator of the process of basal regelation. There is no evidence that completely debris-free basal regelation ice forms at the bottom of ice sheets.

The characteristics of basal debris-rich ice
Figure 3.31 illustrates some of the characteristics of basal debris-bearing ice exposed at the margin of Aavatsmarkbreen in western Spitsbergen (Boulton, 1970). At this site we can be reasonably sure of the origin of this ice by basal freezing-on, and thus some of its other characteristics can be examined as possible indicators of this process.

Distribution and character of the debris
The glacier bed was not exposed at the site at which sampling was done, but nearby exposure at the foot of the ice cliff suggest that it lies between 0.5–1.5 m below the sequence shown in Figure 3.31. The basal 6 m contains a relatively high mean debris concentration of 3.5% by volume (if we ignore a mass of frozen marine sediment with intact sedimentary structures which occurs between 0.5–1 m from the base of the sequence, and 9% by volume if we include it). The debris is highly stratified into well-defined bands of high concentration. An attempt to demonstrate the nature of the banding is shown in Figure 3.31 which shows how banding occurs on several scales, thick debris-rich bands themselves being composed of interstratified elements of high and low debris concentration.

The grain size distribution of debris from the basal 6 m contrasts strongly with that from the overlying ice. The basal debris is in general rich in fine sand and silt compared with debris from the overlying ice which is depleted in these fractions. In addition, clasts in the basal ice are often facetted and striated compared with the angular and unstriated clasts above. This contrast is typical of that between debris which has been in traction over the glacier bed and that which has fallen onto the glacier surface in the accumulation area and been subsequently buried by snow and ice (Boulton, 1979).

Table 3.5. *Reported thickness of subglacially derived debris in the basal parts of glaciers*

Polar and subpolar glaciers	Temperate glaciers
Greenland	
15.7 m – Camp Century (Herron & Langway, 1979a)	0.029 m – Blue Glacier, USA (Kamb & La Chapelle, 1964)
>70 m – Thule (Swinzow, 1962)	0.1 m – Glacier d'Argentière, France (Vivian & Bocquet, 1973)
Antarctica	
4.83 m – Byrd Station (Gow et al., 1979)	<0.3 m – Casement Glacier, Alaska (Peterson, 1970)
40 m – Mirny (Yevteyev, 1959)	
Baffin Island	
15 m – Barnes Ice Cap (Holdsworth, 1973)	0.2 m – Breidamerkurjökull, Iceland (Boulton, 1975)
Spitsbergen	
10 m – several glaciers (Boulton, 1970)	0.3 m – Vesl-Skautbreen, Norway (McCall, 1960)
Bylot Island	
Several metres (DiLabio & Shilts, 1979)	

The lithology of debris clasts varies systematically with height above the bed (Figure 3.31). In the debris-rich ice of the basal 6 m, far-travelled debris particles which can be related to their bedrock source occur in upper horizons, whilst progressively more locally derived material occurs in lower horizons. It is for this reason that the hypothesis of progressive basal freezing-on is strongly advocated for this debris mass. The debris in the ice above 6 m is all far travelled, having fallen on to the glacier surface in the accumulation area. On the glacier surface at this point, there are many locally derived rocks which fell on to the surface in the ablation area.

One feature of the debris lithology distribution which cannot be explained by freezing-on alone is the fact that lithologies typical of low horizons are occasionally found very much higher. In the next section, this will be ascribed to flow processes which produce post-inclusion mobility of frozen-in debris.

Much of the coarse fraction of the englacial debris is suspended in the ice as individual particles. However many small debris inclusions comprise aggregates of fine sand and silt grains. This may reflect masses of comminuted debris produced by the abrasion process at the glacier sole (see Boulton, 1979) and subsequently lifted up to a higher level by the formation of regelation ice beneath it and broken up during subsequent flow.

Isotopic ratios
$\delta^{18}O$ measurements change abruptly from values in the range -11 to $-14^0/_{00}$ in the basal 6 m of ice to -14 to $-15.5^0/_{00}$ in the overlying ice. The values above 6 m are similar to those found by Punning, Vaikmyae, Kotlyakov & Gordiyenko (1980) in glacier ice which had accumulated over the last millenium at the source of the Grönfjord and Fridtov Glaciers 110 km to the south, which suggests that the ice above the 6 m level is derived from glacier surface accumulation. The basal ice is considered on other grounds to have been frozen on basally.

Figure 3.31. Some characteristics of the basal ice exposed in an ice cliff at the margin of Aavatsmarkbreen in Spitsbergen. The basal 6 m are thought to be composed of regelation ice.

O'Neill (1968) has shown there to be a fractionation factor for ^{18}O for ice/water of 1.003. Figure 3.32 shows a simple model of how this might affect the ^{18}O values in basal regelation ice. Consider a glacier in which the basal zone has a uniform ^{18}O content. Meltwater from ice will have the same ^{18}O composition as the mean value in the ice from which it was produced. Passing into a zone of freezing-on, the ^{18}O values in the regelation ice are initially higher because of fractionation during freezing. However, the remaining subglacial water will be progressively depleted in ^{18}O because of freezing-on, and after some distance, the increase in ^{18}O content during freezing is not enough to compensate for the depletion in ^{18}O in the subglacial water reservoir. Thus $\delta^{18}O$ values in regelation ice will be relatively higher than overlying ice near to the meltwater source, but lower than overlying ice if far from the meltwater source. Lawson (1979) describes relatively high $\delta^{18}O$ values from presumed basal regelation ice from the Mantanuska glacier, Alaska, compared with overlying ice, whereas Gow *et al.* (1979) report relatively low values from the basal ice at Byrd Station, compared with overlying ice. I suggest that in these cases, as in the case of Aavatsmarkbreen, the distance from the meltwater source is the critical variable. In any case, basal regelation ice is expected to have a different $\delta^{18}O$ value from the overlying ice.

Entrapped air
Rough measurements of the frequency and size of air bubbles in the ice were made. These showed a sharp increase in the amount of entrapped air above 6 m. However, near to the top of the debris-rich ice, many of the relatively debris-free horizons showed high values of air content. The debris-rich ice at Byrd Station was devoid of air whilst the overlying ice had a relatively high air content (Gow *et al.*, 1979). A similar pattern was found by Lorius, Raynaud & Dollé (1968) in East Antarctica. These observations are compatible with the concept of a basal freezing process in which, during freezing, there is rejection of air contained in the water.

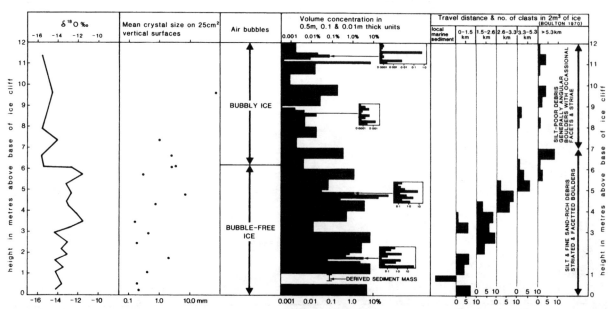

It is suggested that the air-rich, debris-free ice strata near to the top of the debris-rich ice reflect interfolding of overlying ice into it.

Ice crystal size

Below 4 m there appears to be a general decrease in the size of ice crystals, and in the debris-rich basal ice crystals are found to be small where debris concentrations are large. Relatively small crystal sizes have frequently been found associated with basal debris-rich ice (Kamb & La Chapelle, 1964; Lorius, 1968; Boulton, 1970; Anderton, 1974; Herron & Langway, 1979a).

It is to be expected that crystal sizes will be small in zones of high strain rate in the basal parts of ice sheets. However, existing data on the mechanical behaviour of ice with included debris (Hooke, Dahlin & Kauper, 1972) suggests that only ice with very low debris concentrations will have a lower strength than debris-free ice, and that substantial debris concentrations make such ice appreciably stronger. Thus the contrast in crystal size between debris-rich and debris-poor horizons does not support the view that a difference in strain rate between these is the cause of the local differences in crystal size. More likely causes of local crystal granulation in zones of high debris concentration are local stress concentrations produced by interactions between debris particles.

Non-laminar flow in the basal parts of glaciers

Models for the interpretation of isotopic profiles in ice sheets assume laminar flow. Because of the irregularity of glacier beds and the inhomogeneity of basal debris-rich ice, the basal zones, which contain the oldest part of the isotopic record, are most likely to show non-laminar flow. Styles of basal deformation can be observed directly in ice cliffs or subglacial tunnels in debris-rich ice, and the way in which they might evolve during flow can be predicted by using analogues from structural geology.

Non-laminar flow, or flow patterns in which flowlines diverge from parallelism with the bed, may be steady features which are a response to the form of the bed or transient features which reflect temporal changes of shear stress, flow direction or velocity, and the movement of debris in traction on the glacier bed.

While non-laminar flow in basal ice is most readily seen and studied where debris bands are present, there is no reason to suppose that overlying clean ice has not been similarly deformed. This problem is considered further in section 3.11.

Steady features

Decollement planes (Figure 3.33(1a)). There is strong field evidence that planes of decollement can exist roughly parallel to the bed in the basal parts of glaciers. This frequently occurs between basal debris-rich ice and overlying relatively clean ice (Boulton, 1970). It has also been observed in a cold-based glacier frozen to its bed (Holdsworth, 1974) where contact by the moving glacier with the rock bed is only made on the summits of small protuberances, whilst between these, the moving glacier slides over a stagnant ice surface which forms part of an 'effective bed' (Boulton, 1972). The existence of major decollement planes above very large stagnant basal ice masses is suggested by borehole evidence from near the western margin of the Greenland ice sheet at Isua (Colbeck & Gow, 1979), where a 700 m long and 100 m thick ice mass may have become stagnant on the up-glacier side of a subglacial hill.

Fold structures (Figure 3.33(1b)). Stable fold structures in basal ice have been observed to form around the flanks of bed undulations (Boulton *et al.*, 1979). This occurs where ice flowing around the flank of an obstacle flows beneath ice flowing over the obstacle. Thus, three-dimensional obstacles may generate steady folds. However, formation of steady folds by long transverse ridges does not seem to be possible.

Transient structures

Thrust planes (Figure 3.33(2a)). High-angle thrust planes are commonly found near to the margins of glaciers, often in association with folds. Glaciologists

Figure 3.32. A simple model of isotopic fractionation at the base of a glacier in which a zone of net basal freezing-on occurs downglacier of a zone of melting.

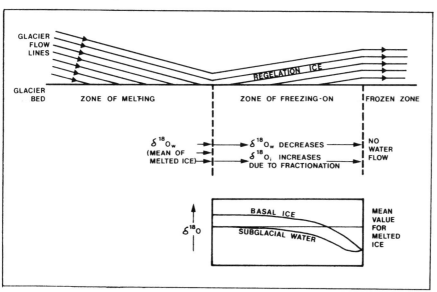

have frequently discussed whether thrust planes exist in ice. Using the displacement of foliation as a criterion, there is no doubt that they do. However, the amount of displacement along them generally seems to be small and there is no field evidence which supports the suggestion (Harrison, 1957) that subglacially derived debris can be introduced into a glacier along them. They develop when principal stresses are near to the horizontal, and thus it is unlikely that they are common features deep within ice sheets. In basal ice, it seems most likely that thrust planes develop in local zones of compression where the glacier flows against obstructions on its bed. Stress builds up until it is released by failure along a thrust plane. When the thrust plane is carried by flow past the obstruction, no further movement occurs along it. Further thrust planes may then develop as stress builds up again against the obstruction, and are likely to form at a high angle to the bed. Subsequent transport of the inactive thrust plane in a regime

of long-term shear strain will then tend to attenuate the plane, and reduce its inclination.

Folds (Figure 3.33(2*b*)). It is common in subglacial tunnels or in natural ice cliffs to find folds which are no longer evolving, but were nucleated elsewhere and subsequently passively transported. In general, folds develop because of local stress inhomogeneities or because of local variations in material strength. The two principle causes of folding at the base of glaciers are likely to be irregularities on the glacier bed and inhomogeneities within the ice. Folds develop as stress builds up on the up-glacier sides of basal obstacles where they are often associated with faults. They relieve the stress, but are then carried beyond the obstacle by glacier flow so that stress again builds up on the up-glacier side of the obstacle and generates further folds. The frequency of folding is controlled by the rate of stress build-up and decay, and factors such as the magnitude and direction

Figure 3.33. A schematic diagram showing styles of tectonic deformation which occur in the basal parts of glaciers. Transient structures undergo an active phase of formation and are subsequently attenuated by shear strain in the ice after they have passed beyond the zone in which they are initiated. The progressive complication of an originally simple isotopic profile is shown in the two sequences in (2). The example of folding shows the results of superimposition of two folding episodes, though many such episodes could of course be superimposed, and could explain the complexity of basal isotopic profiles.

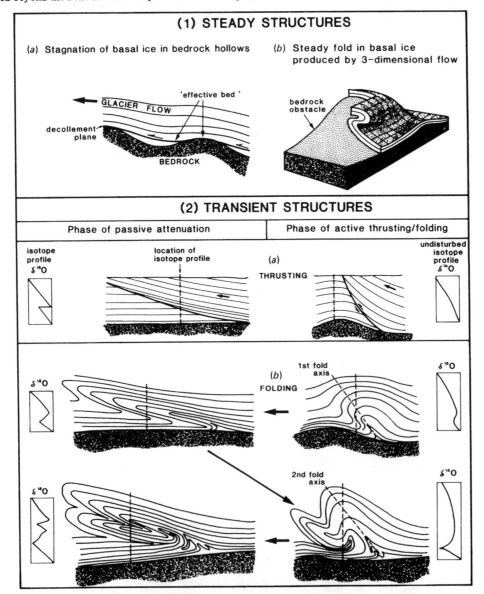

of ice velocity and effective stresses. The precise location of folding (or faulting) is often determined by inhomogeneities within the ice, such as debris-rich masses which are more viscous than cleaner ice and which block deformation so as to produce inclined fold or fault structures.

Long-term evolution of fold and fault structures and their influence on isotopic profiles

Decollement planes may separate active ice above from stagnant ice below, or a lower zone of slowly moving ice from more rapidly moving ice above. If this is a steady feature, the ice above the plane is likely to have had a different source from that below the plane. The plane becomes the 'effective bed' of the glacier and we might expect a close approximation to laminar flow conditions to occur in the ice above it. However, when the plane originally formed, it may have been a cross-cutting feature, and thus we would expect it to cut across isotopically different horizons in the underlying ice (Figure 3.33(2a)).

Steady fold structures such as that shown in Figure 3.33(1b) may originally invert and repeat the isotopic stratigraphy, and this inversion and repetition will be perpetuated during flow, though its thickness may change due to longitudinal extension or compression.

Transient structures may locally invert and/or repeat the isotopic stratigraphy. Once such structures have formed they are progressively attenuated by shear strain in the basal ice, so that high-angle structures are progressively lowered into parallelism with the plane of shear. The effect is shown in Figure 3.33(2). It results primarily in interdigitation of ice from one isotopic level with that from another. The net disruptive effect on the isotopic stratigraphy of a single folding/faulting episode, followed by shear attenuation, depends on the intensity of the original folding/faulting episode. The effect of a series of such episodes will be to produce considerable mixing of originally distinct layers, though retaining a much disturbed remnant of the original isotopic stratigraphy (Figure 3.33(2b)). For instance, the considerable variation in $\delta^{18}O$ values in the basal, debris-rich part of the Byrd Station core (Gow *et al.*, 1979) may reflect this process. It is likely that the mixing of bedrock clasts shown in Figure 3.31 (especially the occurrence of clasts from later episodes of freezing in lower horizons) also reflects this process.

The basal parts of an ice sheet can be expected to show isotopic changes which are a result of non-laminar flow processes. It is thus important to develop techniques by which such processes can be identified. One such technique, though expensive, is the drilling of multiple, closely spaced boreholes. As boreholes on the same flowline may sample the same sequences of non-laminar flow products, they should preferably be sited across flowlines. The character of the radio echo may provide another technique for the identification of zones of non-laminar flow (section 3.11).

3.11

Radio-echo studies of internal layering of polar ice sheets

G. de Q. ROBIN

When discussing the complex deformation of debris-rich layers at the base of polar ice sheets in the previous section, Boulton points out that there is no reason to suppose that overlying clean ice has not been similarly deformed. His models of basal ice flow provide a likely explanation of the 'spikes' in isotopic profiles of the lowest levels of the Camp Century and Devon Island ice cores (see also section 3.4). Within such irregularly deformed ice dating by glaciological modelling techniques will be invalid. We therefore need to know how far this irregularly deformed layer extends above bedrock.

If we can show that deformation of layers initially deposited on the surface is regular and conforms with our models to great depths, we can have more confidence in our method of dating. Visual monitoring of shallow layers due to changes with depth in crystal size, bubble concentrations, ice lenses and other features has been traced reliably to about 100 m depth (Langway, 1970). In Figure 1.9 we saw that annual isotopic layering followed a regular pattern with depth to around 1100 m depth at Camp Century, but ice cores give only a spot sounding and may not be suitable for detection of folded or distorted layering.

Layering within ice sheets, believed to be of depositional origin, is found on radio-echo sounding records. This provides a further means of checking on the continuity of deformation patterns at great depths. Internal layers detected by radio-echo sounding were first reported from north west Greenland by Bailey, Evans & Robin (1964) and have since been found to considerable depths throughout much of the ice sheets of Greenland and Antarctica. These are described in Robin, Swithinbank & Smith (1970), Harrison (1973), Gudmandsen (1975), Robin, Drewry & Meldrum (1977) and elsewhere and are shown in Figures 3.34–3.36. Figure 3.34 from central Greenland, in Gudmandsen (1975), shows traces of layering to within 200 m of bedrock in ice about 3000 m thick around 10 km from Crête, while Figures 3.35 and 3.36 show layering in central Antarctica that appears to reach bedrock on some peaks.

Figure 3.34. Radio-echo profile passing through Crête
(71° 07′ N 37° 19′ W) in central Greenland on a west to east
track. Crête is situated at 0 km, as indicated by the hyperbolic
lines due to reflections from surface structures. Vertical bands
are caused by radio communication interference. The recording
was made in 1974 using a 60 MHz system.

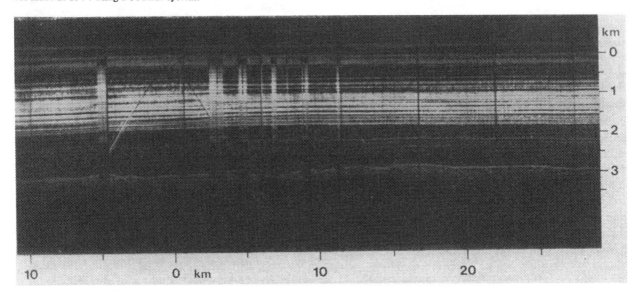

Figure 3.35. Radio-echo profile from central Antarctica running
along a flowline from ice divide 'B' (near right) to a sub-ice lake
near Vostok (left). The velocity of ice movement shown above
the print is estimated from a steady-state flow model.

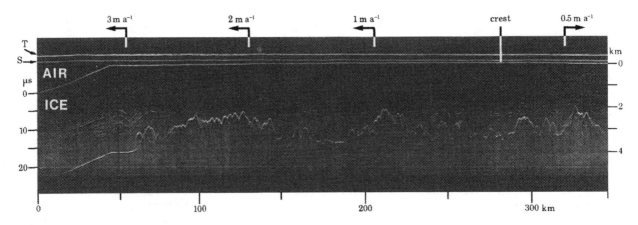

Figure 3.36. Radio-echo profiles from central Antarctica;
parallel to that of Figure 3.35 from 0–95 km, the direction then
changes by 90° to approximately follow a surface contour line.

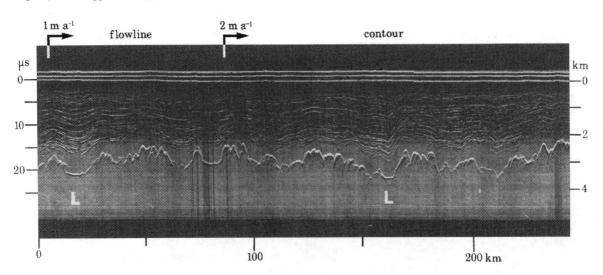

Radio-echo layering in ice sheets is formed as
a result of two different processes. Density contrasts
between depositional layers are believed to be the main
cause of radio-echo layering to some hundreds of metres
depth according to Robin, Evans & Bailey (1969), Paren
& Robin (1975), Clough (1977), Ackley & Keliher (1979)
and Millar (1981*a, b*). However, at great depths the
pressure is too great for density differences to survive.
Paren & Robin (1975) therefore suggested that changes
in conductivity between layers due to varying amounts
of impurities in the snowfall could explain observed
layering found in Antarctic records to depths exceeding
three kilometres. Hammer (1977) showed that the major
impurities affecting the conductivity of melted samples
from ice cores are due to acidity from major volcanic
eruptions that affect snow deposited up to three years
after the eruptions. Hammer, Clausen & Dansgaard
(1980) subsequently confirmed and extended these
results by direct measurements of conductivity of solid
ice cores. Strong peaks of conductivity in ice cores from
Camp Century were found to be correlated with major
volcanic eruptions in the northern hemisphere over the
past 7400 a. Hammer (1980) also reports that Gudmandsen & Overgaard (1978) confirmed that the four most
acid layers in the Crête core at 75, 220, 241 and 306 m
depth are all associated with strong echoes to within the
±5 m uncertainty of the depth estimates. Millar (1982)
measured the power reflection coefficient (PRC) of radio-
layers and from this deduced acidity variations in ice
cores. He confirmed the above results near Crête and
showed that PRCs were some 10–20 dB higher in Greenland than in Antarctica. As a consequence, conductivity
layering in Greenland becomes the dominant mechanism
of internal radio-echo reflections at a much shallower
depth than in Antarctica.

Internally reflecting layers within an ice sheet
provide a record of the total deformation of the ice mass
since the layers were originally deposited. Continuity of
layers over long distances along the ice sheet and
throughout a considerable fraction of the total depth of
ice indicate continuity of deposition to the greatest layer
depth observed. In ice cores obtained from such areas
continuity of time along the ice core can therefore be
accepted.

The converse is not necessarily true, however, since
apparent breaks in the continuity of reflections seen on
some records from undulating or tilted layers, such as
are seen around 290 km on Figure 3.35, are explicable in
terms of the 'optics' of radio-wave reflections from
curved surfaces (Harrison, 1973). No clear case of a break
in the horizontal continuity of layering that cannot be
explained in this way has been observed so far.

Absence of radio-echo layering in a record may be
due to insufficient system performance to detect weak
echoes, to the absence of layers or to the masking of
layers by 'noise', such as long-range reflections from
surface sastrugi and crevasses. The latter cause may
contribute to the absence of layering of the ice streams
in West Antarctica reported in Rose (1978), but noise
appears unlikely to be a limiting factor in the interior of
Antarctica and Greenland. In these areas we are especi-

ally interested in the greatest depth at which layering is
present. This may indicate the depth to which we can be
sure ice deformation has been uniform and regular. At
such depths, the hypothesis that echoes are due to
conductivity changes between layers is not disputed.

Millar's analysis of the PRC of layers at several
sites in Antarctica has shown a general fall in PRCs
between approximately 15 000 and 20 000 a BP. A gap in
layering between about 1.9–2.3 km depth near Crête
station in Greenland is also seen in Figure 3.34. This is
dated by Millar (1981*a*) at around 13 000–23 000 a BP.
It is suggested that conductivity contrasts between layers
are suppressed by a large amount of calcium carbonate
dust neutralising acid layers in the ice sheet. This dust
could come from vast areas of shallow sea bed exposed
at the time of lower sea levels. At Camp Century, radio-
echo layering is seen in the record of Gudmandsen (1975)
to around 1100 m, but it does not reappear around
23 000 a at greater depths as in Figure 3.34. Ice of this
age is expected around 1300 m depth according to
Figure 1.9, that is only 90 m above the base of the ice.
It is not clear whether layering is absent or whether the
system performance was not great enough to detect
layers at this depth.

Millar's studies of PRCs in the Antarctic ice sheet
show reflecting layers to depths around 3500 m where
the ice must be around 200 000 a old. Figure 3.37 shows
results for one such analysis along some 400 m of flightline, around 100 km north of Dome C. On the right of
the diagram we see the normal intensity-modulated (or
Z-scope) profile similar to Figures 3.34–3.36 but at
a considerably expanded distance scale. The horizontal
continuity of reflecting layers is seen clearly, but rapid
fading of echo strength along the flightline of individual
reflecting horizons is also apparent. To obtain a satisfactory mean PRC for an individual layer it is necessary
to take the mean of a large number of measurements of
the PRC. This is done by photographing oscillograph
A-scope displays of the reflected signal. Each exposure
shows an accurate plot of the signal (echo) strength
against time (depth) during an interval of less than 0.01 s
and so records the instantaneous values of the fluctuating
echo strengths. A rapid-run camera taking photographs
of the A-scope at intervals of about 2 m along the flightline (54 frames per second) is used over some 400 m of
flight to obtain the PRCs shown on the left of Figure
3.37. Each PRC measured by A-scope on each layer is
plotted as an individual point. Strong echoes show as
a cluster of points at one depth on Figure 3.37, and the
horizontal spread of points of about 10 dB for an individual layer indicates the amount of fading along the flightline. Fuller details, and a discussion of the significance
of the results in relation to former volcanic activity is
given in Millar (1982).

The maximum depth to which an echo can be
detected on Figure 3.37 depends both on the PRC and
the energy loss due to absorption of radio waves in ice.
Weak echoes of around −80 dB are lost at depths around
2700 m, whereas stronger echoes of −60 dB are still
detected down to 3200 m. It is clear in this case that
absorption of radio wave energy in ice limits the depth

to which weak internal layering can be detected. However, if internal layering is not present below some depths, and our radio-echo system performance is great enough, it should be possible to detect the level at which layering ceases. The best chance of this occurring is near the centre of Antarctica where the ice is so cold that dielectric absorption is low, especially in areas where the bedrock is not too deep (say less than 3 km). Figure 3.38 shows a plot of PRC's against depth obtained at a site at

180 km on Figure 3.35. We see here that echoes with PRC's from −73 to −86 dB all cease at the depth of 2850 m, in contrast to those of Figure 3.37. Also our calculations indicate that the normal absorption figures should have let us see echoes of around −80 dB to depths 500 m greater than those observed, as indicated on Figure 3.38. Both effects indicate that the absence of echoes is due to absence of effective layering below 2850 m depth.

Figure 3.37. (*a*) Power reflection coefficient (PRC) at a site 100 km at a site 100 km north west of Dome C, Antarctica from many measurements with calibrated equipment. Solid line shows maximum variations expected from possible fluctuations of density at different depths. Shaded zone is that beyond the

calculated detection limit.

(*b*) Corresponding 'Z'-scope profile of layering over the same flightline. This shows a record with a much expanded distance scale compared with Figures 3.34–3.36.

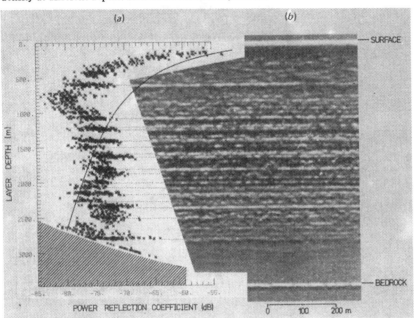

Figure 3.38. Plot of PRC of internal layers against depth for site at approximately 180 km on Figure 3.35. The sharp cut-off at 2800 m indicates the maximum depth of continuous near-horizontal layering, and hence the maximum depth to which vertical deformation can be assumed to follow a regular pattern.

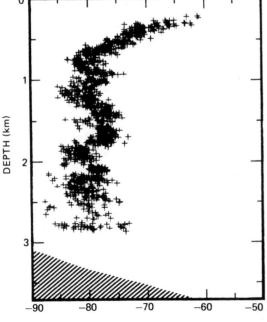

POWER REFLECTION COEFFICIENT (dB)

In Figure 3.39 we show the depth of the deepest layering observed along Z-scope profiles in central East Antarctica. Figure 3.39(*a*) is the same profile as in Figure 3.35, while the profile of Figure 3.39(*b*) crosses that of 3.39(*a*) at right-angles. We also show depths at which we would expect to detect effective layering echoes of given PRCs. It appears that over all of Figure 3.39(*b*) and over much of Figure 3.39(*a*) the deepest echoes indicate the boundary between regular layering and ice that must be assumed to be considerably deformed. This 'non-reflecting zone' appears to be of a thickness of the same order as, but thinner than, the total elevation range of the bedrock relief along the profiles. Robin & Millar (1982) suggest that weak seismic reflections at a similar height above bedrock on West Antarctica reported in Bentley (1971*b*), the change of ice fabric at around 350 m above bedrock in the ice core from Byrd Station, and the cessation of radio-echo layering, may all be due to irregular deformation of ice

flowing over rough terrain. If this is correct it provides an answer to the question raised in the opening paragraph of this section. The deepest layering does give confidence that to that depth deformation of ice and hence age scales are continuous. Table 3.6, from Millar (1981*b*), gives depths and computed ages of the deepest observed radio-echo layering at the main drilling sites in Antarctica. In Greenland corresponding ages will be less than in central Antarctica owing to higher accumulation rates, even in the vicinity of Crête where layering is seen within 200 m of bedrock.

Confidence in the continuity of age scales to the deepest reflecting layer does not imply that the age scales from modelling are correct at all depths, since the vertical strain rates used in our model may be in error. However, Robin *et al.* (1977) and Millar (1981*b*) showed that spacing between some individual layers shown in Figure 3.35 varied uniformly to a rough approximation along the flowline down to the deepest layer, while the gap between the deepest layer and bedrock did not. This gives some confidence in our method of dating where layering is present.

For boreholes, discussed in chapter 4, that penetrate only a limited fraction of the ice depth, radio-echo layering confirms the continuity of the time scale to the maximum depth of coring.

Below the level of the deepest layer, ice will be older, and in some locations perhaps several times the age of ice from higher levels. It will contain information about climate at a much earlier period, possibly to several times that shown in the age column of Table 3.6. Although its value is limited by uncertainty over dating, such information can still be of use. Its value will be greatly enhanced, and the irregularity of deformation better understood, if new methods make age determinations possible at many points along the ice core.

Apart from helping with a critical assessment of ice cores that have been studied, this section draws attention to the value of radio-echo sounding methods for selection of future sites for deep drilling in ice sheets.

Table 3.6. *Maximum age of regular layering in Antarctic ice sheet*

Station	Depth (m)	Estimated age (a BP)	Remark
Byrd Station	>1800	>34 000	Absorption limit
South Pole	2650	≈90 000	Layering ceases
Vostok Station	>3000	>225 000	Absorption limit
Dome C Camp	≈2900	≈170 000	Layering ceases

Figure 3.39. Depths of layering in central Antarctica: thin lines show observed continuous layers and dashed line maximum depth of observed layering; dotted lines show maximum calculated depth at which layers of indicated PRCs should be seen. (*a*) is profile shown in Figure 3.35. (*b*) is profile that crosses (*a*) at right-angles near the mid-point.

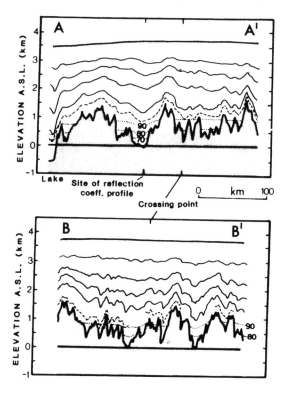

PRESENTATION OF DATA

4.1

General glaciology

G. de Q.ROBIN

The original intention was that this chapter would present data covering all major profiles of temperature and δ values in ice sheets. However, the rapidly increasing volume of such data made this a difficult, if not impossible, task. We have therefore drawn on the conclusions reached in chapters 5 and 6 to present information only at that level of detail that is useful for the climatic interpretation of isotopic profiles.

The most important parameters for this interpretation are the distribution of the temperature and accumulation (mass balance) which are summarised in map form in Figures 4.1–4.4. For wider summaries of other glaciological data, including surface form and ice depth, reference should be made to the American Geographical Society *Antarctic Map Folio Series*, especially to Bentley *et al.* (1964) Folio 2, Weyant (1967) Folio 8, and Heezen, Tharp & Bentley (1972) Folio 16; also to the Tolstikov (1966) Soviet *Atlas of Antarctica*, and to the Scott Polar Research Institute (in press) *Antarctic Glaciological and Geophysical Map Folio*. A useful list of all known ice cores is given in the report by McKinnon (1980) published by the World Data Center A for Glaciology.

All measurements of isotopic, as well as temperature, profiles are presented as a function of depth. This contrasts with many earlier reports of isotopic measurements which were presented (for climatological reasons) as a function of age. The latter presentation involves the conversion of a depth scale to an age scale; this in turn involves modelling assumptions, the validity of which may be questioned, especially for lower levels in an ice sheet. Our presentation in this chapter draws the attention of climatologists to the need to assess the reliability of the age scale.

We are greatly indebted to a number of authors, to whom acknowledgement is made in the course of the chapter, for provision of more detailed isotopic data than has previously appeared in published form.

Figure 4.1. Isotherms of mean annual temperature and surface elevation contours for the Greenland ice sheet, together with measured temperature–depth profiles. The ice thickness in metres (Z) is shown at each profile site except Camp Century where data to bedrock was obtained.
Temperature–depth profiles:
Camp Century (see also Table 4.1): B. L. Hansen (personal communication); Weertman (1968).

Site2: Hansen & Landauer (1958).
Camp VI: Heuberger (1954).
Milcent: B. L. Hansen (personal communication).
Station Centrale: Heuberger (1954).
Crete: B. L. Hansen (personal communication).
Jarl Joset: Philberth (1970).
Dye 3: B. L. Hansen (personal communication).

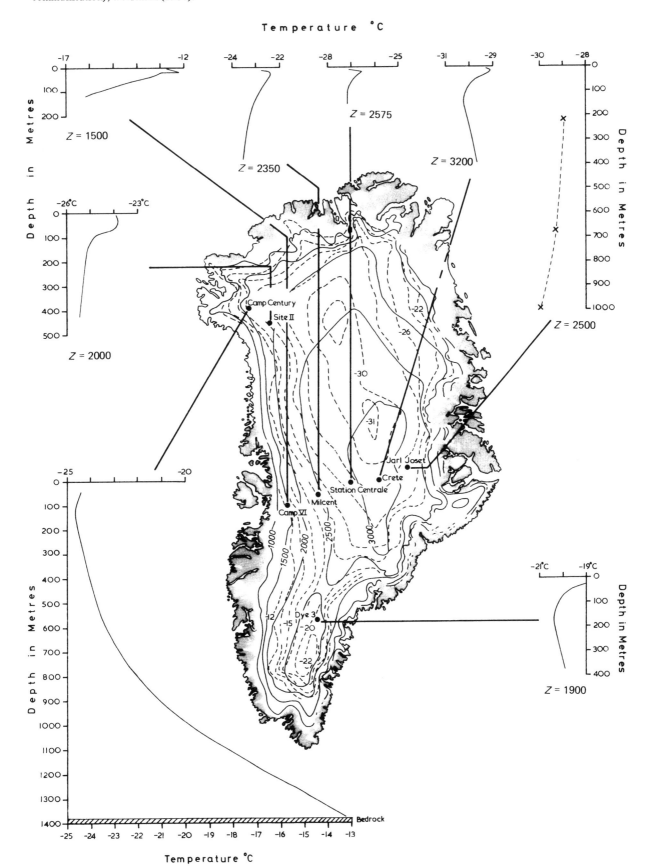

Greenland

Figure 4.1 shows the general contours of ice elevation and isotherms of mean annual surface temperatures. About four-fifths of the total area of Greenland is covered by the ice sheet which extends over an area of 1 726 000 km^2, has an average elevation of 2135 m and, over the central region, is based at or slightly below sea level (see section 2.5). Unlike Antarctica, the Greenland ice sheet is separated from the sea along most of its boundary by ice-free strips of land, one such strip extending 200 km inland; only along limited stretches, the longest being about 400 km, does the inland ice reach to the sea.

Borehole temperature profiles are printed round the periphery of the map with lines linking each profile to its site, for which the ice thickness (Z) is also given. The sites are designated by station names. The distribution of surface temperature measurements in this figure, and of accumulation data in Figure 4.2, is largely the

same since the two measurements were often made at the same sites.

Figure 4.2 shows the pattern of accumulation rate over inland areas of Greenland based on Schytt (1974). We have added dots on the map to show measurement sites on the main ice sheet in order to indicate the extent of interpolation between stations. Thirty-five of these station sites date from 1913, most of the others from the period following 1952. Mean values from the snow-pit studies cover periods ranging from 10 years to a single year, so the value for annual accumulation rate at any given site is not necessarily a good mean value; in addition, the geographical distribution of data points is not uniform. The contours therefore provide a good indication, rather than an exact prediction of regional accumulation patterns. The coastal ice-free regions are not covered in this map.

Antarctica

Figures 4.3 and 4.4 present the corresponding glaciological data for Antarctica. Because of the scale of the continent and the range of temperatures involved, the temperature and depth scales for the temperature profiles are less than half those for Greenland for boreholes on the inland ice. On ice shelves the temperature scale is reduced by a further factor of three in order to show changes of curvature more clearly.

Figure 4.3 shows ice elevation and surface isotherms. Almost the entire bedrock of Antarctica is covered by the ice sheet, about 10 per cent of which extends out to sea in the form of floating ice shelves. The total area covered by inland ice sheet and floating ice shelves is 13 800 000 km^2. Ice-free areas, such as exposed rock around the coast, inland dry valleys and mountain peaks, total less than 200 000 km^2. A more detailed description of the surface form of the ice sheet is given in section 2.4.

The density of measurement sites of mean annual surface temperatures is similar to that for accumulation measurements shown in Figure 4.4. This shows the distribution of net annual accumulation over the Antarctic ice sheet, along with dots that indicate the location of measurement sites. Although reasonable cover exists over much of the ice sheet, in some areas the density of measurements is very inadequate.

Figure 4.2. Accumulation rate over the Greenland ice sheet (adapted from Schytt, 1974, Fig. 2). Dots indicate the measurement sites, with heavy dots showing location of borehole sites on Figure 4.1. The distribution of temperature measurements used to compile Figure 4.1 is similar since both were measured at many of the same sites.

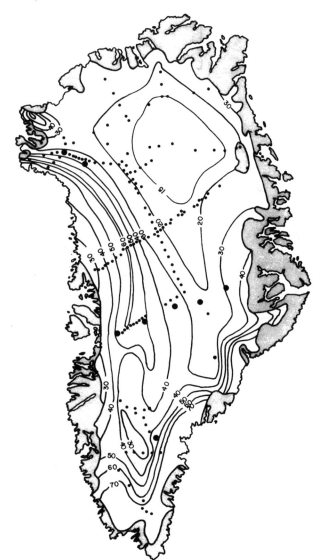

Figure 4.4. Accumulation rates over the Antarctic ice sheet. Dots indicate measurement sites with heavy dots showing the location of borehole sites on Figure 4.3. Shaded areas indicate accumulation zones feeding the three ice coring sites for which isotopic profiles are presented in this chapter. Adapted from *Antarctic Glaciological and Geophysical Map Folio* (Scott Polar Research Institute, Cambridge, in press).

Figure 4.3. Isotherms of mean annual temperature and surface
elevation contours of the Antarctic ice sheet. Temperature-
depth profiles are shown at sites indicated, the ice thickness in
metres (Z) being shown for ice sites where the borehole did not
penetrate to the base of the ice.
Temperature–depth profiles:
Byrd Station (see also Table 4.5): Ueda & Garfield (1970).
J9: Clough & Hansen (1979).

Little America V: Bentley *et al.* (1964).
D10: Gillet, Donnou & Ricou (1976).
Dome C: L. Lliboutry (personal communication).
Mirny (2 holes): Bogoslovskiy (1958).
Amery: W. F. Budd (personal communication).
Maudheim: Schytt (1960).
Vostok: Barkov, Vostretsov & Putikov (1975).

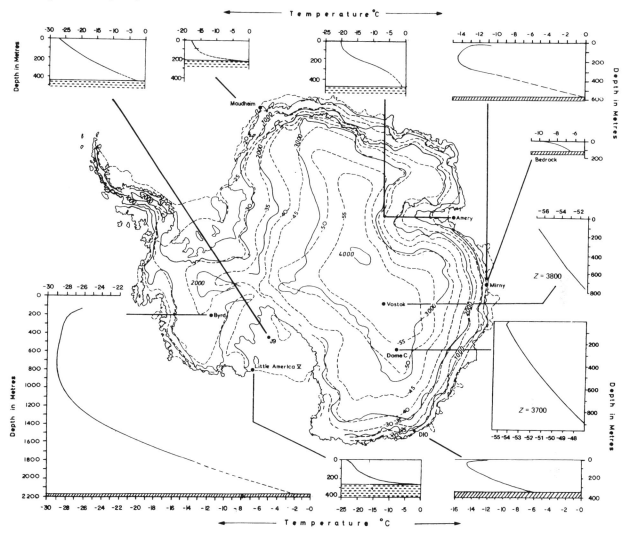

Figure 4.4. (see opposite page).

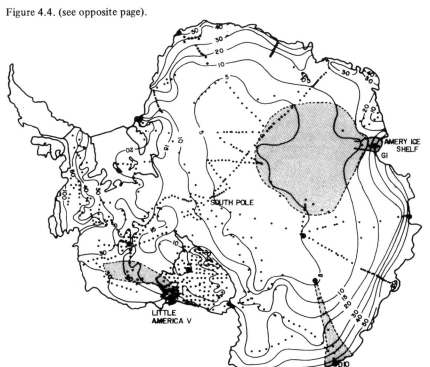

4.2

Profile data, Greenland region

G. de Q. ROBIN

Camp Century

77° 11′ N 82° 08′ W

Elevation   1885 m

Ice thickness   1388 m

Accumulation rate   0.38 m a^{-1} of ice (see
                      Table 3.3)

10 m temperature   −24.4 °C (1963)

Velocity of ice movement   5.5 m a^{-1}

Drilling dates   1963–6

Camp Century is situated on the southern side of
a broad ridge running westward from the main dome of
the ice sheet along a broad peninsula. In Figure 4.5 we
see that the flow of inland ice tends to diverge to either
side of this ridge, to Inglefield Bredning to the north and
Melville Bugt to the south. Bedrock is reasonably
level at around 500–600 m above sea level for at least
50 km to the north, south and west of Camp Century
in profiles of Robin, Evans & Bailey (1969). A radio-echo
profile along the approximate flowline down the ridge
through Camp Century (Figure 4.6) by Gudmandsen
(unpublished) shows that bedrock rises towards Camp
Century by some 400 m between 65–20 km upstream of
Camp Century, then falls by 200 m over the last 20 km.
(These changes are not clear in Figure 4.5.) Possibly
this rise, as well as the proximity of the outlet to
Inglefield Fjord to the north west, helps divert flow and
results in the relatively low surface velocity of 5.5 m a^{-1}.

It is difficult in these circumstances to define
flowlines, and especially the divergence between flowlines upstream of Camp Century, with any accuracy. If
there were no divergence, a mean columnar velocity of
3.8 m a^{-1}, roughly equivalent to a surface velocity of
5.5 m a^{-1}, could be maintained in a steady state by a flowline only 14 km long with the same accumulation rate
as Camp Century. Such flow in the vicinity of ice divides
(see section 2.4) seems likely. However, we see in
Figure 4.5 that the crest of the ridge further inland
from Camp Century is broad, so some contribution
to the flow from inland ice is possible. Both local and
general flow analyses are therefore presented in chapter 5.

Uncertainty over flowlines upstream of Camp
Century and their past variation makes it difficult to
date the lowest 280 m of the ice core and to separate
climatic effects from changes in size of the ice sheet.
Above this level (1100 m depth) sufficient measurements of annual stratigraphy are available (Figure 1.9)
to date the ice core back to about 8500 a BP with an
estimated error of around 2 per cent. Over this time
span climatic interpretation of the isotopic data can
therefore be made with more confidence.

Our discussion in section 2.5 indicates that the
inland ice advanced seawards around Greenland when
sea level fell during the last ice age. This will have had
a considerable effect on the surface contours at Camp
Century. In particular, if the grounded ice advanced
into Inglefield Bredning and Melville Bugt to the 500 m
depth contour as suggested in section 2.5, the ice surface
contours would have become less curved near Camp
Century and divergence of ice flow would have decreased
considerably. This would bring a great increase in the
amount of inland ice flowing over the Camp Century
site during the last ice age.

Flowline parameters

More modelling studies have been carried out on
the ice sheet at Camp Century than at any other location.
The parameters used are set out fully in sections 5.2,
5.3 and 5.5, and they cover a wide range of input data,
from steady-state stationary column (Figure 5.1) to
steady-state flowline data (Figures 5.18–5.34) while
the main variable parameter for non-steady-state solutions is the temperature input derived from δ values
and thickness changes, taking account of either δ value
or both δ value and gas content changes (sections 5.3
and 5.5).

Ice movement

Since survey data was collected before satellite
Doppler fixing methods were available, the methods
available for measurement of ice movement were limited
to conventional survey methods over long distances or
borehole survey methods, the accuracy of which is also
limited. Mock's tellurometer–theodolite traverses in this
region around 1960–3 are reported to give a velocity of
surface movement of 9.9 m a^{-1} in a direction close to
the borehole results, (Hansen, personal communication)
but these results have not been published.

Hansen (personal communication) has reworked
borehole survey data for the completed hole from
1966, 1967 and 1969 using a 'balanced tangential
method'. When data from the 1967 and 1969 surveys
only are used, a surface velocity of 8.0 m a^{-1} was
obtained, but when data from all three surveys were
included the surface velocity obtained was 5.5 ±
0.3 m a^{-1} in a direction 200° true with an uncertainty
of several degrees. The last figures are our preferred
velocity.

We show a comparison of the relative variation
of velocity with depth in Figure 4.7 for the periods
1966–9 and 1967–9 from which we calculate that the
mean column velocity is close to 3.8 m a^{-1}.

Figure 4.5. Topography and flow of ice sheet in north west Greenland. Surface contours are shown by continuous thin lines, bedrock contours by dashed lines. Heavy arrows show direction of ice flow deduced from contours, heavy dashed lines show position of ice divides along which ice flow is low. Topography based on Weidick (undated) *Quaternary Map of Greenland*.

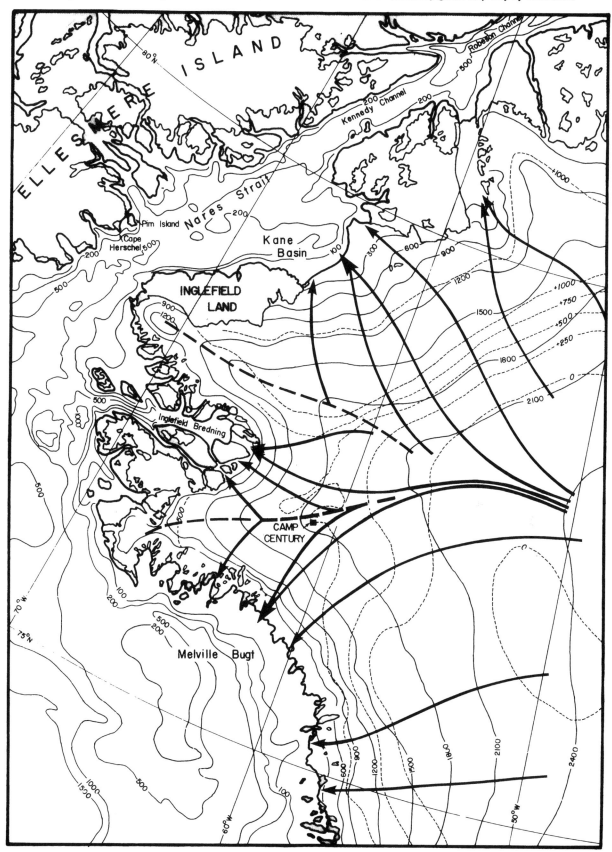

Ice core

Since Camp Century was the first site at which an ice core was obtained to depths greater than 500 m in an ice sheet, various methods were tried during the three seasons of drilling from 1963–6. Splettstoesser (1976) gives general details of drilling methods in ice sheets.

Hansen (personal communication) has given the following summary of progress of drilling operations.

1963 Drilling from the surface to 285 m depth was accomplished by use of a thermal coring drill in a dry hole.

1964 Thermal coring in a liquid-filled hole was attempted; core quality was poor down to the 550 m reached.

1965 Changed to mechanical drill in the middle of the season with marked improvements in core quality.

1966 Drilling completed this season with exceptional quality core, probably due to warmer ice at the bottom of the ice cap.

Core recovery. The top 285 m and bottom 300 m of core are excellent quality, the rest being of fair to good quality.

Table 4.1. *Temperature profile Camp Century –*
1 August 1966

Depth* (m)	Temperature (−°C)	Depth (m)	Temperature (−°C)
14.8	23.98	457.12	23.87
30.48	24.38	487.67	23.77
60.96	24.57	518.16	23.66
91.44	24.68	533.40	23.599
106.68	24.69	609.60	23.255
121.92	24.69	685.8	22.813
137.16	24.68	762.0	22.243
152.40	24.66	838.2	21.533
167.64	24.63	914.4	20.672
182.88	24.59	990.6	19.661
198.12	24.56	1066.8	18.418
213.36	24.52	1219.2	15.977
243.84	24.45	1371.6	13.000
274.32	24.37	1388.3	BOTTOM
304.80	24.29		
335.28	24.22		
365.76	24.13		
396.24	24.04		
426.72	23.96		

* Depths below 1966 reference level.

Figure 4.6. Radio-echo profile along approximate flowline from north central Greenland through Camp Century from airborne radio-echo sounding (P. Gudmandsen, unpublished data).

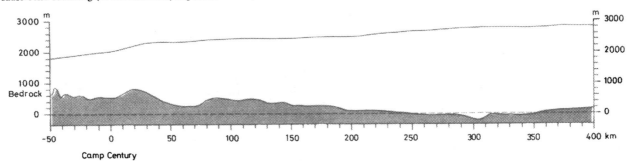

Figure 4.7. Horizontal ice movement velocity profile against relative depth in Camp Century borehole (B. L. Hansen, personal communication).

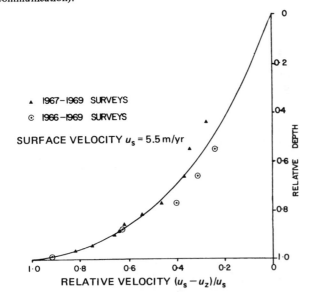

Temperature profile

This was measured on 1 August 1966 after comple-
tion of drilling. In addition to the profile shown on
Figure 4.1 we give details of measurements in Table 4.1
(Hansen, personal communication).

Isotopic profiles

Three profiles are shown with different sample
spacing in Figures 4.8–4.10. Figure 4.8 presents mean
δ values over 4 m lengths for the whole ice core. The
reference marks for layer depths covering 100 a or more
of accumulation are based on Hammer *et al.* (1978),
and these figures lead to calculation of the 95 per cent
confidence limits shown by horizontal bars. Determina-

tion of the magnitude of these confidence limits is
discussed fully in section 6.2. The 95 per cent confidence level indicates the probability that the deviation
of a plotted δ value from the mean trend of values can
be taken as indicating a similar temperature fluctuation
from the mean. For a deviation from the mean lasting
over a number of adjacent levels, the 95 per cent confidence level falls in inverse proportion to the square root
of the number of levels. Thus, although deviations of
most plotted points are of similar magnitude to the
95 per cent confidence level, there are a number of
deviations lasting over several points (some decades)
that clearly indicate climatic perturbations. It will be
seen that from around 250–900 m depth variations

Figure 4.8. Isotopic profile to bedrock at Camp Century,
showing mean δ values over 4 m sections along ice core, except
for some sections between 250 m and 900 m where longer
sections were used. Vertical bars show thickness of 100 a layer
at different depths. Horizontal bars show 95 per cent confidence

level for use of an individual δ value over 4 m as an indicator of
temperature at the time of deposition. At greater depths, the
bars indicate possible measurement error. Age scale from
Dansgaard & Johnsen (1969a): shown also as curve 2 in Figure
5.22. Isotopic data supplied by S. J. Johnsen.

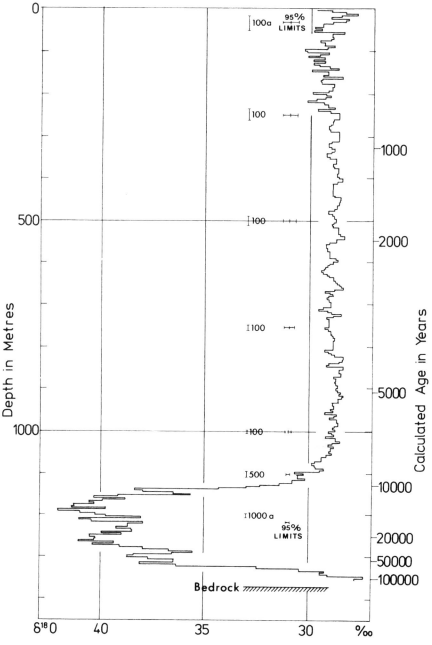

CAMP CENTURY

δ value between adjacent samples appear to be markedly less than would be expected from the 95 per cent confidence limits. Along this portion of the ice core, and especially for the poor quality section from 280–550 m depth, sampling is not continuous and the δ values have been smoothed.

Figure 4.9 presents mean δ values over 0.5 m lengths from 960 m to bedrock. It will be noted that at 1000 m depth the 95 per cent confidence limits on individual δ values are similar to near-surface values with 4 m sampling in Figure 4.8.

Figure 4.10 shows δ values over 0.01 m lengths for core from around 25 m–27 m above bedrock. At this level, diffusion in solid ice between adjacent 1 cm samples is dominant. The type of information that can be deduced from δ values at this level is different to the relatively direct climatic record given by δ values

at higher levels. Section 3.4 should be read for a more detailed discussion.

Basal layer

The bottom 15.7 m of the core contains over 300 alternating layers of clear and debris-laden ice which is described in detail in Herron & Langway (1979*a*). They conclude that the source of the debris is from till at the base of the ice sheet rather than surficial deposits, and that some type of freezing-on process led to incorporation of debris within the ice. The lowest 10 m consists of extremely fine-grained, highly oriented ice that implies a zone of high deformation. This is consistent with large changes of $\delta^{18}O$ values between clear and dirt-laden ice layers reported by Johnsen (personal communication), which suggests that rapid shear may be taking place in the basal ice similar to that shown by isotopic profiles at the base of the Devon Island ice cap (section 3.4).

General comments

Although most δ value shifts between adjacent samples above 1000 m depth are little greater than

Figure 4.9. Isotopic profile from 1061 m to bedrock at Camp Century, Greenland, showing mean δ ¹⁸O values over 0.5 m sections along ice core. Vertical and horizontal bars as for Figure 4.8. Data supplied by S. J. Johnsen.

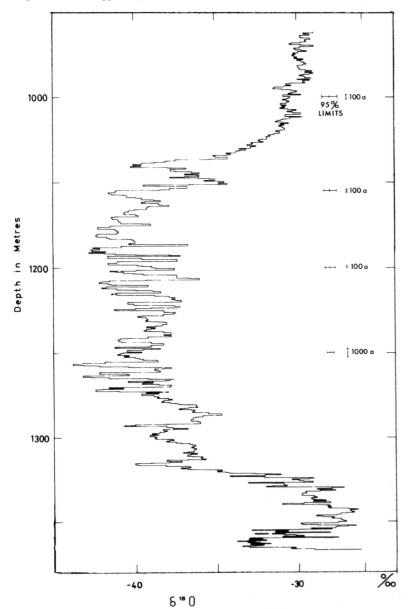

noise levels, many shifts that persist over a run of three or more samples indicate δ value changes that appear to be climatically significant.

The most detailed climatological assessments of this ice core are given in sections 5.3, 5.5 and 6.3.

Central and southern Greenland

The Greenland Ice Sheet Programme (GISP), a combined project of the US National Science Foundation and institutions in Denmark and Switzerland, is carrying out broad studies of the Greenland ice sheet. One aim which was achieved in summer 1982 at Dye 3 was to obtain an ice core to bedrock at a suitable location in central Greenland to study climatological changes over a period back into the last ice age. Since 1971 a series of boreholes have been drilled in different locations, of which extensive results from three cores in central Greenland are now available. We present the results for the three locations together in Table 4.2. All boreholes were drilled by thermal coring techniques developed by the US Army CRREL. Holes were air cored, and consequently the depth was limited by the closure rate of the boreholes at depth. Core recovery was 100 per cent at all sites.

Table 4.2. *Station data*

	Dye 3	Crête	Milcent
Latitude	65° 11′ N	71° 07′ N	70° 18′ N
Longitude	43° 50′ W	37° 19′ W	44° 33′ W
Elevation	2479 m	3172 m	2450 m
Ice thickness	1900 m	3200 m	2350 m
Accumulation rate	0.50 m a^{-1}	0.29 m a^{-1}	0.532 m a^{-1}
Surface velocity	7 m a^{-1}	(4 m a^{-1}?)	48 m a^{-1}
10 m temperature	−18 °C	−30.0 °C	−22.2 °C
Mean surface slope and direction	0.0044 towards 060° (T)	<0.001	0.0043 towards 250° (T)
Drilling date	1971	1974	1973
Core depth (max.)	377 m	404 m	400 m
Ice age (max.)	740 a	1420 a	800 a
$\alpha v \lambda / A$	0.000 62 °C m^{-1}	0.0	0.003 89 °C m^{-1}

Figure 4.10. δ ^{18}O values over 0.01 m sections at Camp Century, from approximately 25–27 m above bedrock. Data supplied by S. J. Johnsen. (See also section 3.4.)

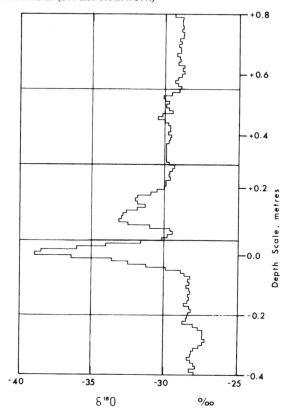

Dye 3

The station lies about 30 km to the east of the main ice divide of central Greenland, at just over 100 m lower elevation than the divide. Reeh *et al.* (1977) showed that the accumulation rate increases from around 0.32 m a^{-1} ice at the divide to 0.50 m a^{-1} at the station. The effect is seen in the accumulation map (Figure 4.2), so it is clear that much of the precipitation at this site comes from the south and east, and consequently depends on different weather systems to that at Milcent on the western slope. Ice movement has not been measured directly at this station. The tabulated figure is calculated on a steady-state basis using data in Reeh *et al.* (1977).

Crête

The station is situated on the north–south ice divide of central Greenland. The divide shows a slight slope towards the south at this site that averages around 0.0005, whereas slopes average 0.0008 over the first 10 km to either side of the station. Consequently, the component of flow along the ice divide will be small, and for practical purposes the ice may be considered as stationary in regard to horizontal motion. If this is correct, the velocity of 4 m a^{-1} towards the north reported in Hofmann (1974) is due to errors of triangulation over the 500 km net from the west coast of Greenland. Since the station is on the ice divide, weather systems reach the stations from all directions, and in particular the station is equally exposed to systems from either west or east.

Milcent

This station is sited on the west-facing slope of the ice sheet, some 150 km in from its western boundary, about 280 km from and 600 m lower than the central ice divide. It is thus well exposed to weather systems approaching with a component of motion from the west, but protected from any surface weather systems from the east. With an ice movement of almost 50 m a^{-1}, the lowest part of the ice core will have been deposited some 35 km inland at around 140 m greater elevation.

Isotopic profiles

δ value–depth profiles for 2 m samples are shown in Figure 4.11. Depths are plotted as equivalent depth in solid ice, so the maximum depth appears less than in Table 4.2. In addition to showing the thickness of 100 a layers and the 95 per cent confidence limit of plotted δ values (section 6.2), sloping lines show the calculated effect of the change of elevation of ice on δ values with depth in the ice core. These lines have been calculated from equation (1.10), except that instead of using a value of the mean surface temperature–altitude gradient of 1° per 100 m, we use a moist adiabatic Rayleigh gradient of 0.65°/$_{oo}$ per 100 m (see section 3.2). The general agreement between the trend of this line and data at Milcent supports our interpretation of the processes involved. The slightly greater noise levels at Milcent and Dye 3 compared to Crête are due to heavier accumulation that results in fewer years of accumulation in each sample. Also seen in the upper layers at Dye 3 and Milcent is a 'pseudo'-cyclic

Figure 4.11. Isotopic profiles in upper levels of the Greenland ice sheet giving mean δ values over 2 m intervals, using ice equivalent depths. At Crête, depths are plotted in relation to the 1974 surface, Milcent uses the 1973 surface and Dye 3 the 1976 surface. Depths of 100 a of accumulation and confidence levels are shown as in Figure 4.8. The dashed lines show the calculated trend of values due to outward flow of the ice sheet. Data supplied by S. J. Johnsen.

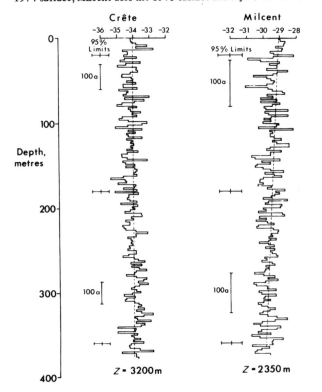

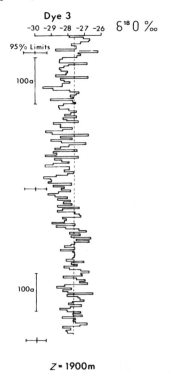

variation of δ value with depth, which is due in part to sampling varying fractions of the odd part of the annual cycles of δ values. This effect is referred to as 'aliasing' in computing techniques and is discussed more fully in section 6.2.

The very detailed record of past accumulation at Milcent given by isotopic profiling is shown in Figure 4.12, from Hammer et al. (1978). In this case the accumulation rate is large enough to prevent ambiguities through the effects of diffusion in surface layers (see section 3.4). Similar records have been obtained at Crête and at Dye 3, but these have not been published in full. The records are marked with the date of each summer layer on the left, while true depths are shown on the right (sloping figures in heavier type). Also shown on the right, in upright figures, is a depth scale corrected for variations of vertical strain since deposition, so that annual accumulation may be calculated directly from this scale. On the left of the figure are plotted variations of total β activity which are due primarily to nuclear explosions of known date. These variations provide a good check on dating over the last two decades.

With a somewhat lower accumulation rate at Crête, occasional ambiguities in identifying summer layers can arise due to diffusion in the firn layer. By use of deconvolution techniques (section 3.4) and cross-checking with the corresponding micro-particle profile, ambiguities of dating can be largely removed. Hammer et al. (1978) estimated that errors of dating of their Crête core did not exceed three years at the base of the core (AD 548).

Temperature profiles

Temperature profiles for the three sites shown in Figure 4.1 have been supplied by L. Hansen (personal communication), and have not been published.

Greenland – other drill sites

Results from several other sites in Greenland are of importance, although more limited in scope than data from the four major locations already covered. We summarise work at four such sites in Table 4.3.

Temperature profiles at the four sites are shown in Figure 4.1.

Site 2

At this site, pioneering studies of deep drilling techniques were carried out by the US Army, Snow, Ice and Permafrost Research Establishment (later CRREL), in preparation for deep drilling in Antarctica during the International Geophysical Year. A wide range of stratigraphic studies on the ice cores from this site are described in Langway (1970). In addition to showing that annual layering was well preserved by isotopic stratigraphy at 300 m depth, he carried out a thorough range of physical and geochemical studies of the ice core that laid the foundations for later work. Temperature profiles at Site 2 are given in Hansen & Landauer (1958) and their interpretation is discussed further in Robin (1970). Core recovery during 1957 was nearly continuous from 19–110 m, about 75–85 per cent was recovered from 110–305 m and two 5 m lengths were obtained at 360 and 411 m. A hand-drilled ice core was obtained from 0–25 m to cover the near-surface layers.

Station Centrale, Camp VI

These holes were drilled in 1950–1 by Expédition Polaires Francaises, and while little analysis of ice core material has been published, the temperature profiles of Heuberger (1954) for the two sites (Figure 4.1), provided a major stimulus that helped to further the study of glacier thermodynamics (Robin, 1955).

Jarl Joset

Use of the Philberth thermal probe at this site as part of the EGIG programmes in 1968 has provided spot temperatures at 220, 600 and 1000 m depth, as shown in Figure 4.1. The probe uses electrical heat to melt its way down in the ice, uncoiling the electrical leads behind it as it sinks. The leads freeze into the ice behind the probe, and are used to send temperature and other information to the surface. To measure ice temperature in situ it is necessary to stop the probe at one depth for around four days to ensure that all meltwater around the probe is fully frozen, and the

Table 4.3. *Station data*

	Site 2	Camp VI	Station Centrale	Jarl Joset
Latitude	76° 59′ N	69° 42′ N	70° 55′ N	71° 20′ N
Longitude	56° 04′ W	48° 16′ W	40° 38′ W	33° 30′ W
Elevation	2000 m	1598 m	2961 m	2865 m
Ice thickness	≈2000 m	≈1374 m	≈3020 m	≈2430 m
Accumulation rate	0.36 m a^{-1}	32 m a^{-1}?	32 m a^{-1}	32 m a^{-1}
10 m temperature	−24.0 °C	−13 °C	−27 °C	−28.5 °C
Ice velocity	17 m a^{-1}	100 m a^{-1}	14 m a^{-1}	17 m a^{-1}
Surface slope	0.0043	0.0075	0.0024	0.0034
Drilling date	1956, 1957	1950	1950	1968
Drilling method	Rotary air-filled	Rotary air-filled	Rotary air-filled	Thermal probe
Depth drilled (max.)	411 m	126 m	151 m	1000 m

Figure 4.12. Continuous δ¹⁸O profile along the 398 m long GISP ice core from Milcent. Dating (cf. AD numbers to the left of the curves) is accomplished by counting summer peaks downward from surface, the interpretation in the upper strata being

supported by the specific β activity profile shown to the outer left. The δ values are plotted along a linear depth scale (normal figures) corrected for varying density, varying accumulation rate and ice thickness upslope, and for total vertical strain as calcu-

lated by two-dimensional ice-flow modelling. The sloping figures are true depths in metres. (Hammer *et al.*, 1978, Fig. 4. Reproduced from the *Journal of Glaciology* by permission of the International Glaciological Society.)

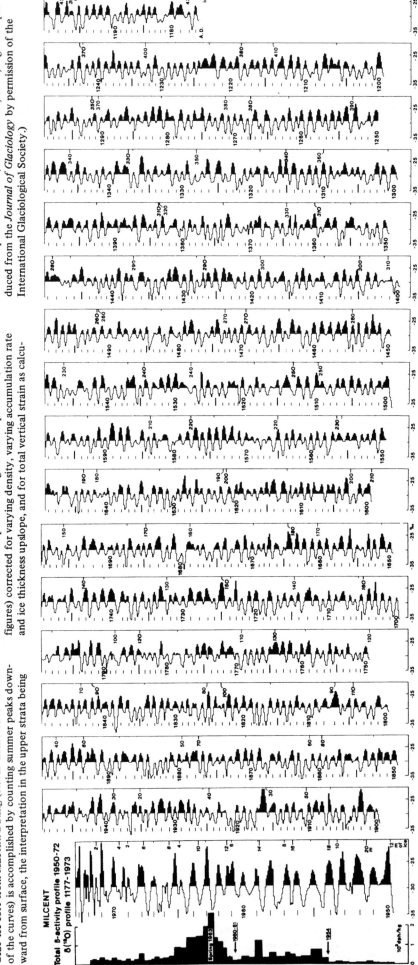

temperature–time curve is then extrapolated to give the in situ temperature. Although the technique provides a way that is simple in principle for obtaining ice temperatures at considerable depths, technical problems such as failure below 1000 m at this site have discouraged further development.

Central Greenland: ablation area

Although readily accessible in comparison with the central region of an ice sheet, little attention has been given to isotopic measurements in the ablation zone of ice sheets. Raynaud (1976) presents isotopic measurements from the western edge of the Greenland ice sheet, in line with the end of the EGIG line. These can be considered in relation to equations (1.16)–(1.18) and the accompanying discussion. Near the outer margin samples were about 30 cm in length, and further inland the sample size was up to 1.5 m. With similar reasoning to that used in chapter 1, it appears that each sample will cover at least 10 a accumulation at the deposition site and possibly up to 100 a. δD values were obtained both in the marginal zone and along the EGIG profile, as shown in Figures 4.13 and 4.14. Over the outermost 500 m, samples were collected every 25 m along two parallel lines 100 m apart, starting from the lateral surface moraine that probably marks the boundary between inland ice and ice of relatively local origin. The mean value of δD for each pair of values is shown in the figures, the mean difference between each pair of δD values being $8.5^{0}/_{00}$ ($1.1^{0}/_{00}$ $\delta^{18}O$).

The significance of the results can be estimated approximately on a steady-state model by linking sites in the ablation zone with the corresponding point on the inland EGIG profile at which surface firn has the same δD value. A series of such links is shown by dashed lines in Figure 4.14. We can check the model of Figure 1.20 described in chapter 1, by use of equation (1.18), taking the equilibrium line at 1400 m

and at 50 km on Figure 4.14 and using velocities and thicknesses from Figure 2.17 and values of accumulation A and ablation N as shown in Table 4.4. Some figures in Table 4.4 are rough approximations since no measured data is available, especially for Z_N, the ice thickness in the ablation zone, and to some extent for mean ablation rates N. In spite of these approximations, the calculated width (δX_N) of strips in the ablation zone show a broadly similar variation to the observed values. The latter are the sections of the ablation zone shown in Figure 4.14 with the same δ values as the sections of the accumulation zone, δX_A, used for calculations of δX_N.

For the strip from the ice divide (Crête) to Station Centrale, the calculated δX_N is 1.30 km compared with an observed value of 0.2 km, a ratio of 6.5 : 1. This is probably a minimum value as $\bar{Z}_N$ may be over-estimated and the observed value of δX_N may be high as the δD curve bends at this point. This suggests that the theory is inadequate, perhaps through use of a vertical column model. Alternatively, the basal ice may be more readily channelled into outlet glaciers or lost by bottom melting in the outer part of the ice sheet.

Comparison of the observed values of δX_N with calculated values for the other three cases shows the observed values to be greater by factors of 2.16, 3.19 and 4.5 as one moves to the equilibrium line. Although large errors may be present, the pattern nevertheless suggests that much more ablation is taking place in this zone than is necessary to keep the ice sheet in balance. To bring about a balance, it appears necessary to lower the equilibrium line by 200–250 m, thus moving it 20–25 km towards the ice margin. This, together with matching adjustments to ablation zone parameters of say halving the ablation rate and adding 60 per cent to the product of ice thickness and ice movement, would produce a rough overall balance for each of the three sections from the equilibrium line to Station Centrale.

Figure 4.13. Variation of δD values with distance inland from the moraine at the western edge of the Greenland ice sheet at 69° 43' N. Each point is the mean of corresponding values along

two parallel sampling lines 100 m apart. The right-hand scale shows corresponding $\delta^{18}O$ values. Data from Raynaud (1976).

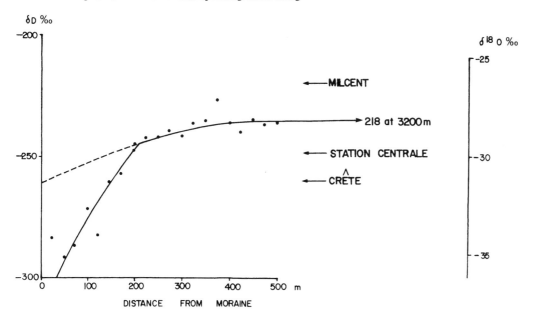

Our general conclusion from Table 4.4, that there is excess ablation over that for a steady-state model, matches the results of Bauer (1968) who measured a mean lowering of the surface of the outermost 35 km of the ablation zone of 0.33 m a^{-1} between 1948 and 1959.

One further link between isotopic evidence from the marginal zone and other glaciological parameters deserves mention. We can calculate from the measured surface velocities shown in Figure 2.17 the travel time of our vertical column along the EGIG profile and through the marginal zone. As pointed out by Raynaud (1976), the sudden fall in δD values, 200 m from the moraine (Figure 4.13), to below the present-day δ values

at Crête could be explained if the marginal ice had been deposited during the last ice age. We take the change of slope of the δD curve to correspond with the esti-mated date before which the δ values fell rapidly at Camp Century, which Hammer *et al.* (1978) put at about 9500 a BP. Figure 4.15 indicates that in this time, a vertical column moving with the present-day surface velocities would have travelled from a location close to Station Centrale to the ice margin. By comparison with present δD values at Crête, Raynaud suggested that δD values were around 35⁰/oo (4.4⁰/oo of δ^{18}O) lower in central Greenland during the ice age. This figure is increased to about 43⁰/oo (or 5.4⁰/oo of δ^{18}O) if the ice originated at Station Centrale. The figures could be

Table 4.4. *Calculated and observed widths of ablation zone strips corresponding to accumulation zone sections, observed values being given by identical isotopic δD values*

	Accumulation zone					Ablation zone					
							Corresponding points			δX_N	
Site	Distance from moraine (km) (= X_A − 50)	δX_A (km)	A (m a^{-1})	U_A (m a^{-1})	Z_A (m)	Distance from moraine (km) (= X_N + 50)	N (m a^{-1})	U_N (m a^{-1})	Z_N (m)	Calculated from equation (1.18) (km)	Observed (km)
Crête	510.8					(0?)					
Station		122.7	0.32	6	3000		2.7	18	100	1.31	0.2
Centrale	388.1					0.2					
Milcent	226.8	161.3	0.45	27	2700	3.5	2.7	20	250	1.84	3.3
Carrefour	104.6	122.2	0.50	70	2050	12.5	2.5	27.5	500	2.34	9.0
Equilibrium		54.6	0.40	104	1400		1.0	70	800	8.4	37.5
line	50.0					50.0					

Figure 4.14. Variation of isotopic δD values with distance from the edge of the Greenland ice sheet as for Figure 4.13. (*a*) shows variations from 0–50 km over the ablation zone, while (*b*) shows variations from 0 to about 500 km on a different distance scale, but using the same δ value scale. Values inland of the equilibrium line were measured at the sites indicated. Horizontal lines link locations with same observed (surface) mean δ values in the accumulation and ablation zones. Data from Raynaud (1976); Merlivat, Ravoire, Vergnaud & Lorius (1973).

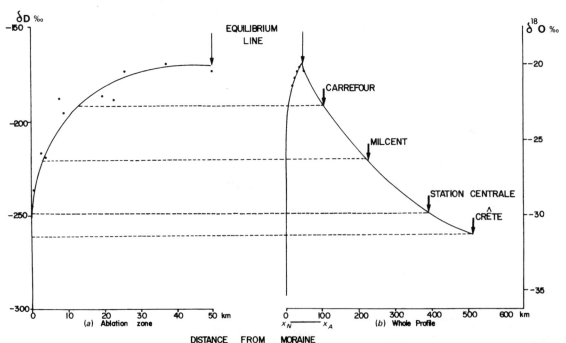

further increased if allowance is made for a decreasing
ice velocity with depth below the surface and hence
a point of origin closer to the coast. Changes of ice
thickness would also affect δD values of ice in central
regions during the last ice age, but the approximate
agreement with δ shifts in Devon Island suggests that
the ice thickness in central regions was not very different
from the present-day value towards the end of the ice age.

West Greenland iceberg study

A pioneering project carried out in 1958 made
a study of radio-carbon ages and $\delta^{18}O$ values of ice from
eleven icebergs off West Greenland between 61.5° N and
75.0° N as reported in Scholander et al. (1962). Since
both techniques were still under development, the
results are not accurate in comparison with present
techniques, but they did show that ice with the lower
δ values was of the greatest age. If we allow for ice in
a trunk glacier reaching the sea through a major glacier
about 1000 a sooner than it would reach the ice margin
in Figure 4.15, we find the times shown in Figure 4.15
are about double those estimated by Scholander et al.
(1962) from their field evidence. A further study along
the same lines would be interesting, as any major discrepancy between travel times deduced from icebergs and
from measurements of inland ice movement would
indicate irregular movement and changes of dimensions
of the ice sheet.

Figure 4.15. Travel time for surface point on the ice sheet to
move along the EGIG profile from the ice divide in central
Greenland to the edge of the ice sheet at the location of Figure
4.13. Times are integrated from the velocities of Hofmann
(1974) which are also shown in Figure 2.17.

Devon Island

75° 20′ N 82° 30′ W

Elevation   1800 m

Ice thickness   (1) 300 m   (2) 298.9 m
                 (3) 299.4 m

Accumulation rate   0.22 m a^{-1}

10 m temperature   −23.1 °C

Ice movement   ≈1 m a^{-1}

Date of drilling   (1) 1971   (2) 1972   (3) 1973

Method   Thermal coring

Ice coring depth   (1) 220 m   (2) 298.9 m
                    (3) 299.4 m

Age of basal ice   (2) & (3) ≈125 000 a

Ice cores were obtained from three sites, of which holes
72 and 73 are 27 m apart on the same flowline, about
900 m north of the ice divide (Figure 4.16). Hole 73 is
upstream of hole 72. Hole 71 is about 600 m north of
the ice divide in the same locality. Comparison of results
between the 1972 and 1973 boreholes shows to what
extent results from a single borehole can be considered
as representative of the area around the drill hole. Results
from this site also provide a useful comparison with data
from Camp Century about 600 km distant at a similar
elevation (section 6.3).

Comprehensive studies down the drill hole and of
the surrounding ice cap provide much detailed information that makes it possible to check the validity of
glaciological models that are in common use. Figure
4.17 shows the time scale down the ice core from
Paterson et al. (1977), based on the measured vertical
motion of irregularities in the diameter of the drill hole
(Paterson, 1976) and on ^{14}C and ^{32}Si dating of ice
samples extracted from an adjacent borehole. Except

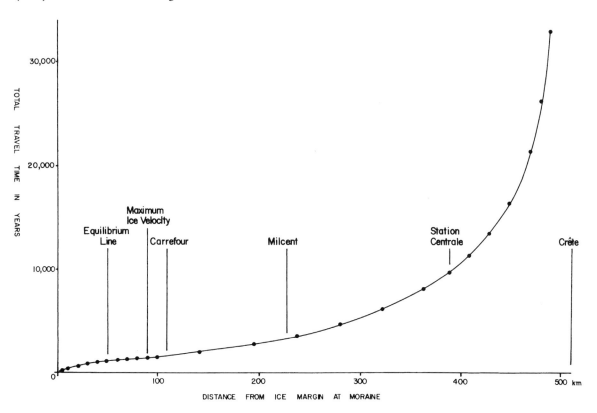

for one ^{14}C measurement that may have had some contamination present in the ice, agreement between dating methods and that of the usual steady-state calculation is reasonable over the top 95 per cent of the borehole.

The Devon Island ice cap is situated on a plateau that falls steeply to Jones Sound to the north and Lancaster Sound to the south (Figure 4.16). During the last ice age these sounds must have contained large drainage glaciers from the main ice sheet over the Arctic Islands. It is likely that the level of these drainage glaciers was so far below the surface level of the ice cap that the latter continued to exist during the ice age as an independent ice cap or ice divide

Figure 4.16. Map showing location of boreholes on the Devon Island ice cap and its position relative to Camp Century, Greenland. From Paterson *et al.* (1977, Fig. 1).

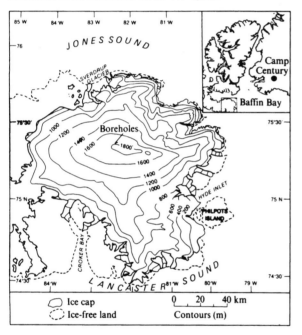

Figure 4.17. Time scale for the core, based on measured vertical velocity, and absolute dates. The vertical lines besides the ^{32}Si dates indicate the depth interval over which each sample was taken. Each ^{14}C sample covers a depth of 1.5 m. Crosses, ^{32}Si date; dots, ^{14}C date. Bars are 1 standard error on each side of the mean. From Paterson *et al.* (1977, Fig. 2).

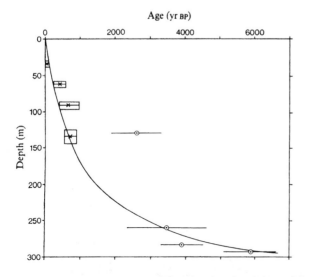

running out from the main ice sheet. There is no evidence that ice from the main ice sheet rode over the Devon Island ice cap during the Wisconsin ice age, so it appears that the basal layers in the ice core must be of local origin. The ice cap may well have been thicker at this time, but this is limited by the size of the plateau on which it is sited. Fisher (1979) puts this increase at a maximum of 150 m, or only 100 m if allowance is made for isotopic subsidence. The effect of this change on the δ^{18}O value is put at $-1^0/\!{}_{00}$.

The full δ^{18}O depth profile of the 1973 ice core from Devon Island is shown in Figure 4.18 using equivalent ice depths. Although there are various significant trends over a few decades above 200 m depth, these are of the order of $1^0/\!{}_{00}$ only, but below 240 m depth a more pronounced trend is seen down towards the ice

Figure 4.18. Isotopic profile to bedrock on Devon Island (1973 core) showing mean δ^{18}O values over 2 m sections along the ice core, except over the bottom 10 m which show mean values over one metre. Depths shown are the ice equivalent depth. The thickness of layers containing 100 a accumulation (vertical bars) and 95 per cent confidence levels (horizontal bars) are plotted as on Figure 4.8. (From data supplied by W. S. B. Paterson.)

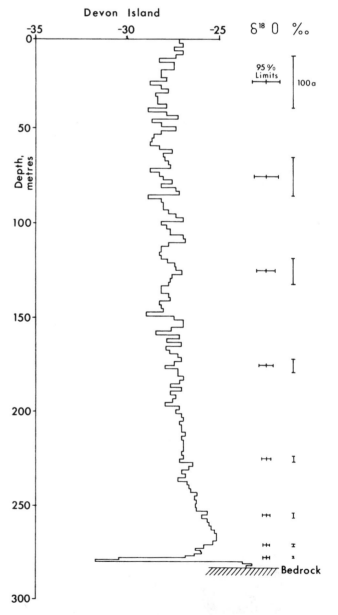

of glacial age which is shown by two points at 278–280 m. This ice-age information is very compressed and close to bedrock compared to profiles from Byrd Station in Antarctica and even to Camp Century in Greenland.

Figure 4.19 shows mean δ values over 10 and 50 a intervals from data averaged over both the 1972 and 1973 cores. This doubles the number of annual values used in each mean, hence the 95 per cent confidence limits are reduced by √2 over other estimates. Paterson *et al.* (1977) determined correlation coefficients between δ values measured on the two ice cores. Comparison of 50 a mean values between the two cores gave correlation coefficients of 0.965 down to 13 m above the bed. This figure is in line with evidence in section 6.2. The correlation then decreased to 0.449 between 13 and 5 m. Measured δ values of core segments of length equal to the mean annual accumulation were compared by matching individual sets of data covering 50 a from the two cores so as to give the best fit. This gave correlation coefficients that varied from 0.45–0.72.

Figure 4.20 (from Paterson *et al.*, 1977) presents

Figure 4.19. (a) 10 a-mean δ ¹⁸O values for past 800 a. Values from 1972 and 1973 cores are combined.

(b) 50 a-mean δ ¹⁸O values for past 10 000 a using values from both cores. Beyond 5250 a BP the time scale is unreliable.

95 per cent confidence levels shown by horizontal bars are slightly less than for a single core. From Paterson *et al.* (1977, Fig. 4) with addition of confidence limits.

Figure 4.20. Values of δ, for samples 10 mm long (a, core 73; b, core 72), against distance above bedrock for lowest few metres of each core. Sections and features marked with the same number or letter are considered to be equivalent. The 'cold peak' at 2 m in core 72 is doubtful; it may result from a sampling error. Gaps up to 50 mm long are apparent in both cores. From Paterson *et al.* (1977, Fig. 3).

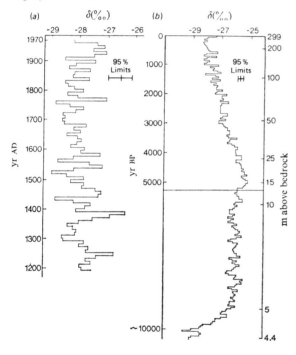

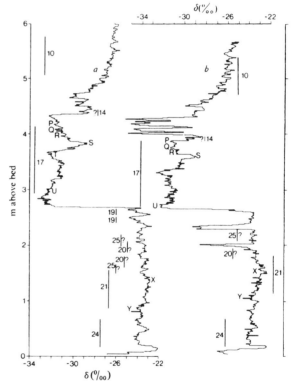

Figure 4.21. Temperature–depth profiles in 1971 and 1972 boreholes on Devon Island ice cap. Redrawn from Figs 2a & b of Paterson & Clarke (1978).

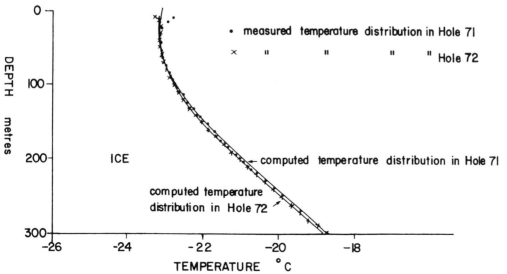

a detailed study of the lowest 6 m of both the 1972 and 1973 ice cores, showing mean δ values for individual 10 mm sections. We do not attempt to show the depth of 100 a sections as there are time gaps in the core, which is discussed in some detail in section 3.4, nor are 95 per cent confidence limits shown. If we assume the ice age lasted 50 000 a, each 10 mm section will cover a time span of about 250 a on average. Paterson *et al.* (1977) have combined both 1972 and 1973 profiles of Figure 4.19 into one, but gaps in the record are still considered probable, while discussion in section 3.4 suggests that some sections could be duplicated within each profile. Sharp discontinuities of δ values indicate that very recent shearing (<1000 a) has occurred at lower levels and could even be taking place at present (Koerner & Fisher 1979, and section 3.4).

Temperature distribution

The temperature profiles measured in hole 72 and a temperature profile measured in a hole drilled to 220 m in 1971 that did not reach bedrock due to difficulties with drilling are shown in Figure 4.21. Hole 71 penetrated 220 m out of an ice thickness of about 300 m. It lies about 300 m closer to the ice divide, at an elevation 25 m above that of the other two holes. Temperatures were measured at intervals of 2–5 m in both boreholes one year after drilling to ensure that any temperature perturbations had died out. Calibration of thermistors used was accurate to ±0.01 °C, which is also the r.m.s. difference between readings by two thermistors, and was taken as the standard error of each temperature.

A detailed analysis of the temperature profiles in Paterson & Clarke (1978) takes account of past variations of the latent heat released by refreezing of meltwater within the upper firn layers, of past surface temperatures as indicated by $\delta^{18}O$ variations, of the accumulation rate of snow and ice and of the geothermal flux. They show that in this location, where ice lenses in the firn are very pronounced as a result of summer melting, variations of the amount of refrozen meltwater with time have a greater influence on the temperature profile than do the variations of $\delta^{18}O$ values over the same period. The effect is to raise the mean ice temperature a few metres below the surface by varying amounts of up to a few degrees above the mean surface temperature. There is probably no runoff of meltwater as on Meighen Island ice cap (section 3.4), so unlike the latter the mean δ values still provide a complete record of δ values of precipitation. Thus the isotopic profile gives a reasonably satisfactory record of past surface climate, although the seasonal cycles in isotopic profiles have been largely obliterated through the action of meltwater. At stations on Greenland of similar latitude and elevation (Site 2 and Camp Century) there is relatively little ice lensing, so effects of meltwater percolation are not significant there.

4.3

Profile data, inland Antarctica

G. de Q. ROBIN

Byrd Station (new)
80° 01′ S 119° 31′ W

Elevation   1515 m
Ice thickness   2164 m; ≈2200 m (radio sounding)
Ice core   to bedrock
Accumulation rate   0.13 m a^{-1}
10 m temperature   −28.0 °C
Ice movement   12.8 m a^{-1} towards 140° True
(see section 3.7)
Drilling date   1968
Drilling method   Thermal coring to 227 m then electro-mechanical rotary drill
Deepest radio-echo layering   ≈1800 m possibly absorption limited

Byrd Station is 300 m lower and 160 km downslope to the south west of the ice divide that separates ice flowing northwards to the Amundsen Sea from that which flows westwards to the Ross Ice Shelf. Section 2.4 includes a discussion of the ice sheet of Marie Byrd Land and its past changes. Very detailed studies of ice flow upstream of the borehole along the Byrd Station Strain Network (BSSN) are described in sections 3.7 and 5.5, and in Whillans (1973, 1976, 1977, 1979).

The top 65 m of the drill hole was lined with casing and isotopic data are not available for this section. Thermal coring continued to 227 m, after which drilling was done in a liquid-filled hole with an electro-mechanical rotary drill lowered by wire as described in Ueda & Garfield (1970). Core recovery exceeded 99 per cent. With the exception of brittle and fractured core, between 400 and 900 m depth, the overall condition of the core varied from good to excellent (Langway, 1970).

The value of climatic information based on ice core analysis from the Byrd ice core has been queried because of controversy over the stability of the ice sheet of West Antarctica (Hughes, 1973; Thomas, 1976; Weertman, 1974; and others). Whillans (1976) showed that no major changes of the flow regime have occurred during the past 30 000 a. Slow but limited changes of ice thickness with time are however indicated from changes in total gas content (sections 3.9 and 5.3), by mass balance studies (section 3.7; Whillans, 1976, 1979) and by comparison of isotopic profiles of Byrd and Dome C (section 6.3), but these changes appear smaller than the corresponding changes affecting the Camp Century ice core (section 5.3). The Byrd ice core is of particular value because its depth range means that ice deposited at the culmination of the last ice age occurs 800 m above bedrock and that at the end at 1100 m above bedrock. At these levels any irregularity of deformation due to flow over and around rough bedrock relief will be small.

In addition to the temperature profile shown in Figure 4.3, Table 4.5 gives details of a set of temperature measurements made down the borehole between 2nd and 7th February 1968, after completion of drilling on 29th January 1968. Comparison of measured temperature with computed curves in Radok, Jenssen & Budd (1970) show random fluctuations of around $\pm 0.1\,^{\circ}$C in the lower half of the profile where temperature gradients are strong, and less than half this amount between 200–800 m depth where gradients are smaller. Since measurements were made shortly after drilling it seems unrealistic to expect the temperatures to have an accuracy better than $\pm 0.1\,^{\circ}$C above 1000 m depth, or $\pm 0.2\,^{\circ}$C or more at lower levels. No detailed temperature measurements are available for the lowest 350 m of the ice column because water encountered at the base of the ice first upwelled some 55 m into the hole; this water then convected into ethylene glycol above this level creating a heavy slush in the lowest 350 m of the hole which prevented lowering of the temperature probe. Therefore the basal temperature is calculated from the overburden pressure and the observed presence of water at the base of the ice. Thus, although we know the mean temperature gradient over the lowest 350 m, we lack knowledge of any variations of gradient near the base of the ice sheet.

Isotopic profiles

The isotopic–depth profile in Figure 4.22 shows mean δ values over 4 m sections from 68 m depth to bedrock. Figure 4.23 shows more detail from 1400 m to bedrock by presenting mean values over 1.5 m intervals to about 2060 m and 1.0 m intervals for the bottom 100 m. Because of the relatively low accumulation rate, our 95 per cent confidence limits for interpreting isotopic data as indicators of temperature change are shown as 50 per cent higher than at stations where the accumulation rates are higher (see section 6.2). The absence of sudden changes of δ values of 3–5^{0}/$_{00}$ between adjacent samples down to a depth of 2100 m contrasts with some records from Camp Century and Devon Island. This indicates that, even during the ice age, changes of climate in Antarctica took place gradually rather than catastrophically.

While the deepest radio-echo layers at Byrd Station are seen at 1800 m depth, the absence of sharp jumps in δ values down to 2100 m depth suggest that deformation of this part of the ice column may have been relatively uniform. Below 2050 m sudden changes

Table 4.5. *Temperature profile of Byrd Station, 2–7 February, 1968 (Hansen, personal communication)*

Depth (m)	Temperature ($-^{\circ}$C)
133.5	26.10
163.7	26.93
194.2	27.44
224.3	27.77
254.8	28.02
285.0	28.15
315.2	28.26
345.6	28.33
375.8	28.37
406.3	28.40
436.5	28.48
467.0	28.53
497.1	28.58
527.3	28.62
557.8	28.65
588.0	28.69
618.1	28.74
648.6	28.79
679.1	28.84
709.0	28.85
739.1	28.84
769.0	28.84
798.9	28.84
829.1	28.82
858.9	28.76
889.1	28.70
919.0	28.61
949.1	28.50
979.0	28.38
1009.2	28.20
1039.1	28.12
1068.9	27.87
1098.8	27.63
1128.4	27.33
1159.2	27.04
1187.8	26.68
1217.7	26.31
1247.2	25.96
1277.1	25.56
1306.7	25.10
1336.5	24.57
1366.4	24.03
1396.0	23.44
1425.9	22.91
1455.7	22.36
1485.6	21.82
1515.2	21.05
1545.0	20.39
1574.6	19.58
1603.9	18.90
1633.7	18.13
1663.3	17.39
1811.1	13.10
2164.4 (bottom)	1.6 (calculated for 197 bar pressure)

in the δ profile could be due either to discontinuities in the age–depth scale caused by folding or shearing of basal layers or to a very high compression of layers covering a considerable time span.

Other studies

Although dating of ice cores, particularly at great depths, may be subject to considerable error, when several different studies are made on the same ice core relative dates between the different types of study are closely defined by distance along the ice core. The significance of parameters such as the dust content of the atmosphere can be determined in relation to the climatic record shown by the $\delta^{18}O$ profile. Thompson, Hamilton & Bull (1975) report such a study of the Byrd ice core.

The internal structure and crystal fabric of the Byrd ice core have been studied by Gow & Williamson (1976) who consider its implications on the flow of the ice sheet in this region.

Measurements of the tilt of the borehole down to a depth of 1482 m, and hence of the amount of internal shearing in the ice mass, reported in Garfield & Ueda (1976) are shown in Figure 3.25.

Studies of the impurity content of the Byrd ice core to establish the relative concentrations and variations with depth (time) of the major cationic constituents,

Figure 4.22. Isotopic profile of the Byrd borehole giving mean δ values over 4 m intervals from the surface to bedrock. Bars show thickness of 100a or 1000a of accumulation at different depths and 95 per cent confidence levels as in Figure 4.8. The age scale on the right is that shown in Figure 5.50. Isotopic data supplied by S. J. Johnsen. (See also Figure 1.2.)

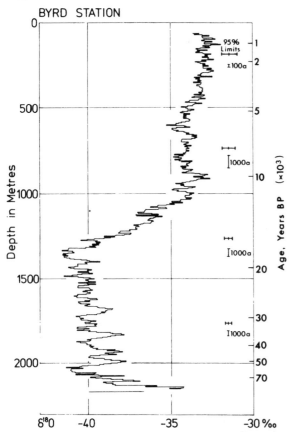

Na^+, K^+, Ca^{2+} and Mg^{2+}, have been reported in Ragone & Finelli, (1972).

Basal ice

The bottom 4.83 m of the ice core contains abundant stratified debris ranging from silt-sized particles to cobbles, as reported in Gow, Epstein & Sheehy (1979). This ice layer is almost air free, in contrast to the ice immediately above. It is concluded that the layer consists of refrozen meltwater that has incorporated basal debris during freezing. The δ values indicate that the source of basal water has come from melting of ice of δ values close to those found above this level, but some fractionation effects during refreezing have occurred.

Dome C
74° 39' S 124° 10' E

Elevation   3240 m
Ice thickness   3700 m
10 m temperature   −53.5 °C (1978)
Accumulation rate   0.037 m a⁻¹
Ice movement   Nil (from topography)
Depth of ice core   906 m
Drilling method   Thermal coring
Drilling date   1977–8

The drilling site on an ice divide (dome) was chosen for simplicity of interpretation of past ice motion. It lies at point C on Figures 2.9 and 2.10. Although a repeated fix by satellite Doppler technique has not yet confirmed absence of movement, it is unlikely to exceed 1 m a⁻¹. Some uncertainty remains over past changes of ice thickness, especially prior to 10 000 a BP, and there is a possibility that the position of the ice divide changed from these earlier times (see section 2.3), but the shift is unlikely to be large.

Apart from some doubt over past variations of ice thickness, climatic interpretation of the isotopic profile is straight-forward since this is the only site chosen on glaciological grounds for ice coring to depths greater than 500 m. Elsewhere ice coring has been carried out at stations whose siting was decided on other criteria. Comparison with the results of the Byrd ice core in section 6.3 helps to confirm the value of this site for ice core studies.

Drilling was by thermal coring in a dry hole, so the depth attained was limited by hole closure due to pressure of overlying ice. Due to the thermally induced stresses in drilling the ice core was cracked extensively and was unsuitable for determination of total gas content. Core recovery was 98 per cent complete. Isotope values in Figure 4.24 are plotted for 4 m increments of core, corresponding to a time interval varying from about 100 a near the surface to about 135 a at the lowest level. The age scale shown on the left, from Lorius, Merlivat, Jouzel & Pourchet (1979), makes allowance for a decreased accumulation rate during the ice age.

The temperature profile shown on Figure 4.3 shows a similar profile to that at Vostok Station, with a proportion of the basal heat flux still being conducted upwards at the surface. An analysis of this temperature profile has been prepared by Ritz, Lliboutry & Rado (in press).

Geochemical studies from pits and ice cores have also been made and are discussed in *Trace metals in Antarctic snows since 1914* (Boutron & Lorius, 1979), *Sulphate in Antarctic snow: spatio-temporal distribution*, (Delmas & Boutron, 1978), *Crystal size and total gas content: two indicators of the climatic evolution of polar ice sheets* (Raynaud et al., 1979) and *Polar ice evidence that atmospheric CO_2 20,000 yr BP was 50% of present* (Delmas, Ascencio & Legrand 1980).

Studies of isotopic deposition noise made by Benoist et al. (1982) in the 1980–1 season are mentioned in section 3.3 and are discussed more fully in section 6.2 (see Figure 6.2). These studies showed that the 95 per cent confidence limits for interpretation of changes of δ values as showing temperature changes in the Dome C ice core (Figure 4.24) are about three times as large as in regions of high accumulation rate.

Vostok

78° 28′ S 106° 48′ E

Elevation   3500 m

Ice thickness   3800 m

10 m temperature   −57 °C

Accumulation rate   0.027 m a⁻¹

Ice movement   ≈3 m a⁻¹

Depth of ice cores   952 m (1972–3) and 905 m (1974)

Drilling method   Thermal coring

Vostok lies around 300 km to the east of an ice divide and at around 300 m lower elevation in an area of low precipitation. Movement of ice is estimated at about 3 m a⁻¹ from steady-state calculations, and a similar figure has been obtained from astronomical measurements (Liebert & Leonhardt, 1974). The ice core results come from the top 26 per cent of the ice sheet, so corrections for deformation are not large. Flowline data, such as accumulation rate, ice movement and temperatures, upsteam of Vostok Station have to be

Figure 4.23. Detailed isotopic profile of Byrd borehole below 1400 m showing mean δ values over 1.5 m intervals to 2060 m depth, and over 1 m intervals over the lowest 100 m. Accumula- tion and confidence limits shown as on Figure 4.23. Data supplied by S. J. Johnsen.

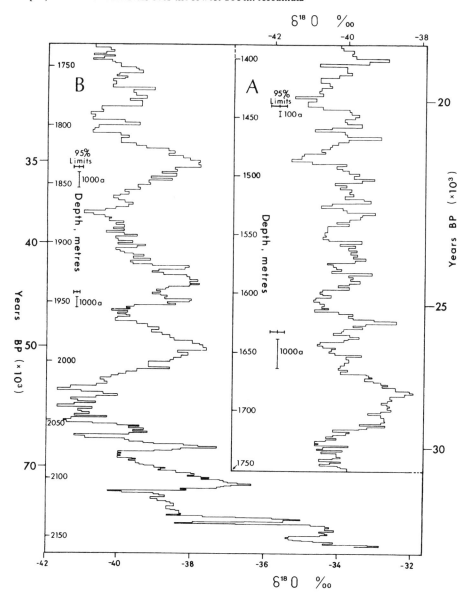

estimated from a few traverse measurements which
indicate average conditions in this region. The data
used for modelling are shown in Figures 5.58 and
5.66. Due to its remoteness from the coast, to the
thickness of the ice and to a low accumulation rate,
changes of ice thickness with time are likely to be
slow. Voronov (1960) estimated the elevation increase
at Vostok to be possibly 100 m during the Wisconsin,
and not more than 300 m during the Quaternary
maximum.

Drilling by thermal coring in an air-filled hole was
limited in depth by hole closure. The resultant thermal
stresses in very cold ice produced considerable cracking in
the ice core. Core recovery was up to 99 per cent over
many sections (Korotkevich & Kudryashov, 1976).
Isotopic data is not available for the top 50 m, and
a continuous core was not obtained over several sections
up to 40 m in length. Core recovery appears to be almost
complete from 330–800 m depth.

The isotopic profile in Figure 4.25 is based on
data from 300 samples from Barkov, Korotkevich,
Gordiyenko & Kotlyakov (1977) who state that the
measured accuracy for not fewer than three measurements on each sample reached $\pm 3^0/_{00}$. Sample lengths
ranged from 0.5 m to 5.0 m for the main analysis but
δ values from some samples of 0.05 m to 0.5 m from
earlier measurements above 500 m appear to be included.

Since sampling intervals are irregular, we cannot indicate
95 per cent confidence limits as on other isotopic
profiles. We have therefore grouped the data in Figure
4.25 into 50 m intervals and have added a heavy line
showing the mean $\delta^{18}O$ value over each 50 m interval,
as well as providing a shaded zone showing the standard
deviation as an indication of the error of the mean. In
Figure 5.65 a smoothed curve from a running mean of
20 measurements is shown. This was used for calculating
temperature–depth profiles in section 5.5.

Detailed studies of shorter sections of ice core,
made in Copenhagen on samples of 0.005–0.02 m and
reported by Dansgaard, Barkov & Splettstoesser (1977),
do not show the rapid fluctuations of δ value with
depth of Figure 4.25 from Barkov *et al.* (1977) whose
values are generally some $3^0/_{00}$ lower than the corresponding measurements of Dansgaard *et al.*

The temperature–depth profile to a depth of
480 m is shown in Figure 5.61. From the deviations of
measured points from the computed curve it appears that
the relative accuracy of measurements was around
$\pm 0.02\,^\circ C$, and the absolute accuracy is believed to be
better than $\pm 0.1\,^\circ C$.

Figure 4.25. Isotopic profile from Vostok. The profile is not
continuous but based on 300 samples of varying length. The
heavy vertical line shows the mean δ values over 50 m sections,
and shaded areas are the standard deviation as a guide to errors.
Depths of 1000 a of accumulation are shown by vertical bars.
Data from Barkov *et al.* (1977).

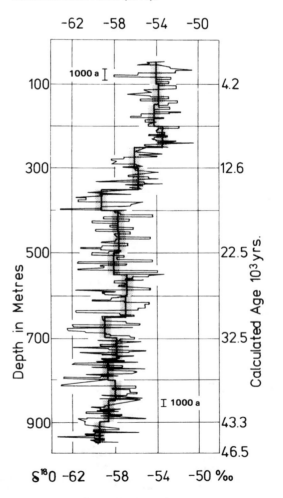

Figure 4.24. Isotopic profile at Dome 'C', Antarctica (adapted
from Lorius *et al.*, 1979, Fig. 2). Although measurements were
made of δD ratios, data is presented as the equivalent of $\delta^{18}O$
mean values over 4 m sections of ice core equivalent to about
100 a accumulation near the surface and increasing to about
135 m at 900 m depth, so no vertical 100 a layer bars are shown.
95 per cent confidence limits are shown by horizontal bars as in
Figure 4.8. The age scale makes allowance for a decreased
accumulation during the ice age.

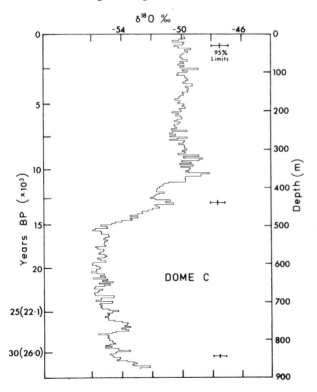

Law Dome – Strain Grid F

$\approx 66^\circ\ 09'$ S $\approx 111^\circ\ 00'$ E

Elevation   375 m

Ice thickness   380 m

10 m temperature $\approx -11.5\ ^\circ$C

Accumulation rate ≈ 0.10 m a^{-1}

Ice movement   9 m a^{-1}

Depth of ice core   348 m

Year of drilling   1974

Although Law Dome is linked to the main ice sheet of East Antarctica, its surface form and measurements of ice movement show that it has an independent flow regime at present. Since it is only around 100 km across, a comprehensive range of glaciological studies spanning most of the Dome have been carried out and reported in Budd (1970*b*), Budd & Morgan (1973, 1977), Budd, Young & Austin (1976) and Russell-Head & Budd (1979). The temperature and related studies of the 1976 paper are shown in Figure 1.17 as a good example of the way in which steady-state temperature theory fits observed profiles. The Law Dome isotopic profiles now available compress the climatic record during the last ice age in much the same way as in ice cores on Devon Island. Unlike Devon Island, where the size of the ice cap will have been little different during the last ice age, the Law Dome may have expanded in area and thickened considerably. This makes quantitative climatic interpretation during the last ice age difficult. On the other hand the comprehensive range of measurements of physical parameters on Law Dome makes possible certain studies of ice flow on a practicable

scale as opposed to the high cost of tackling similar problems on a continental scale.

Figure 4.26 from Budd & Morgan (1977) shows the measured isotopic profile at strain grid F compared with a steady-state profile and estimated age scale. The latter two parameters have been calculated from ice movement, strain grid measurements (see Figure 1.17) and near-surface isotopic values over the Dome upstream of the boreholes. The additional drop of about $7^0/_{00}$ in δ^{18}O values around 300 m at age over 8000 a BP corresponds approximately in date and magnitude with falls on the major ice core profiles from inland Antarctica, and is particularly close to that at Byrd Station where the corresponding fall in δ values is attributed mainly to climatic cooling, with an added effect due to the regional surface elevation being 200–400 m higher at the culmination of the ice age (section 5.3). A similar explanation seems likely on Law Dome, although the possibility that the low isotopic values are due to flow of ice from further inland is also discussed at some length in Budd & Morgan (1977).

Since the summit of Law Dome is only 1400 m above sea level, much of the Dome lies below 1000 m, the level at which the isotopic–elevation relationship seemed to break down in Adélie Land (section 3.3). Figure 4.27 shows the variation of isotopic ratios with surface elevation for Law Dome and for points further south on the main ice sheet. The former points show a different δ value–elevation gradient to the latter, which is similar to that at higher elevations in East Antarctica. The gradient of total gas content against elevation of about 15 ml kg^{-1} km^{-1} is similar to that found by Raynaud (section 3.9) over a wider range of elevations, but low values of total gas content below

Figure 4.26. The isotopic δ^{18}O profile at Site F near the coast of Law Dome, Antarctica, together with a steady-state profile and estimated age scale. Adapted from Budd & Morgan (1977, Fig. 4).

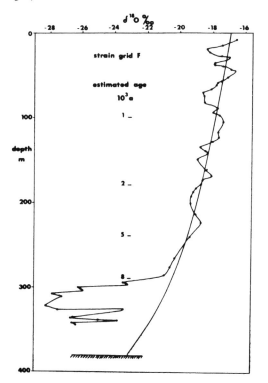

Figure 4.27. Variations of δ^{18}O values with elevation for surface samples from two lines on the Law Dome and for the main ice sheet to the south. From Budd & Morgan (1977, Fig. 5).

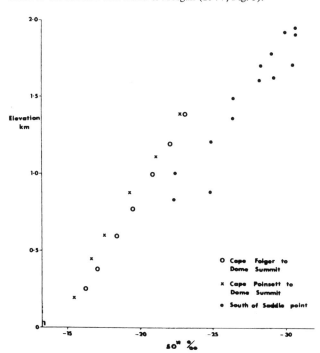

300 m depth in Cape Folger A core suggest a much higher elevation of origin of this ice than appears possible on Law Dome. Reference should be made to the papers quoted for fuller details. The range of glaciological information coming from the studies on Law Dome is of great value to our general understanding of ice sheets. We shall not, however, present more details here as the major records of ice age climate are to be found in the deeper ice cores from further inland.

4.4

Coastal sites, Antarctica

G. de Q. ROBIN

At these stations, the lower level of ice may have been deposited far inland and carried to the coast by the outward motion of the ice sheet. Although precise tracing of the trajectory of the past motion of ice at different depths is not possible, we can nevertheless estimate the region in which the ice at any depth was deposited by use of a steady-state model that assumes that the form, flow and climate of the ice sheet are constant at present-day levels and have not varied with time. The simple theory involved is described in chapter 1 (p. 13). We find that this approach explains the main features of isotopic profiles in coastal regions, and gives some indication of the loss of basal ice by melting beneath ice shelves. Such information helps to confirm our model of the flow of the Antarctic ice sheet, rather than to contribute directly to our interpretation of past climate. We therefore discuss our conclusions regarding ice flow from coastal stations in this section, rather than try to integrate these conclusions in the final chapter.

Terre Adélie D10
66° 42′ S 139° 55′ E

Elevation   270 m
Accumulation rate $\approx 0.18 \, \mathrm{m \, a^{-1}}$
Ice temperature $-14 \, ^\circ \mathrm{C}$
Surface velocity $5.4 \, \mathrm{m \, a^{-1}}$
Depth of ice   303 m
Ice core length   303 m
Drilling method   Thermal coring
Date of drilling   1973–4

On Figure 4.4 the shaded zone bounded by two flowlines diverging from Dome C shows the accumulation zone feeding a section of the coastline in the vicinity of the borehole D10. The accumulation data and area between the flowlines have been used to calculate the steady-state isotopic profile near the coast according to equation (1.15).

D10 borehole is situated 5 km inland from the coast and 1 km from the Astrolabe Glacier which drains the main outflow of inland ice in this locality. Earlier ice cores obtained closer to the coast in this region (Figure 4.28) show continuity of the drainage pattern of inland ice to the coast in spite of the presence of Astrolabe Glacier.

In Figure 4.29 the curve of elevation of origin based on our steady-state model is compared with the measured isotopic profile and with measurements of the total gas content from the same ice core. The latter two scales are plotted to correspond with observed isotopic–elevation (section 3.3) and total gas content–elevation relationships of the present day (section 3.9), so that departures from steady-state conditions stand out. From 50 to 100 m depth, departures of δ values and total gas content from the elevation curves are considered to be of relatively local origin, caused by the lack of an isotopic–elevation relationship below 1 km in this area, as mentioned in section 3.3. From 100–200 m depth there is reasonable agreement between the steady-state curve and observed values, but from 210–250 m both isotopic and total gas content measurements indicate considerable departures from steady state and suggest that a colder and thicker ice sheet was present at that time. The presence of morainic debris in the ice from 227–237 m depth, as well as in the lowest 2 m, shows that a break must be present in the time–depth relationship in the ice core at this level. The results above 225 m indicate that the isotopically coldest period occurred before the maximum ice thickness was reached and that the decrease in ice thickness took place relatively rapidly. Raynaud, Lorius, Budd & Young (1979) have modelled isotopic and gas content measurements along similar lines to those of sections 5.3 and 5.5, which involves estimating ice velocity over the shaded area in Figure 4.4 in addition to the parameters used in our steady-state approximation. They conclude that over the past 5000 a conditions have been approximately those of a steady state, and prior to that the maximum increase of ice elevation around 250 km from the present coastline would have been around 400 m. This is around the figure suggested in section 2.4 for a seaward advance of the coastline of about 100 km.

Below 225 m at least two explanations of the profile are possible, since the presence of moraine between 227–237 m depth indicates that ice that has been near the base of the ice sheet has overridden clean ice. The processes discussed in section 3.10 suggest that overfolding could result in ice from 270–303 m being of the same age as ice from around 210–235 m,

Figure 4.29. Isotopic profile at D10 in Terre Adélie, showing mean δ D values over approximately 5 m intervals (vertical lines), measurements of total gas content (X, or X with horizontal line for mean of several measurements) and calculated elevation of origin (continuous curve) plotted against proportional depth. Isotopic and total gas content data from Raynaud (1976).

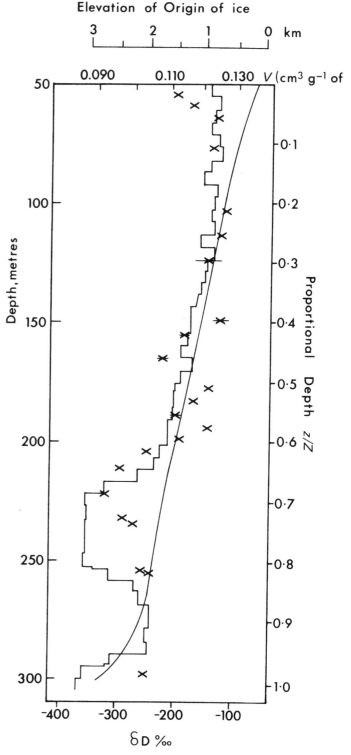

Figure 4.28. Cross-section of coastal ice of Terre Adélie showing deuterium content of two ice cores and in the surface ablation zone. From Lorius (1968, Fig. 1).

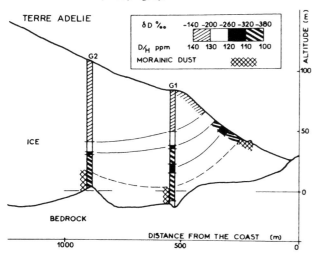

while the time scale could be inverted from say 270–237 m. Alternatively, sub-horizontal shearing of ice over a subglacial peak could have produced a time gap in the ice column, as occurred near the base of the ice cap on Devon Island according to Paterson *et al.* (1977). In this case the ice from 270–303 m would be from the last interglacial period. Points can be made in favour of both explanations, so no firm conclusion is possible.

It would be useful to apply equation (1.18) to see how the surface isotope values in Figure 4.28 fit a steady-state model. Lorius (1968) indicates that the accumulation zone starts about 1 km from the coast, that is in the vicinity of hole G2. In the outermost kilometre, the surface mass balance is stated to be practically zero and the ice movement to be very small 'if not practically zero'. This agrees qualitatively with the predictions of chapter 2, since Figure 4.24 shows that ice of $\delta D = -260^0/_{00}$ reaches the surface about 0.6 km from the accumulation zone (equilibrium line?), while similar ice is deposited 245 km inland at present. Using mean values of ice thickness, accumulation rate and velocity to this inland point predicts that, for ablation of $1\,\mathrm{m\,a}^{-1}$ in the ablation zone, we would expect movement of $5\,\mathrm{m\,a}^{-1}$. For decreased ablation the predicted movement would decrease in proportion. Thus movement of say $0.2\,\mathrm{m\,a}^{-1}$, which might be consistent with the phrase 'if not practically zero' quoted above, would correspond with an ablation rate of $0.04\,\mathrm{m\,a}^{-1}$, which again is consistent with the 'very low intensity' of melting rate reported for this area. We can therefore say that, qualitatively, the surface isotopic data appears consistent with a steady state for ice deposited since the last ice age. Raynaud *et al.* (1979) have used flowline studies to show that steady-state calculations can produce a close match to the temperature profile at D10 below a depth of 60 m. Again this result suggests that steady-state conditions could have existed in this region over the past few thousand years, but does not prove that this has been the case.

Little America V
78° 11′ S 162° 10′ W

Elevation   44 m flat ice shelf
Ice thickness   258 m
Surface velocity $400\,\mathrm{m\,a}^{-1}$
Surface accumulation rate $0.25\,\mathrm{m\,a}^{-1}$
Bottom melting rate $0.80\,\mathrm{m\,a}^{-1}$
10 m temperature $-22.3\,^{\circ}\mathrm{C}$
Ice core length   255 m
Drilling method   Rotary with compressed air
Date of drilling   1958–9

This is one of two ice shelf cores for which both isotopic and temperature profiles are presented. Studies of the temperature profile and of mass balance in Crary (1961) indicate local bottom melting of around $0.80\,\mathrm{m\,a}^{-1}$ in this locality. From the surface profile and ice velocities of Thomas, MacAyeal, Eilers & Gaylord (in press) it is estimated from a steady-state model that the ice at Little America V will have been afloat for some 1300–1800 a.

It is generally assumed that bottom melting falls off rapidly with distance from the ice front, but Robin (1979) has pointed out that considerable melting or deposition of ice should be expected where the thickness of a floating ice shelf varies in locations far from the ice front. Furthermore, due to the change of freezing point of sea water with depth (pressure), melting should continue throughout the year where water is moving beneath an ice shelf from the open ocean, even when the ocean surface is freezing. This suggests that the amount of bottom melting beneath ice shelves may have been underestimated in many studies.

In Figure 4.30, showing the isotopic–depth profile from Dansgaard *et al.* (1977), we also show, against calculated proportionate depth, the estimated elevation of origin of ice reaching this site from the accumulation zone shaded in Figure 4.4. The δ scale in Figure 4.30 is plotted to make δ values at Little America and at Byrd Station correspond with elevations at those stations. We also assume that the δ values

Figure 4.30. Isotopic profile at Little America V Station showing mean values over approximately 1.5 m intervals, from Dansgaard *et al.* (1977, Fig. 3). The continuous curve shows the calculated elevation of origin plotted against proportional depth (see text).

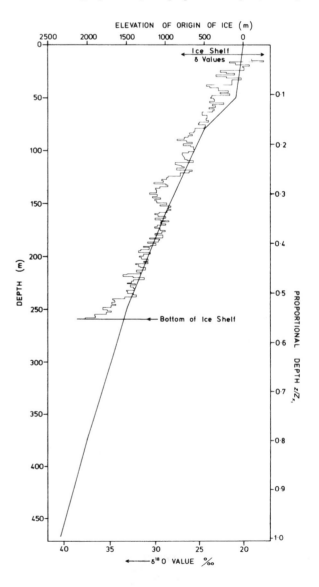

and elevation of origin profiles should correspond (as was seen between 100 and 200 m in the Adélie Land D10 core), so we match the metric depth scale to the proportionate depth scale to bring the mean δ value plot into approximate coincidence with the elevation of origin curves. This is not possible for the upper 90 m, since δ values of ice deposited on the ice shelf fall off with distance from the coast (section 3.3). Below 90 m depth there is moderate agreement between the calculated elevation of origin and the observed isotopic curve, but it appears that about half of the ice column expected from the plot of elevation of origin is missing. Although other matches of the isotopic and elevation scale are possible, it is difficult to reduce the missing portion of the ice column to less than one-third of that calculated on the steady-state model.

Absence of the lower part of the expected ice column can readily be explained by melting off the bottom of the ice shelf. The rate of melting at Little America V could melt the equivalent ice thickness at Little America V (255 m) in 320 a. Alternatively, a mean rate of melting of 0.3 m a^{-1} would melt 450 m of ice in some 1500 a, the approximate time the ice has been afloat. This is approximately half of the thickness at the flotation line (≈ 700 m) plus surface accumulation since flotation (≈ 250 m). A more precise calculation would take account of thinning due to spreading of the ice shelf after the ice column started to float, but to be effective we would also need to know the distribution of bottom melting. The two simple calculations bracket the limits of models that could explain disappearance of half the ice column by bottom melting.

If ice from 235–255 m was deposited at the end of the ice age, as is suggested by lowering of δ values, we could date it around 10 000 a BP. The mean ice velocity required to transport this basal ice from the present 1500 m contour line (the estimated elevation of origin) to the grounding line would then be around 30 m a^{-1}, which appears reasonable on steady-state calculations. Additional thinning due to the final stage of the retreat of the ice sheet at the end of the ice age would also be likely, but this need not be large. In contrast, Dansgaard et al. (1977) tentatively suggest that if some apparent annual layering of around 0.02 m near the base of the ice shelf is real, the age of the basal ice would be around 3500 a. This is likely to be an underestimate due to disappearance of layers. Such an age would still imply considerable bottom melting, in line with the calculations of Crary (1961). However, the more reliable satellite measurements of Thomas et al. (in press) show higher ice velocities than Crary and a floating time of about half his earlier model. This in turn requires a higher contribution of inland ice in the column; this is confirmed by the isotopic profile. Whether this has resulted from the rapid disintegration of the West Antarctic ice sheet, as implied by Dansgaard et al.'s estimate of age of basal ice, or whether it could be explained by limited variations from a steady-state model, as suggested by our analysis, cannot be decided without further information. For this purpose we need further surface glaciological and isotopic data along the flowline to Little America as

well as knowledge of the total gas content in the lower levels of the ice shelf.

Amery Ice Shelf
$\approx 69^{\circ}\,00'\,\text{S}\,71^{\circ}\,00'\,\text{E}$
Elevation ≈ 50 m
Ice thickness   428 m
Surface velocity   800 m a^{-1}
Surface net accumulation rate   0.39 cm a^{-1}
Bottom accumulation rate $\geqslant 0.30$ m a^{-1}
10 m temperature $-20\,^{\circ}$C
Ice core depth   315 m
Drilling method   Thermal coring
Drilling date   1968

Although this ice core also comes from an ice shelf, the isotopic profile appears very different from that of Little America. The ice shelf is fed primarily by the Lambert Glacier that drains a large basin stretching back to the major ice divide of East Antarctica, as shown by the shaded area in Figure 4.4.

The drilling project was carried out during five months under primitive conditions, and led to substantial improvements of technique for work on Law Dome. In general the core quality was good, although there were problems due to shattering when cores were exposed to the thermal shock of $-40\,^{\circ}$C and 10 m s^{-1} winds. About 40 per cent of the cores were returned for analysis (Bird, 1976).

Figure 4.31 presents the isotopic measurements of Morgan (1972) with some revised values (Budd,

Figure 4.31. $\delta\,^{18}$O values (crosses) plotted against depth at Station G1 on the Amery Ice Shelf. Data from Morgan (1972) and W. F. Budd (personal communication). Also shown is the calculated elevation of origin plotted against proportional depth (see text).

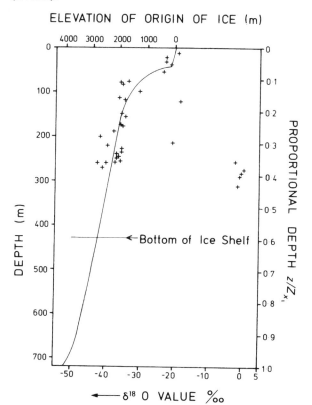

ELEVATION OF ORIGIN OF ICE (m)

personal communication) plotted against depth, along
with a steady-state curve of elevation against proportional ice depth calculated according to equation (1.15).
The isotopic scale has been matched to the elevation
scale from δ values in this region shown in Figure 3.10,
while the proportional depth scale has been matched
to the metric depth scale to bring the δ values and
elevation curve into approximate coincidence. The
form of these curves differs markedly from that of other
coastal profiles.

In the upper layers we see the limited amount of
accumulation that comes from a low elevation on the
ice shelf and Lambert Glacier, while at greater depths
converging flow brings large quantities of ice from
higher elevations, in contrast to Adélie Land which is
dominated by diverging flow. The isotopic and elevation
data in Figure 4.31 show reasonable agreement down to
a depth of 270 m, but our matching of scales indicates
that about 60 per cent of the ice column has been
melted off. About one-third of this has been replaced
by basal deposition of 160 m of ice with δ values characteristic of frozen sea water, as was the case for one
sample at only 255 m depth. Morgan (1972) has also
published conductivity figures for melted ice samples
that lead to the identical conclusion, since high conduc-
tivities typical of ice frozen from sea water were also
measured at 255 m depth and below 270 m depth.

The freezing of substantial quantities of ice to the
base of this ice shelf was first suggested by analysis of
strain measurements in Budd (1966*b*), which indicated
that a basal freezing rate of $0.90 \, \mathrm{m \, a^{-1}}$ was needed for
mass balance in this area. The unusual form of the
temperature–depth curve for this site shown in Figure
4.3, when analysed by Radok and others (personal
communication), suggested basal freezing was taking
place at about $0.70 \, \mathrm{m \, a^{-1}}$. The isotopic and conductivity
measurements of Morgan (1972), when combined with
measured ice velocity data and the assumption that
basal freezing took place beyond the apparent grounding
line at the outlet of Lambert Glacier into the ice shelf,
suggest a mean basal freezing rate of about $0.30 \, \mathrm{m \, a^{-1}}$.
Although the three different estimates of basal freezing
rate indicate some uncertainty, it is clear that the rate
is substantial and is as large as, or more than, the surface
accumulation rate. This amount of basal freezing cannot
be explained by conduction of heat through the ice
shelf to the surface, and must be due to thermal processes
in the sea water beneath the ice shelf. The results at this
site were the first to show convincingly that oceanographic processes were dominant in such situations.
Robin (1979) reviews evidence of these processes,
especially beneath the Ross Ice Shelf, while radio-echo
strengths measured by Neal (1979) show that the
pattern of freezing and melting beneath ice shelves is
complex and related to bottom topography.

Measurements of ice movement and thickness at
the entry to, and exit from, Lambert Glacier have been
used by Allison (1979) to determine the mass budget
of the region. He finds that only half the amount of ice
falling on the basin inland of Lambert Glacier is at
present discharging into the head of the glacier. It
also appears that little more than one third of the ice
entering the glacier is being discharged into the Amery
Ice Shelf, which means that the mean level of the
Lambert Glacier should be rising by about $0.19 \, \mathrm{m \, a^{-1}}$.
He suggests that this pattern indicates that there has
been a surge of ice flow from the whole basin in the
past and that the present lack of balance shows that
ice levels are now building up towards their former
level as the ice sheet recovers from the surge.

An alternative process to the surging has been
put forward by Robin (1979) who suggested that most
of Lambert Glacier is afloat on sea water (see Figure
2.13*b*). At greater depths heavy melting could be
taking place at temperatures down to $-3.0 \, ^{\circ}\mathrm{C}$, and
the less saline and less dense water produced would rise
under the Amery Ice Shelf towards the open sea. The
decreasing pressure in the rising water results in the
formation of ice which forms the saline ice layer at the
base of the ice shelf. The imbalance of the mass budget
studies on the Lambert Glacier would then be explained
by the heavy basal melting and only partial replacement
of ice by refreezing. The imbalance between the interior
basin and the Lambert Glacier inflow might be due to
overestimation of accumulation rates. Although Allison
has taken what are considered low values to avoid this
criticism, microwave radiation data (Zwally & Gloersen,
1977) suggest that the whole basin is one of unusually
low accumulation. More data from the drainage basin
is needed to sort out this question. An isotopic profile
through the ice sheet near the entrance to Lambert
Glacier would be of great help.

ANALYSIS OF TEMPERATURE, ISOTOPIC AND TOTAL GAS CONTENT PROFILES

5

5.1

Synopsis

W.F.BUDD

Introduction

The aim of the work described in this chapter has been primarily to interpret the data from ice core isotope measurements in terms of past surface temperature changes, through analyses of the present temperatures in the ice sheet which are observed in the boreholes. This work may be divided into the following parts:

(1) to examine whether isotope-derived temperatures, when used with the ice sheet dynamics regime (to the extent that it is known), give rise to the observed temperature profile;

(2) to examine what the measured gas volumes in the ice may imply concerning past elevations of the origins of the ice;

(3) to consider the effect of such elevation and possible ice thickness changes on the subsequent temperature profile;

(4) to examine the uniqueness of the above derived results and the consequences of plausible variations in the unknown factors involved;

(5) to determine what resultant climatic temperature and ice thickness changes can be inferred from the data.

Results of previous studies

The application of Robin's (1955) analysis of heat conduction for steady state to the Camp Century temperature profile gives a first-order approximation to the measured profile with a maximum discrepancy of about $3\,^{\circ}C$. The most interesting aspects of such an analysis are the reasons for the discrepancies. Weertman (1968) made a careful analysis of the problem and showed that if this steady-state model is improved to take account of the downstream advection, the maximum discrepancy is increased rather than decreased. The addition of the effect of internal heating in the ice also was examined and found unable to account for the discrepancies. Weertman also noticed that a closer matching would result if a lower accumulation rate applied. These features suggested that some degree of non-steady state may be relevant.

Dansgaard & Johnsen (1969*b*) obtained some slight decrease in the discrepancy by using a different form of approximation to the velocity profile, but still were not able to account for the bulk of the discrepancy. Nevertheless, the isotope data also provided by Dansgaard, Johnsen & Møller (1969) indicated for the first time that substantially non-steady-state conditions probably occurred prior to about 10 000 a BP. In addition, the isotope data for the annual layers presented by Johnsen, Dansgaard, Clausen & Langway (1972) suggest that the accumulation rate has been reasonably close to its present value since the large isotope change.

Although Philberth & Federer (1971) were able to achieve closer matching to part of the profile, there was still a discrepancy of about 1 °C between the temperature differences from the 150 m depth to the base. Again this raises the question of whether the remaining discrepancies reflect non-steady-state effects.

Non-steady-state analyses of the type described by Budd (1969, section 4.10) and Robin (1970) indicate that the upper part of the Camp Century temperature profile should be governed by the relatively uniform conditions of the last 10 000 a, except for the reversed gradient in the upper 150 m which could be simply accounted for by the fluctuations in climate over periods of about the last century.

Radok, Jenssen & Budd (1970) showed how the various elements in the heat conduction equation affect typical temperature profiles and indicated that computer techniques could be used to scan the fields for each variable to find the best-fit values.

Such a study for Camp Century was carried out by Budd, Jenssen & Radok (1971*b*) which showed that close matching could be achieved over a wide range of different values of accumulation rate, A, and surface temperature change rate, S, with a fixed rate of A/S. In fact, the closest fits corresponded closely to the observed accumulation rate with a net surface *cooling* rather than the net warming advection, due to the down-slope motion, required for steady state. Such a slight net cooling, of the order of 0.2-0.4 °C $(10^3 \text{ a})^{-1}$, could be due to either climatic change or to a surface rising, of about 20–40 mm a^{-1}, which would be a very small degree of net imbalance. A slight net cooling of about this magnitude (depending on the temperature–isotope factor) is also apparent in the isotope record. However, from the steady-state, or steady warming-rate, analyses of Budd *et al.* (1971*b*), the time period over which the deduced best-fit temperature change applies is not clear because of the large diffusion of the time variations.

For the workshop, Robin suggested using the isotope data as the source for specifying temperatures as a function of time as input to the models of heat conduction, to examine whether the measured temperature profile results. This led directly to the analyses and results described in this chapter.

Cambridge workshop analyses

Two different approaches were undertaken as part of the workshop. These two approaches were complementary and, although they differ in concept and output,

it is shown in the summary that the most important conclusions were confirmed by both methods. The first approach, as described in subsequent sections by Jenssen & Campbell in section 5.2 and Jenssen in section 5.3 is as follows. A vertical column fixed in space at the Camp Century location was considered and the heat conduction equation for ice moving through this column was modelled. A two-layer velocity profile of the Dansgaard & Johnsen (1969*a*) type was used and the effects of variable density and compaction in the upper layers were considered. To begin with a steady-state temperature profile is calculated with the present surface temperature and the present basal temperature gradient. The surface temperature is held constant and the normal steady-state horizontal advection associated with the down-slope flow of the ice is neglected. The effects on the temperature profile are examined for a number of variables including accumulation rate, variable density, lower layer velocity gradient, variable thermal parameters, initial temperature profile, grid spacing, and so on. For the transient, or non-steady-state study, the surface temperatures which were fed into the model were derived with a linear equation from the $\delta^{18}O$ isotope ratios provided by Johnsen in a digitised form as a function of time from 20 000 a BP. (BP – at the time of the temperature measurement was 1966.) Averages over 500 a, 50 a, and 5 a were used up to 20 5000, 2000 and 500 a BP respectively. The time scale used was that of Dansgaard & Johnsen (1969*a*), and for these calculations the base gradient was fixed at its present value, as were the ice thickness and the accumulation rate at 0.37 m a^{-1} (of ice).

A number of different runs were carried out using different digitised forms of the temperature input data. Finally, further runs were carried out using, in addition to the temperature changes inferred from the isotopes, the elevation changes deduced from the gas volumes as described below and incorporated into the model simply as ice thickness changes of the column.

In section 5.3 Jenssen describes the technique for using the gas volume data of Raynaud & Lorius (1973) for Camp Century, together with the isotope data, to derive elevations of origin of the ice and past climatic temperature changes after allowance has been made for effects of changing elevation. Measurements of Raynaud & Lebel (1979) and section 3.9 shows that the total gas content of ice is mainly a function of elevation of origin, with a weak dependence on temperature at the time the gas is trapped in the ice. The theory developed to handle this problem also derives past gradients of surface temperature–elevation for a given surface temperature–isotope conversion factor. The numerical results approximate to present-day values. Similar calculations are also presented for Byrd data using the gas volumes of Gow & Williamson (1975).

The second approach, adopted by Budd & Young, is primarily based on flowline type analyses; however, single-column steady-state and transient analyses are also considered for special purposes. The theory is set out in section 5.4 and the applications to the Camp Century, Byrd and Vostok regions are given in section 5.5.

The main differences from the first approach are as follows.

(1) The temperature variations in the upper 150 m are not dealt with, as this region is strongly influenced by small fluctuations (which are damped out at greater depths) and possibly also by the drilling activities. Hence the temperature at the 150 m depth rather than an arbitrary 'surface' temperature is used as the upper boundary condition.

(2) The velocity profile was taken to vary smoothly with depth in a manner similar to the measured velocity profile. The vertical strain rate and velocity were varied similarly with depth.

(3) For the steady state, non-zero horizontal heat advection was studied over a wide range of positive and negative values.

(4) Likewise the model's reaction was also studied to wide ranges of variations of accumulation rate, ice thickness, basal gradient, and thermal parameters; and with varying degrees of sophistication of the basic model.

(5) A detailed study of the effects of conduction deep into the bedrock was made for the single column model to determine the effect of artificially keeping the base gradient fixed at its present value, as was done in the first approach and in the flowline calculations.

(6) For the flowline model, a time–depth scale was obtained for the isotopes using velocities varying with depth, as in (2) above, and upstream along the flowline. These calculations provided information on the elevation and temperature of origin and other features such as accumulation rates, ice thickness and strain rates upstream.

(7) The gas volumes were used to determine past ice thicknesses, and the effects of using these in the flowline calculations was also examined.

(8) Ranges of the unknown factors were examined and the values giving the best fit were determined, including the basal gradient when conduction in the rock is considered.

(9) The uniqueness of the past history thus derived was confirmed by an examination of a series of alternative histories resulting in exact fits to the measured profile.

(10) The resultant contributions to the total change in temperatures (or isotopes) for the Camp Century ice from the following three independent effects were estimated:
   (i) the climatic temperature changes,
   (ii) the downslope flow of the ice,
   (iii) the ice thickness changes.

Finally similar steady-state column and flowline, and transient column and flowline, analyses were likewise carried out and are described for the profiles of isotopes and temperatures at Byrd and Vostok. The final section presents a summary and conclusions drawn from the results of the whole chapter.

5.2

Heat conduction studies

D.JENSSEN & J.A.CAMPBELL

5.2.1 Introduction

The aim of the computations here described is twofold: firstly, to develop a numerical model of the thermodynamic processes within large ice masses which will incorporate more physical parameters and processes than hitherto, and, secondly, to determine the effects of defining the ice surface temperature changes with time from the oxygen isotope data.

The first aim is, of course, a prerequisite to the second, although it is the latter which in this study was given the greater importance. In the following pages the model used will be briefly described, the nature of its differences from previous models detailed, and then its application to reality, utilising oxygen isotope data, will be analysed.

5.2.2 The heat conduction model
5.2.2.1 *The equation*
The model is closely based on that described in detail in Budd, Jenssen & Radok (1971*a*), which takes as the governing heat conduction equation

$$\frac{\partial \theta}{\partial t} = \frac{K}{\rho c} \frac{\partial^2 \theta}{\partial z^2} - \left(w - \frac{1}{\rho c} \frac{\partial K}{\partial z} \right) \frac{\partial \theta}{\partial z} + Q \right) \tag{5.1}$$

where θ is the temperature of the ice at depth z and time t, w the vertical velocity, K the thermal conductivity, ρ the density, c the specific heat and Q is the internal (frictional) heating. The quantity $K/\rho c$ is the thermal diffusivity.

Equation (5.1) has been derived from the fuller heat conduction equation in two dimensions

$$\frac{\partial \theta}{\partial t} + U \frac{\partial \theta}{\partial x} + w \frac{\partial \theta}{\partial z} = \frac{K}{\rho c} \left(\frac{\partial^2 \theta}{\partial x^2} + \frac{\partial^2 \theta}{\partial z^2} \right)$$
$$+ \frac{1}{\rho c} \left(\frac{\partial K}{\partial x} \frac{\partial \theta}{\partial x} + \frac{\partial K}{\partial z} \frac{\partial \theta}{\partial z} \right) + Q$$

(where U is horizontal velocity and x the horizontal coordinate) by assuming:

(1) $\dfrac{\partial^2 \theta}{\partial x^2}$ and $\dfrac{\partial K}{\partial x} \cdot \dfrac{\partial \theta}{\partial x}$ are both negligibly small;

and *either*

(2) columnar motion for the ice (that is, $U \neq \mathrm{f(z)}$ and is constant with depth) so that $\partial\theta/\partial t$ in equation (5.1) is a local derivative in a coordinate system *moving with* the ice column, and this implicitly *incorporates* the term $\partial U/\partial x$ in it;

or

(3) $U(\partial\theta/\partial x)$ is also negligibly small.

Of (2) and (3) it is the former which is more exact, but it must be noted that equation (5.1) is equally well applicable to either a moving column or a fixed geographical vertical column with ice flowing past it, depending on whether assumption (2) or (3) is chosen.

It is possible to regard equation (5.1) in a third manner, using it to follow the motion and temperature distribution of an *initially* vertical ice column, but which becomes distorted with time if $U = \mathrm{f(z)}$. If this function is such that U changes rapidly only near the base of the ice, say in the lower 10%, then the temperatures found can be assumed to hold for the top 90% or so and the lower portion neglected. This third physical interpretation of equation (5.1) is possibly the least attractive since the longer the time in which the ice is studied, the greater is the region of the distorted ice, and hence the less accurate is the entire calculation.

In reality, U is a function of z so that equation (5.1) describes an ideal situation (or, as just stated, holds for the upper 90% of the ice), but it is possible to incorporate the effect of a non-constant horizontal velocity through a correction term applied to the vertical velocity. This is done below in section 5.2.3.3(a).

Although it is a relatively simple matter to incorporate the effect of a non-constant horizontal velocity this study does not. Such heating has its largest effect near the bedrock, whilst the incorporation of oxygen isotope data modifies the upper portion of the ice. Since emphasis is placed here on the isotopes, the heating Q is neglected. It is, however, a small effect, and some discussion is given by Budd, Jenssen & Young (1973). The basic equation is thus

$$\frac{\partial\theta}{\partial t} = \frac{K}{\rho c}\frac{\partial^2\theta}{\partial z^2} - \left(w - \frac{1}{\rho c}\frac{\partial K}{\partial z}\right)\frac{\partial\theta}{\partial z} \qquad (5.2)$$

It is assumed that this equation applies to an ice column moving along a flowline, wherein horizontal temperature gradients are negligibly small.

5.2.2.2 *Computational considerations*

(a) Finite differences. In order to apply equation (5.2) to a computer, the derivatives must be replaced by finite difference analogues. Examples may be found in Jenssen & Straede (1969). These analogues operate on data equally separated in space and time. Thus the continuous ice column of equation (5.2) is replaced by a series of vertical points.

(b) Axes transformation. To avoid computational problems incurred by ice and bedrock surfaces which vary (perhaps randomly) in elevation, the vertical axis (z) is transformed to a 'relative coordinate' (ζ) equivalent to the transformation on p. 92 of Budd *et al.* (1971*a*),

except that the origin is kept at the ice surface, by

$$\zeta = \frac{\mathrm{z} + \mathrm{M}}{Z} \qquad (5.3)$$

where M is the total depth of meltwater, and Z is the ice thickness. This transformation implies that the heat conduction equation which is used to produce temperatures at some future time is

$$\frac{\partial\theta}{\partial t} = \frac{K}{\rho cZ^2}\frac{\partial^2\theta}{\partial\zeta^2} - \frac{1}{Z}\left(w - \frac{1}{\rho cZ}\frac{\partial K}{\partial\zeta}\right)\frac{\partial\theta}{\partial\zeta} \qquad (5.4)$$

where w is discussed in the section following. Note that now the ice has (in the ζ-system) a horizontal boundary at both the ice/air interface ($\zeta = 0$) and the ice/bedrock interface ($\zeta = 1$). Fuller details of such a transformation are given in Budd *et al.*

(c) The variables of the heat conduction equation. In order to utilise equation (5.4), information on the following variables must be supplied.

θ An initial distribution is required. This is obtained from equation (5.4), but by using that equation in a 'steady state' mode. From this distribution, the finite difference form of the equation will yield $\partial\theta/\partial t$, and hence θ at some future time – at $t + \Delta t$, say.

Equation (5.4) for a steady state is

$$S = \frac{K}{\rho cZ^2}\frac{\partial^2\theta}{\partial\zeta^2} - \frac{1}{Z}\left(w - \frac{1}{\rho cZ}\frac{\partial K}{\partial\zeta}\right)\frac{\partial\theta}{\partial\zeta} \qquad (5.5)$$

where S is constant at all depths and is interpreted as the surface time rate of temperature change. This change is compounded of a climatic warming and cooling (warming) due to a rising (sinking) ice surface. No attempt is necessary (or is made) to separate these components for steady states.

When $S = 0$, no temperatures can change with time; when $S \neq 0$, (the realistic case) temperatures vary, but temperature gradients remain constant.

A treatment of steady-state computations is given in Budd *et al.* (1971*a*). As will be seen later, the exact values of the initial temperature distribution are not critical and two profiles differing by up to 3 or 5 °C will lead to the same final transient state when equation (5.4) is integrated over 20 000 a or more.

w Assuming columnar flow, it can be shown that the vertical velocity in the ζ-system is simply

$$w = \frac{1}{Z}\left(A(1 - \zeta) + \frac{d\mathrm{M}}{dt}\zeta\right)$$

See Budd *et al.* (1971*a*). Other forms of w and w are discussed in section (i) 6–8. Here A is the accumulation, and M is the total amount of meltwater under the ice.

K This is a function of density and temperature (known at every grid point in the ice), and is given by Weller & Schwerdtfeger (1970). Thermal diffusivity (k) is also taken from that paper.

ρ This is prescribed (at 5 m intervals from the ice surface to 500 m: after that depth it is assumed constant at $0.918\,\mathrm{g\,cm^{-3}}$). It can be any arbitrary (real) set of values.

c Specific heat is a function of temperature only, and from the Smithsonian Meteorological Tables (List, 1949) the function $c = 0.50027 + 0.0015\theta$ has been derived.

Z The ice thickness is prescribed initially, and is either prescribed or computed at later times. See section 5.2.2.2(h).

M The total melt is computed; the process is discussed in the section following and in section 5.2.2.2(h).

(d) Phase changes. As is clear from equations (5.4) and (5.5), melting may play an important role in the determination of temperatures. So, therefore, may freezing. Phase changes, in turn, are determined by both the lower boundary (ice/bedrock) conditions and by the heat input here.

If the base temperature exceeds the pressure melting point (that is $-0.00077Z\,^\circ\mathrm{C}$), or if meltwater exists under the ice, the base temperature gradient before and after resetting the base temperature is found. If their difference is I, then the amount of melt/freeze is

$$\frac{\mathrm{dM}}{\mathrm{d}t} = \frac{K}{\rho LZ\Delta t}\,I$$

where L is the latent heat of fusion of ice. Note I/Z has units of $^\circ\mathrm{C\,m^{-1}}$ since I is dependent on derivatives in the ζ-system. The quantity $K/\rho L$ has been chosen to be constant at $0.2310\,\mathrm{m^2\,a^{-1}\,^\circ C^{-1}}$. Fuller details may be found in Budd *et al.* (1971*a*).

Melting occurs if the initial base temperature is greater than the pressure melting point; and freezing occurs in the presence of basal water when the temperature is less than the pressure melting point.

(e) The lower boundary heat input. For the integration of equation (5.4) a lower boundary condition must be specified. This is of two main forms:

(1) the base temperature (θ_b) is kept at a specified (input) value, *or*

(2) θ_b is computed, given that $(\partial\theta/\partial z)_b$ is compounded of both the geothermal heat flux and the frictional heating assumed to be only at the ice base. That is,

$$-\left(\frac{\partial\theta}{\partial z}\right)_b = \frac{\Gamma}{K} + \frac{U\tau_b}{JK}$$

where Γ/K is the geothermal heat flux expressed as a temperature gradient, and τ_b is the base stress $(=-\rho g(\partial E/\partial x)Z$ where E is the elevation). The quantity JK is treated as constant, and has the value $6.0087\times10^2\,\mathrm{m^2\,a^{-1}\,^\circ C^{-1}}$ bar (where τ_b has the unit of bar).

For form (2), the lower boundary condition is the prescription of both Γ and U. Alternative (1) is very rarely used, since it is applicable only over water or when melting occurs.

(f) Parameter specification. As already pointed out above, an initial temperature distribution must be specified.

But, as well, there are other parameters which must be given before any computation can proceed.

Since the heat conduction equation is of second order, two temperature boundary conditions must be specified. The lower boundary has already been dealt with above; the upper boundary condition is simply to prescribe θ there.

As well, however, the bedrock elevation, the accumulation, the geothermal heat flux and the surface elevation of the ice, the horizontal velocity, and the lateral divergence must all be given along the entire flowline in which the ice column moves.

Lateral divergence has not been mentioned before, but is simply explained. As the ice moves, the horizontal velocity, the bedrock elevation and ice elevation can all change. In order for these to be prescribed in a consistent manner, it is assumed that the ice may flow out of the flowline, at right-angles to the motion. This divergence (or convergence) is the 'lateral divergence' term. It is discussed in section 5.2.2.2(*h*) below.

All the parameters listed above are assumed to be quadratic functions of space and linear functions of time, so they are defined by

$$\chi_{x,t} = \chi_{x_0,t_0} + a(x - x_0)$$
$$+ b(x - x_0)^2 + c(t - t_0)$$

The constants a, b, c are chosen to fit observations, and may take any values.

(g) Computational stability. Before the calculations begin, the time and space increments of the finite difference analogues must be such that they do not lead to computational instability. A discussion is given in Budd *et al.* (1971*a*). For the present model, it is found that

$$\Delta t < \text{MINIMUM OF } (3(\Delta z)^2/8k,\ 1.2k/A^2)$$

where Δt is the time increment, Δz is the space increment, and $k(=K/\rho c)$ is a representative value of the thermal diffusivity. In fact, the above equation is used to automatically select Δt from the given Δz. This process will effect the integration of the heat conduction equation most efficiently.

(h) Continuity. Briefly, it is assumed that mass is conserved with time. Thus the ice column gains mass from the accumulation, and (perhaps) freezing of basal water, but loses mass by (perhaps) melting. Now, if the ice column is of thickness Z and velocity U, then the continuity equation to be satisfied is

$$\frac{\mathrm{d}((M + Z)\sigma)}{\mathrm{d}t} = A\sigma$$

where σ is the cross-sectional area of the column. Now the change in σ and the horizontal velocity are linked by

$$\frac{1}{\sigma}\frac{\mathrm{d}\sigma}{\mathrm{d}t} = \frac{\partial u}{\partial x} + \frac{\partial v}{\partial y}$$

where v is the velocity *across* the flowline, and $\partial v/\partial y$ is termed the lateral divergence. These two equations thus allow

(1) melt (or freeze) to be computed, and

(2) accumulation, horizontal velocity, ice thickness and surface elevation to be prescribed (from observations)

in a fully self-consistent fashion. Any imbalance of mass which results is explainable in terms of the lateral divergence, which is computed as a residue. (In fact, the program allows two out of horizontal velocity, surface elevation and lateral divergence to be specified, and computes the remaining one. In the present study, lateral divergence has not been prescribed simply because it is less easy to determine with the same accuracy as the other parameters.)

(i) New aspects of the model. The model described above has additional features to that of Budd *et al.* (1971*a*). Those already discussed are:

(1) variable specific heat, c;
(2) variable density, ρ;
(3) variable thermal conductivity, K;
(4) improved, consistent, continuity (section 5.2.2.2(*h*));
(5) optimised time step selection using the stability criterion (section 5.2.2.2(*g*));
(6) the model also allows for a completely arbitrary vertical velocity (consistency with continuity is still assured) of the form

$$w = \sum_{r=0}^{5} a_r z^r \quad \text{for } z = 0 \text{ to } Z^* \quad (5.6a)$$

$$w = \sum_{r=0}^{5} b_r z^r \quad \text{for } z = Z^* \text{ to } Z \quad (5.6b)$$

That is, the vertical velocity (dz/dt) may be specified by two fifth-order polynomials, the change occurring at the arbitrary depth Z^*. This, in effect, allows variable strain rate with depth to be incorporated into the model. A simple example is given in the discussion of Camp Century computations.

The velocity so prescribed is transformed, within the computer program, to its equivalent in the ζ-coordinate system, that is, to w.

(7) With variable density, compaction within the upper portion of the ice column must be considered. This is done simply by amending the vertical velocity ($w = dz/dt$) to be

$$w_c = w(\rho_i/\rho_z) \quad (5.6c)$$

where ρ_i is the density of ice, ρ_z is the density of the column at depth z, w is the vertical velocity of equations (5.6*a*) and (5.6*b*) and w_c is the amended vertical velocity. For further details see Budd *et al.* (1973).

(8) As stated above, the ice will not, in reality, move as a column along the flowline. In order to ease what may be a restrictive assumption (columnar flow), variable horizontal velocity can be treated through the vertical velocity (Budd *et al.*, 1973). Thus

$$w_s = w_c(u_z/u_s) \quad (5.6d)$$

where u_z is the horizontal velocity at depth z, u_s is the surface horizontal velocity and w_s is the

vertical velocity incorporating both compaction and horizontal velocity shear. u_z may be found from the computed temperature profile (of the previous time step) and a flow law. Details of such computations are to be found in Budd *et al.* (1971*a*).

5.2.3 Use of the model: relative importance of parameters

5.2.3.1 Vertical velocity profile

Although equations (5.6) allow almost any arbitrary vertical velocity profile to be fitted analytically, any computation for real data must deal with physically sensible values for the vertical velocity.

Consider the situation which will hold in any actual case, say, for example, at Camp Century. Here, the following assumptions may be made:

(1) the surface vertical velocity is proportional to the accumulation;
(2) the change in vertical velocity with depth, at the surface, equals the measured strain rate;
(3) the vertical velocity is a continuous function of depth (so that equations (5.6*a*) and (5.6*b*) give the same value at $z = Z^*$;
(4) $\partial w/\partial z$ is also continuous at $z = Z^*$;
(5) vertical velocity is zero at the ice/bedrock interface;
(6) $\partial w/\partial z$ is also zero at the ice/bedrock interface.

Under all these assumptions it is found that

$$w = A(1 - 2z/(Z + Z^*)) \quad z = 0 \text{ to } Z^* \quad (5.7a)$$
$$w = A(Z - z)^2/(Z^2 - Z^{*2}) \quad z = Z^* \text{ to } Z \quad (5.7b)$$

where, as before, A is the accumulation rate.

This vertical velocity profile, and the corresponding strain rate $\partial w/\partial z$, is shown in Figure 5.1(*a*). Also shown are the constant strain rate (A/Z) and corresponding vertical velocity of the original heat conduction model (Budd *et al.*, 1971*a*). (For Camp Century, the ice thickness has been taken to be 1400 m, the accumulation to be 0.37 m a^{-1}.) Since layer thickness in a steady-state ice sheet is proportional to vertical velocity, it is of interest to note the close correspondence of the latter with the observed layer thicknesses in Figure 1.9.

5.2.3.2 The basic calculation

In order to study the effects of varying the basic parameter values, and to judge the importance of the improvements (1)–(8) of section 5.2.2.2(*i*) above, a 'basic workshop' computation was performed to act as a reference.

Here the calculations were performed with the following variables.

Ice thickness: 1400 m
Accumulation: 0.37 m a^{-1} ice equivalent
Thermal conductivity, thermal diffusivity, specific heat: all variable
Density: variable, and as given in Figure 5.1(*b*)
Base temperature gradient: 0.018 75 °C m^{-1} (this incorporates geothermal heat flux and frictional heating)

Surface temperature: $-22.5\,^{\circ}$C (This temperature is
arbitrary since any change in it will merely shift
the entire temperature profile. The heat conduction equation (5.4) deals with differences in
neighbouring temperatures, and not the magnitudes.)

Vertical velocity: as in Figure 5.1(*a*) and section 5.2.3.1

Number of points in the column: 29

Initial temperature profile: the steady state obtained for
the parameters listed above

Integration time: 20 500 a (The choice of this particular
time is governed by the length of record available
for the oxygen isotope ratios. Section 5.2.4.1(*b*)
below provides full details.)

The corresponding profile at a time 20 500 a after the
commencement time is shown in Figure 5.2, as curve 1.

5.2.3.3 Variations of the basic calculation

(a) Initial temperature distribution. Rather than use
a steady-state profile for the initial temperatures, a linear
profile was used, with the same temperature as the steady
state at the ice/air and ice/bedrock profile. Thus, at
900 m depth this initial profile is 3.7 $^{\circ}$C warmer than
the steady state at the same depth.

In spite of this seemingly very drastic change in
the initial temperatures, the profile which results 20 500 a
later differs everywhere by less than 0.07 $^{\circ}$C from curve
1, the maximum deviation occurring at the ice/bedrock
interface. This difference is so slight that the temperature
profile has not been plotted in Figure 5.2.

(b) Accumulation. In order to test the effect of
a small (10%) variation in accumulation, a value of
0.402 m a^{-1} was used instead of 0.37 m a^{-1}. This would

simulate observational inaccuracies in measuring that
parameter.

The computation gave a profile whose maximum
difference to curve 1 was less than 0.75 $^{\circ}$C. For this
reason it has not been plotted in Figure 5.2.

(c) Vertical velocity profile. The choice of a depth
of 1000 m for Z* (see Figure 5.1) is slightly arbitrary,
and could, for example, be as high as 800 m. To judge
the effect of varying this parameter, the value of 800 m
for Z* was used. The resulting profile is shown as curve
2 in Figure 5.2.

For a Z* of 800 m, the vertical velocity as given
by equations (5.7) is less at any interior point in the ice
than that for Z* of 1000 m. Thus the downward transport of ice, and the concomitant cold temperature, will
be greater for the basic workshop calculation than for
this variant. Curve 2 bears this analysis out, in which it
can be seen that the temperatures resulting are warmer
everywhere than in curve 1. The differences, however,
are quite small, and do not exceed 0.5 $^{\circ}$C anywhere.

The use of a constant strain rate and linear vertical
velocity, as in the original model, was also made. The
resulting profile is shown as curve 3 in the figure. From
Figure 5.1(*a*) it can be seen that the vertical velocity for
a constant strain rate is greater than that for a variable
strain rate, and hence it is expected that the cold surface
temperatures will be transported more effectively
downward. Curve 3 shows this effect.

*(d) Constant density, specific heat, thermal
conductivity.* If density, specific heat and thermal
conductivity are kept constant, the resulting profile

Figure 5.1. Vertical velocity, strain rate and density profiles used
in the computations for Camp Century. (*a*) shows the vertical
velocity and strain rates used in the model, together with a constant strain rate (roughly the average of that used) and the
vertical velocity corresponding to that average. These last two

profiles are typical of the model most frequently used and are
presented as comparisons for the more realistic profiles used
for the present calculations. (*b*) is the ice density–depth relation.
Below 150 m the density is assumed constant at the value shown.

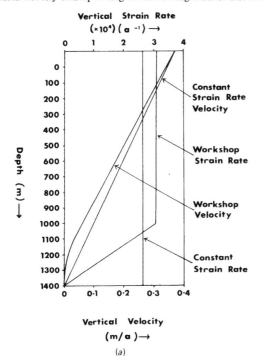

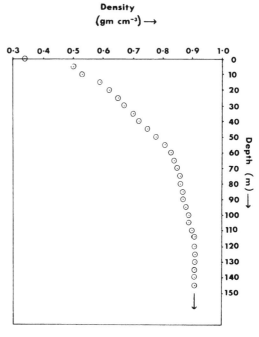

is yet again only slightly changed with the largest deviations from the 'basic' calculation occurring at the bedrock boundary (curve 4). Although density varies most rapidly at the upper boundary, the temperature gradients are greatest near bedrock, with the result that thermal diffusivity ($K/\rho c$) and thermal conductivity (K) vary most significantly there (see equation) (5.4)). Thus the temperature profile is most changed in the lowest few hundred metres, though the magnitude of this change is less than 0.6 °C everywhere. The magnitude of the deviation, of course, depends on the constant values chosen for density, specific heat and thermal conductivity.

(e) Non-columnar flow. The effect of amending the vertical velocity for non-columnar flow is shown in Figure 5.3(*a*), but note that this correction has been applied to the vertical velocity of equations (5.6*a*), (5.6*b*) and not to (5.6*c*); that is, compaction has not been considered. This allows the effect of u_z/u_s to be separated from that of compaction. As expected, the maximum changes from the 'basic' profile occur near the bedrock where u_s varies most rapidly. However, since the vertical velocity is very small there anyhow, the total change is very minor indeed, and reaches a maximum of 0.1 °C at the bedrock.

(f) Compaction. The effect of modifying the vertical velocity to account for compaction was also studied. It would be expected from the form of this correction (see section 5.2.2.2(*i*)) that the temperature profile would be most affected where the density of the column material differs most markedly from the density of ice (here assumed to be 0.915 gm cm^{-3}) that is, in the upper 50–150 m of the profile.

Figure 5.3(*b*) shows the effect of compaction for a specially chosen case in which temperature varies greatly with depth. This profile was selected as an extreme case in which the effect of compaction would be most evident. In spite of the very large temperature gradients in the upper 100 m, compaction is clearly seen to have a small effect. It acts, as one would expect, so as to transfer the temperature changes to a lower depth. It also acts to slightly change the magnitude of the temperature maximum above 100 m in Figure 5.3(*b*). That is, compaction slightly decreases the damping of the downward diffusion of surface temperature changes compared with the case of no compaction. The actual difference in the value of the warmest temperature (at about 60 m) is very small indeed however.

(g) Summary. Changes to the original model of Budd *et al.* (1971*a*), listed in section 5.2.2.2(*i*) and examined above, indicate that they all are 'second-order' effects, and will not grossly affect the temperature profiles produced. Certainly, their effects will be masked, if not overridden, when the surface temperature is allowed to fluctuate with time with an amplitude given by climatic changes over the last few tens of thousands of years.

Nonetheless, while each modification may produce only slight changes, their combined effect, especially over long periods of time, may significantly alter the temperature profile, particularly in the upper regions of the ice where most of them have their maximum effect. This region is also that of major interest in this study.

Consequently, the model as used for calculations involving surface temperature time changes specified by the oxygen isotope ratio data, incorporates all the features described above. Even more, in order to treat the

Figure 5.2. The 'basic workshop' temperature profile, and those resulting from slight variations in the boundary conditions. Curve 1 is the basic workshop profile (see section 5.2.3.2 for definition of the boundary conditions). Curve 2 is the profile which results when the depth at which the vertical strain-rate changes from a constant to a linear function of depth is 800 m (instead of 1000 m as in curve 1). If the strain rate is taken to be constant throughout the entire depth of the ice (see Figure 5.1(*a*)), then curve 3 results. Curve 4 shows the effect of using constant density, specific heat and thermal conductivity.

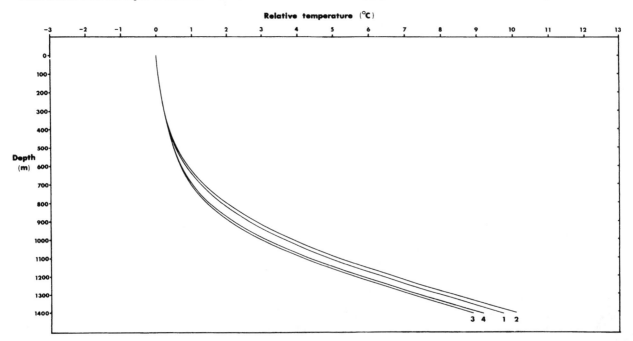

rapid variation of the density in the upper few metres, all calculations have been run using a very fine (12.5 m) vertical point spacing.

5.2.4 The main calculation

5.2.4.1 *Surface temperature time changes*

(a) *The oxygen isotope data.* The most important aspect of this study was to examine the use of oxygen isotope data in prescribing ice surface temperature changes. For this, of course, adequate data must be provided, and this was presented to the workshop by Johnsen. The form of the data was threefold: as 500 a mean values, 50 a means, and 5 a means. These are presented as Figures 5.4, 5.5 and 5.6, in which the isotope ratios (δ) have been converted to temperatures by means of the empirical relation used for these workshop calculations:

$$\theta = 1.633\delta + 25.4 \qquad (5.8)$$

where δ is in units of parts per thousand (and is negative) and θ is in $^\circ$C.

It will be noted that the temperatures obtained at 0 BP (1965 AD) are not the same for the three sets of data. This, of course, is only to be expected when the core data have been analysed independently and are subject to error. Quite apart from these considerations, the 500 a mean, for example, could not be expected to agree with the mean of the last 5 a data. Core sampling techniques and measurement errors will lead to determinations of 5 a means (or 50 a means) which are not entirely consistent with 50 a means (or 500 a means). Thus, for example, if 50 a means are calculated from the given 5 a means and then compared with the given 50 a means, discrepancies of up to $1^0/_{00}$ can occur. Although this figure is not large, consistency between the various means was ensured by computing 50 a means from 5 a data (to 500 BP) and 500 a means from 50 a data (to 2000 BP)

Figure 5.3. The effects of internal deformation of the ice, and compaction. When the horizontal velocity of the ice is not assumed constant with depth, the vertical velocities are changed (see section 5.2.2.2(*i*)) and thus also the temperature profile. (*a*) shows the effect of non-columnar flow of the ice, an effect which is clearly small and confined to the lowest 14 per cent or so of the ice. The warmer curve is that wherein vertical velocities allow for non-columnar flow. (*b*) gives the effect of compaction of the ice, which also changes the vertical velocity. The profile with the larger maximum temperature (at a lower depth) is that for which compaction has been treated. Again, the effect is small, and again, as expected, confined to the upper 18 per cent of the ice. The case shown is an extreme one; very few profiles will show such large temperature fluctuation in the upper 200 m.

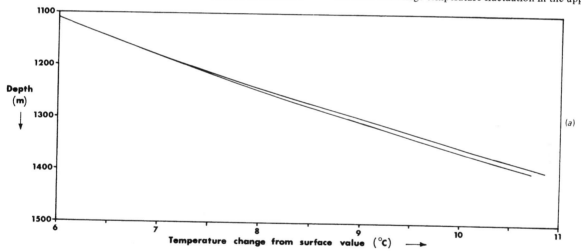

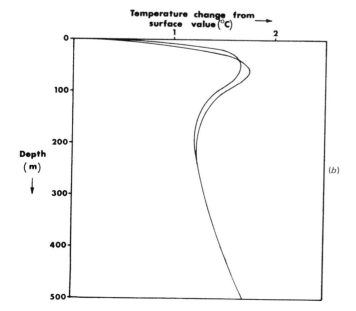

Figure 5.4. Oxygen isotope data used in the heat conduction studies. The data cover the last 21 000 a, and are presented as 500 a mean values in which the oxygen ratios have been converted to temperatures as described in text. It is assumed, therefore, that all the variations in the isotope values are due to climatic changes. The data are for the Greenland Camp Century core.

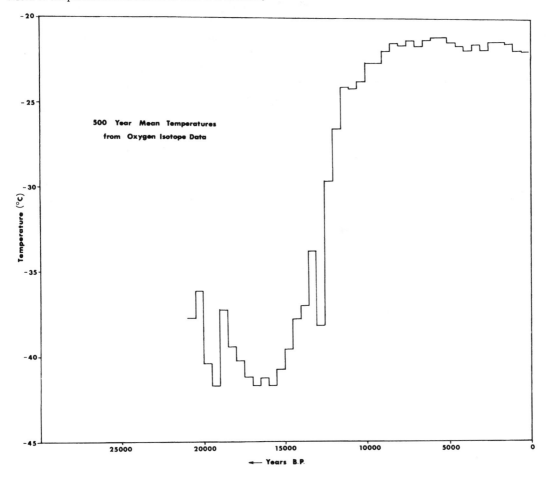

Figure 5.5. As for Figure 5.4 using data for the past 2000 a for Camp Century, as 50 a means.

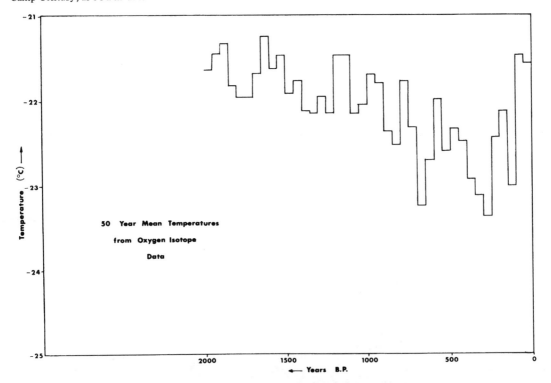

and using these values to replace the given means for those periods. It is these adjusted data which are shown in Figures 5.4, 5.5 and 5.6. It is stressed that the consistency so ensured entails only a very slight modification of the δ values. It should also be noted that the first 500 a means computed from either the 5 or 50 a data are almost identical to the given value.

(b) Deduced time changes. From the data presented in Figures 5.4–5.6 it is a simple matter to compute surface temperature changes per unit time. This information then becomes the upper boundary condition for the integration of the heat conduction equation.

The computational procedure is

(1) set the surface temperature to that given by the δ value at 20 500 BP (remember BP is before 1965 AD);
(2) compute temperatures a short time, Δt, later;
(3) adjust the surface temperature using the computed time change;
(4) repeat from (2) until the time is 500 a later: at this point replace, $(\partial \theta / \partial t)_{\text{surface}}$ by a new value;
(5) continue in this way until 0 BP is reached.

To utilise the 50 a data the process is exactly as above, save that the starting time is 2000 BP with the initial temperature profile being that from the 500 a means integration at that time. New surface temperature time changes are, of course, read in every 50 a.

Finally, for the 5 a data the starting time is 200 BP, and new data is input every 5 a. However, as pointed out above, there remains one inconsistent feature of the temperatures computed from the δ value: the fact that the 0 BP temperature is different for the three data sets. Two possibilities exist: to use the data as it stands, and to accept as unavoidable these inconsistencies; or to force some consistency. The former is preferable, for, due to reasons stated above, some measure of disagreement for 0 BP temperatures must be expected even from perfect data. Nonetheless, the second alternative has also been used. This is simply to adjust the 0 BP temperature for the 5 a means to the observed surface temperature, shifting all other 5 a temperatures by a constant amount; to adjust the 50 a temperature at 0 BP to be that observed, shifting all other 50 a temperatures accordingly; and to adjust the 500 a mean temperature at 0 BP to be that observed also, shifting all other 500 a temperatures accordingly. In effect, this means that the constant (25.4) in equation (5.8) is slightly different for each set of data.

(c) Results using adjusted isotope data. The calculations were performed with the surface temperature specified by the isotope data in three ways:

(1) 500 a mean temperatures from 20 500 BP to 0 BP;

Figure 5.6. As for Figure 5.4 using data for the past 500 a for Camp Century, as 5 a means.

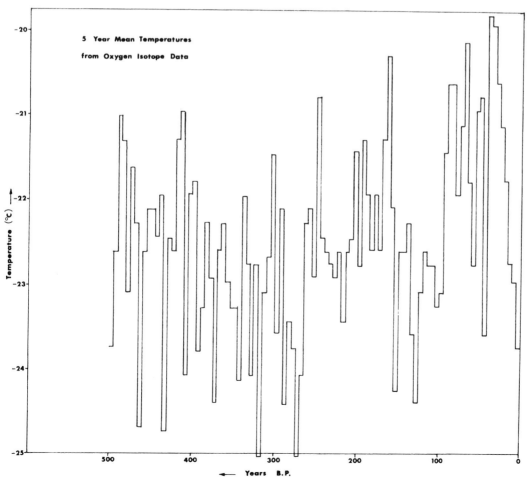

(2) 500 a mean temperatures from 20 500 BP to 2000 BP and 50 a data from 2000 to 0 BP;

(3) 500 a data from 20 500 to 2000 BP, 50 a data from 2000 to 500 BP and 5 a data from 500 to 0 BP.

As well, a steady-state profile was computed.

The temperatures which result directly from the oxygen isotope data and equation (5.8) are called 'raw'; those which use a different equation (5.8) for each means set are called 'adjusted'.

Results for Camp Century and the adjusted data are shown in Figure 5.7. Deviations between the observed profile and calculated temperatures are shown in Figure 5.8. The observed profile was supplied by C. Langway and is reported by Hansen & Langway (1966) (see also Figure 4.1 and Table 4.1). The model used here was that incorporating *all* the amendments discussed above.

The features of this calculation are as follows.

Base temperature: No attempt was made to fix this at any pre-determined value, and the computation differs, therefore, in this important respect from such 'steady-state' model work as that of Budd *et al.* (1971*b*). In spite of this, the lower boundary temperature of the *transient* model agrees remarkably well with that observed, being only one degree different. This is most significant and suggests that the transient model does incorporate a high degree of reality.

Base gradient: This is a combination of geothermal heat flux and frictional velocity. For the calculation it was chosen to be that observed.

Surface to bedrock temperature difference: This is less than that observed, suggesting either a slightly too low value of geothermal heat flux and/or frictional heating,

or surface temperatures in the past which are slightly too cold. This latter point, for reasons given in section 5.2.4.1(*d*), is thought to be the main contributor.

Mid-core temperatures and gradients: The temperature deviations between about 500–900 m depth are of the order of 0.5 °C. The gradients, however, as is clearly shown in Figure 5.8, correspond more closely to that observed, again suggesting that surface temperature fluctuations are of much the right amplitude but in error in absolute magnitude. The worst gradients are those resulting from a steady-state calculation in which, it is stressed, it was the lower temperature gradient and not the temperature which was fixed. Above 500 m depth it is, surprisingly, the 5 a mean curve which is most in error, both in temperature and temperature gradient. Following the argument above, this could be an indication of errors in the 5 a means of both magnitude and amplitude of variations. However, consistency has been forced on the δ values (so all 0 BP temperatures are the same). A glance at Figures 5.5 and 5.6 will show this has meant a relative adjustment of about 2.2 °C between the 50 and 5 a means. (Figures 5.4 and 5.5 show only 0.6 °C adjustment between 500 and 50 a means.) This 2.2 °C is a very large difference, and causes a 'shock' to the temperature profile when switching over from 50 to 5 a means at 500 BP during the integrations. Such shocks, also occurring at 2000 BP, will have the greater effect the closer they occur to the present. Earlier shocks, including the step-wise character of the surface temperatures of Figures 5.4 and 5.5 will be smoothed out and will be of decreasing importance as time passes. Thus, while the variation in the upper 100 m of the 5 a mean curve in Figure 5.8 must be due to the actual

Figure 5.7. Observed and computed Camp Century temperature profiles when the integration of the heat conduction equation uses the 'adjusted' oxygen isotope data (see sections 5.2.4.1(*a*)– (*c*)). The curve labelled 1 is the observed profile. Curve 2 is the steady-state profile with boundary conditions specified by the existing values of the observed parameters. Curve 3 is the profile obtained from a transient state temperature calculation after

20 500 a of integration from the temperatures of curve 2: the surface temperature changes during this time are specified by the data of Figure 5.4 (500 a means). Curve 4 is the 20 500 a temperature profile (from curve 2) using both the 500 a and 50 a mean isotope data (Figures 5.4 and 5.5) for the surface temperatures. Curve 5 is the 20 500 a profile using 500, 50 and 5 a mean isotope data (Figures 5.4–5.6).

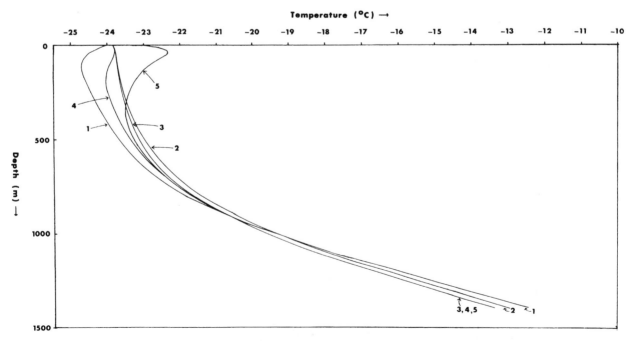

δ values, the lower variations are attributed to the shock effect. This hypothesis is substantiated in the next section.

Surface gradients: No surface gradients are correct; the upper temperature profiles, moreover, are all in error. This could be due to errors in the δ values, in the δ/θ relation, in the measurement of the observed profile, or in the model. It is thought that the profile measurement is correct, and that the discrepancies are due largely to a physical inadequacy of the model, coupled with possible sources of errors in the isotope ratios.

(d) Results using raw isotope data. The three computed curves in Figures 5.9 and 5.10 are for the three sets of means. The use of these means follows the description given in the section above, but here the raw data has been employed so no adjustment to the temperatures of Figures 5.4–5.6 was made. Surface Camp Century temperatures therefore vary by 2.2 °C, from -21.56 °C for 50 a mean data to -23.74 °C for the 5 a mean data. The 500 a means give a temperature of -22.12 °C.

Immediately obvious, now, is the agreement between the three sets of data from 100 m depth to bedrock, in complete distinction to the data of Figure 5.7 where agreement within 0.4 °C did not occur until about 350 m below the surface. Since the only difference between the two sets of curves lies in the forcing of consistency, it can be seen how important is the accurate specification of surface temperature, and the need to accept the inconsistencies in δ values arising from taking averages over different times.

Other main points of the 'raw' computations are as follows.

Base temperature: For the calculations, this is 0.8 °C too warm. Again this is felt to be a remarkable agreement with observation considering the calculations are from a transient model, which leaves the base temperature entirely free, and determined only during the integration by the prescribed base temperature gradient. Even if the upper 100 m or so are ignored and the computed profile superimposed over the observed Camp Century profile (see Figure 5.11), the lower temperature is still only 1.2 °C different from that observed. This time it

Figure 5.8. Deviations of the curves labelled 2, 3, 4, 5 from curve 1 of Figure 5.7. Thus, curve 2 here is the difference between curves 2 and 1 in Figure 5.7 (the steady-state and the observed temperature profiles).

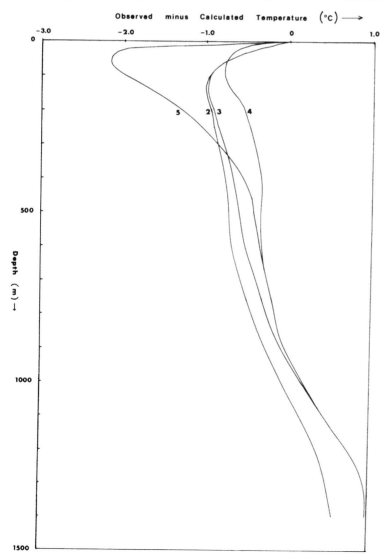

is colder. Now the magnitude of the base temperature is determined, in these calculations, mainly by three factors: the base gradient, the surface temperature, and surface temperature fluctuations with time. A small change in base gradient, say 10%, will affect the temperatures, but not by more than a few tenths of a degree. Temperatures at the surface some 10 000 a ago, if warmer/cooler than those from the given δ values, would warm/cool the lower boundary temperature; and if the 5 a means over the past 30 a or so (1935–65) were ignored, the curve for the 5 a means of Figure 5.9 would not

show the marked fluctuations present in the upper 50 m. Thus, neglect of the very last few years' temperature changes, coupled with a slightly different set of surface temperatures over the past 10 000 a could change the gradients of all the curves of Figure 5.9 sufficiently to provide excellent agreement with the observed profile. What is really being said here, of course, is that the temperatures from the δ values may need a further correction, such as a small change in the coefficient (1.633) in equation (5.8).

Figure 5.9. Observed and computed Camp Century temperature profiles using 'raw' oxygen isotope data (see also sections 5.2.4.1(*a*)–(*c*)). Curve 1 is the observed profile: curve 2 is that which results from an initial steady-state profile (curve 2 of Figure 5.7) after 20 500 a of integration in which surface temperatures are prescribed by the data of Figure 5.4; curve 3 is the profile obtained using the data of Figures 5.4 and 5.5, and curve 4 is the profile generated using the data of Figures 5.4–5.6.

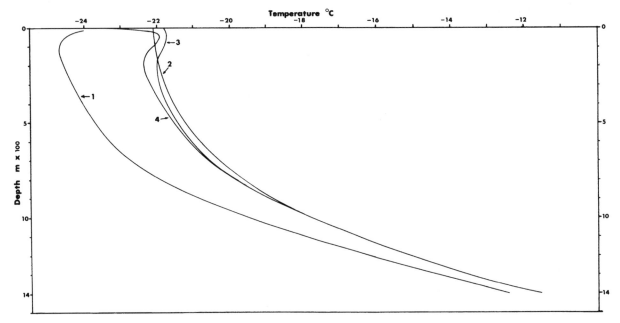

Figure 5.10. A comparison of the observed profile and curve 4 of Figure 5.9. The observed profile is curve 3. Curve 2 is simply this profile shifted to warmer temperatures (by about 2.6 °C). Curve 1 is curve 4 of Figure 5.9. The very close matching of the observed and computed temperature gradients between about 250 and 750 m is obvious. This matching is also brought out in Figure 5.11.

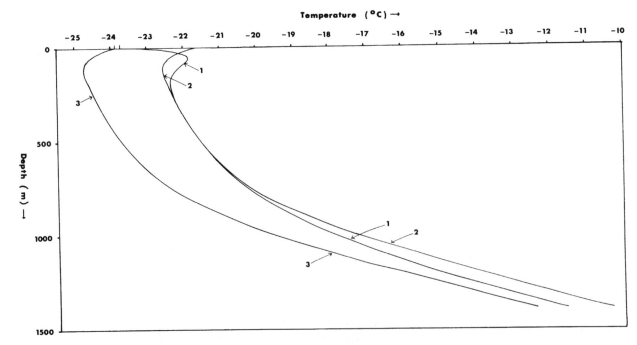

Surface to bedrock temperature difference: Except for the 5 a means calculation, this is about 1.3 °C smaller than that observed. As just stated above, this difference could be partially due to slightly too low a value for geothermal heat flux and/or frictional heating, but it is felt to be mainly the result of the surface temperatures of the past. (This might seem to be an argument based on a steady shape of the ice cap with time. The different thickness of the ice in the past, with concomitantly changed surface temperature due to climate and higher elevation, would also lead to different horizontal velocities of the ice. The changed frictional heating, thickness and surface temperature all must affect the present-day profile. However, surface temperature appears to be the most significant variable.) As for the 5 a means result, the surface to bedrock temperature difference agrees with that observed much more closely than the other curves only if the anomalous upper 50 m are accepted. Otherwise the surface to base amplitude is also about 1.3 °C too small.

Mid-core temperatures and gradients: The absolute magnitudes of the computed profiles are, from Figure

Figure 5.11. Deviations of the curves 2, 3 and 4 from curve 1 of Figure 5.9. Thus, curve 2 here is the difference between curves 2 and 1 of figure 5.9 (the profile obtained using 500 a mean isotope data and the observed profile).

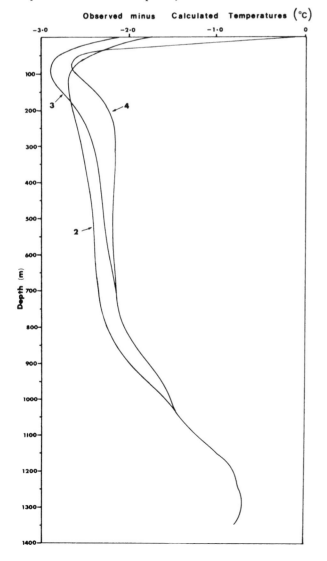

5.11, about 2.5 °C too warm. But, also from that diagram, gradients are in very good agreement, as is shown by their vertical character. In fact, from 200–750 m, the 5 a means curve is almost exactly that of the observed profile. The 500 and 50 a means also agree well, with the latter being the most deviate. This is felt to be due to the warm surface temperature at 0 BP for that data set. This high temperature will warm the upper portion of the profile, thus producing poorer gradient agreement than for the 500 a curve.

Surface gradients: Again, in the upper 200 m, for all curves, these are seriously in error. It would seem that the physical reality of the model, in regions where the ice properties are changing, is deficient.

(e) Surface gradients from mean annual isotopic δ values. Although mean annual δ^{18}O values for surface layers at Camp Century are not available to make more precise calculations of near-surface temperature gradients than those in this section, such data from the borehole at Crête in central Greenland have been used by Johnsen (1977b). His methods were similar to those presented in this section. His conversion factor of 1.62 °C($^0/_{00}$)$^{-1}$ is close to that of our equation (5.8), and his constant term in this equation (not given) was chosen to bring calculated and observed curves together. It is therefore the comparison of temperature gradients that is significant. His calculated and observed profiles are shown in Figure 5.12. This shows small temperature gradients with 'maxima at the same depth but the gradients in the lower part are different. This is probably due to too high temperatures in the second half of the last century, as shown by the δ^{18}O profile.'

Although Johnsen's analysis lacks the control of the Camp Century study in that it does not extend far back in time or to great depth, it does provide higher resolution by use of mean annual δ values instead of 5 a means. This would be expected to improve the matching of calculated and observed profiles at shallower depths, as is shown by agreement of the depth of the temperature maxima. As with the Camp Century results, the matching of the Crête results could be improved by changing the isotopic–temperature conversion factor. However, to produce very close agreement in Figure 5.12, we should have to approximately double the coefficient of 1.62 °C($^0/_{00}$)$^{-1}$ used by Johnsen, as opposed to a change of around only 20% needed to match the profiles in Figure 5.9 for Camp Century. This difference may be due to the relatively large effect of noise in the isotopic–temperature relationship when using 1 and 5 a mean δ values (see section 6.2).

5.2.5 Summary

The development and application of the transient heat conduction model described in section 5.2.2 has proved successful. Results of the computation for Camp Century suggest the following.

(1)   Oxygen isotope ratios may be successfully employed to provide information on surface temperature changes, except as noted in (3) below.

(2)   There is some physical inadequacy of the model in dealing with the upper portion of the column.

(3)   5 a means of isotope ratios fluctuate too rapidly over the past 60 a to provide meaningful inputs to the model.

(4)   Due to either or both the effects of the last two points, the upper 250 m of the computed profile is considerably in error.

(5)   From 250–850 m the oxygen isotope ratios give extremely good agreement with observation as regards temperature gradients. Agreement with temperatures would also be excellent if the anomalies of the upper 250 m were discounted.

(6)   From 850 m to bedrock calculated temperature gradients do not fit observed values as closely as at higher levels. This may be due to changes in the basal temperature gradient with time as discussed in sections 5.4 and 5.5.

It is therefore felt that the incorporation of oxygen isotope and total gas content data now allows the transient heat conduction model described here to have potentially the same ability to match observed temperature profiles as a steady-state model, such as that of Budd *et al.* (1971*b*). Thus, the philosophical disadvantages of using steady-state assumptions in a transient situation can now be overcome.

Figure 5.12. Measured temperature profile from Crête (1974) compared with a calculated profile based on the Crête 1974 δ ¹⁸O record. From Johnsen (1977*b*, Fig. 1).

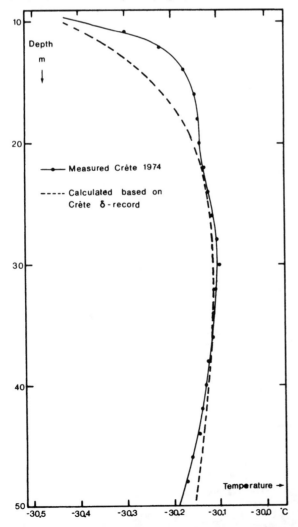

5.3

Elevation and climatic changes from total gas content and stable isotopic measurements

D.JENSSEN

Introduction

The value of the relative concentration within ice masses of the isotopes of oxygen (^{16}O and ^{18}O) depends to a large extent on the temperature of formation of the initial precipitation as outlined in chapter 1. The isotopic ratio (δ value) is thus a measure of the climatic conditions existing at that time in the past when the ice was accreted. Many factors, of course, play their part in obscuring the climatic record thus laid down, but principal among these is that due to the elevation of the surface of the ice mass at formation.

If, for example, a δ value hints that the temperature of formation was 5 °C colder than the currently existing ice surface temperature, this may in fact be due not to a climatic temperature change but to a lowering of the ice surface since deposition, because such a drop will entail a warming of the air. It becomes necessary to separate climatically induced temperature changes from those due to elevation rises or falls. This section describes a scheme for so doing.

A major part of the elevation change described above may be due to motion of the ice from higher elevations further inland, where the ice was deposited, to the site of the borehole. It is therefore necessary to distinguish between the effect of such motion and of general changes in the size and thickness of the ice sheet. This is done in sections 5.4 and 5.5 where flowline models are developed. Continentality, that is the distance from the outer boundary of the ice sheet, may also affect δ values as well as elevation, and although a theory has been developed to cover this factor, we confine this section to separation of effects of altitude and climatic changes only. The further discussion of other factors affecting δ values, such as latitude and various meteorological factors given in section 6.2, suggests that at present we need study only the major variables affecting δ values, namely elevation and temperature.

The theory depends on the fact that for a particular region of the Earth's surface, the relation between the δ value and surface temperature is linear in form, thus:

$$\theta = f\delta + f*$$

where f and $f*$ are determined from observed data but are not the same over the whole Earth, as discussed by Dansgaard *et al.* (1973) and in sections 3.2 and 3.3.

In order to obtain more accurate and realistic estimates of past temperatures than we get from simple application of the above equation, we must separate out climatic from elevation effects. For this, additional data must be used. This comes from the gas content of the ice, that is from the amount of gas trapped in bubbles in the ice as these close off under the weight of overlying firn. The volume of gas per gram of ice depends on temperature, pressure (section 3.9) and on the density of ice at close-off.

Earlier measurements (Langway, 1958; Bader, 1965) suggested that this close-off density was remarkably constant, even for different geographic regions, so that only the atmospheric pressure and temperature determined the total gas content during deposition of ice. Since past temperatures can be estimated from δ values, measurement of total gas content should enable us to determine the past pressure and hence elevation at which the ice was originally deposited. Recently, careful measurements on polar ice samples from a wide range of localities have shown that the pore close-off volume varies linearly with temperature by a small amount (Raynaud & Lebel, 1979, and section 3.9). The original theory for a constant density of close-off (Jenssen, unpublished; used in Raynaud's thesis, 1976) has been modified to take account of the later studies of Raynaud and Lebel. The elevation changes derived with the later studies are reduced by a factor of around 0.7, compared with the original theory, at both Camp Century and Byrd, while the past temperature changes are not significantly affected.

Since we use the total gas volume to determine atmospheric pressure at close-off after taking account of temperature from δ values, this gives sufficient data to determine the past climate and surface level changes of the ice.

While the later studies improve the theoretical approach, it should be stated that the early total gas volume measurements that we use are to be remeasured to derive more accurate values of V. These remeasurements may affect the results significantly, so the results obtained should be considered as preliminary rather than final values. Nevertheless, these are presented, as in principle the method provides the most effective way of deriving both climatic change and elevation change from measurements on a single ice core.

A relation between total gas content and elevation

The expression for total gas content (V) (proportional volume at standard temperature and pressure), in terms of pressure (P_c) and temperature (θ_c) at close-off, and standard pressure (P_0) and temperature (θ_0) is

$$V = V_c \frac{P_c}{\theta_c} \frac{\theta_0}{P_0}$$

where V_c is interpreted as the proportional gas volume

at close-off. It is a deduced quantity and is derived from the observed V and θ_c and from the observed, but partly empirical, P_c.

Further observational work (section 3.9) has shown a strong linear relationship between V_c and the temperature θ_c. Since V_c is a derived quantity, this relation is better expressed as

$$\frac{V\theta_c P_0}{\theta_0 P_c} = c\theta_c + d$$

The values of c and d may be determined by linear regression from data published by Raynaud & Lebel (1979) and are $c = 7.446 \times 10^{-4}\,\text{cm}^3\,\text{g}^{-1}\,^\circ\text{C}^{-1}$ and $d = -0.0589\,\text{cm}^3\,\text{g}^{-1}$. Rearranging the above equation gives

$$V = \frac{\theta_0 P_c}{P_0}(c + d/\theta_c) \tag{5.9}$$

What is necessary now is to find P_c in terms of the more readily observable θ_c.

Consider Figure 5.13 and assume a constant lapse-rate within the inversion layer. Integration of the hydrostatic relation in the vertical then leads to

$$P_c = P_a(1 - T\gamma_i/\theta_c)^{-g/\gamma_i R} \tag{5.10}$$

where P_a is the atmospheric pressure at the upper boundary of the inversion, T is the inversion thickness, γ_i is the lapse-rate $(= -\partial\theta/\partial z)$ within the inversion, and R is the specific gas constant for dry air. (The quantity $\theta_c - T\gamma_i$ is simply the atmospheric temperature (θ_a) at the top of the inversion.)

This equation contains P_a, a quantity which will vary as the ice surface elevation changes, so it will now be replaced by a function of some standard pressure. From Figure 5.13 and the reasoning which led to equation (5.10),

$$P_g = P_a(1 - (E + T)\gamma_a/\theta_g)^{-g/\gamma_a R}$$

where P_g is main sea-level pressure (equivalent to P_0), θ_g is the mean temperature at sea level at the edge of the ice (and is close to 0 °C and so is equivalent to θ_0), E is the elevation above sea level, and γ_a is the lapse-rate of the atmosphere outside the inversion

Figure 5.13. The relations between various temperatures (θ), pressures (P) and elevations at a particular location. Further details are given in the text.

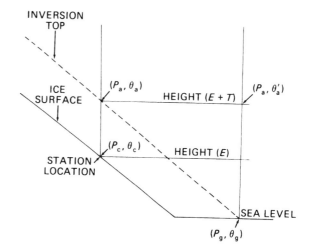

near sea level. Substitution into equation (5.10) gives

$$P_c = P_0 (1 - T\gamma_i/\theta_c)^{-g/\gamma_i R}$$
$$\times (1 - (E + T)\gamma_a/\theta_0)^{g/\gamma_a R} \qquad (5.11)$$

Thus P_c is now expressed as a function of surface temperature, elevation of the site and inversion thickness. Substitution into equation (5.9) shows that

$$V = \theta_0 (c + d/\theta_c)(1 - T\gamma_i/\theta_c)^{-g/\gamma_i R}$$
$$\times (1 - (E + T)\gamma_a/\theta_0)^{g/\gamma_a R} \qquad (5.12)$$

Consider now the approximate expansions of the power terms.

$$V \approx \theta_0 (c + d/\theta_c)\left(1 + \frac{gT}{R\theta_c}\right)\left(1 - \frac{g(E + T)}{R\theta_0}\right)$$

Writing

$$\alpha = \theta_0 (c + d/\theta_c)(1 + gT/R\theta_c)$$

this becomes

$$V = -\frac{\alpha g}{R\theta_0} E + \alpha \left(1 - \frac{gT}{R\theta_0}\right) \qquad (5.13)$$

Now, since θ_c lies within the limited range 0 to $-70\,^\circ$C, the maximum change in $1/\theta_c$ about its mean value is little more than 10%. It is therefore assumed that it is permissible to replace $1/\theta_c$ by the constant q where

$$q = \frac{1}{\theta_2 - \theta_1} \int_{\theta_1}^{\theta_2} \frac{1}{\theta_c} \, d\theta_c$$

With $\theta_1 = -70\,^\circ$C, $\theta_2 = 0\,^\circ$C, it is seen that $q = 4.2324 \times 10^{-3}\,^\circ$C^{-1}.

Taking $T = 100$ m, R = 287 m^2 sec^{-2} K^{-1}, $\theta_0 = 273$ K, and the values of c and d quoted above, equation (5.13) is

$$V = -1.7158 \times 10^{-5} E + 0.1355 \qquad (5.14)$$

Thus it has now been shown that the definition of total gas content, together with the empirical relation between V and θ_c (expressed by Raynaud & Lebel (1979) as between V_c and θ_c) and standard meteorological theory, leads to a linear relationship between total gas content and surface elevation. This relation has, of course, already been established empirically by Raynaud & Lebel. In their paper, Raynaud & Lebel find that by a least-squares linear regression fit to their observed data,

$$V = -1.66 \times 10^{-5} E + 0.138$$

The agreement between theory and observation is satisfyingly close.

The theoretical form of $\Lambda(= -\partial\theta_c/\partial E)$

Should the ice elevation (E) change with time at the location being considered, P_c will also change. The rate of change of P_c with E is given by differentiation of equation (5.10), and is

$$\frac{\partial P_c}{\partial E} = \frac{P_c}{P_a} \frac{\partial P_a}{\partial E} - \frac{gP_c T}{R\theta_a\theta_c} \frac{\partial\theta_c}{\partial E}$$

where $\partial P_a/\partial E$ is the change in pressure at the top of the inversion per unit height change in the surface of the ice. To a first approximation, this is given by the hydrostatic relation for the free atmosphere:

$$\frac{\partial P_a}{\partial E} = -\frac{P_a g}{R\theta_a'}$$

where $\theta_a' = \theta_0 - \gamma_a(E + T)$.

Thus

$$\frac{\partial P_c}{\partial E} = -\frac{P_c g}{R\theta_a'} - \frac{gP_c T}{R\theta_a'\theta_c} \frac{\partial\theta_c}{\partial E} \qquad (5.15)$$

This equation, of course, can also be found simply by differentiation of equation (5.11) with respect to elevation.

Now take the derivative of equation (5.9) with respect to elevation:

$$\frac{\partial V}{\partial E} = \frac{\theta_0}{P_0}\left(c + \frac{d}{\theta_c}\right)\frac{\partial P_c}{\partial E} - \frac{\theta_0 P_c d}{P_0 \theta_c^2} \frac{\partial\theta_c}{\partial E}$$

Making use of equation (5.15), this is

$$\frac{\partial V}{\partial E} = -\frac{\theta_0}{P_0}\left(c + \frac{d}{\theta_c}\right)$$
$$\times \left(\frac{P_c g}{R\theta_a'} + \frac{gP_c T}{R\theta_a'\theta_c}\frac{\partial\theta_c}{\partial E}\right)$$
$$- \frac{\theta_0 P_c d}{P_0 \theta_c^2}\frac{\partial\theta_c}{\partial E} \qquad (5.16)$$

The theory developed in the preceding section, which gives a physical basis for the empirically determined relation between elevation and total gas content, is based on the observed correlation between V_c (or V) and surface temperature θ_c. There seems little reason to assume that this latter, well-supported relation of Raynaud & Lebel (1979) will be time dependent. Thus it will be treated as holding for the time span covered by the core data. It follows then, that the linear dependency of V on E will also hold in the past, and, since the range of surface temperature will still be within $0\,^\circ$C to $-70\,^\circ$C, equation (5.14) will hold. Thus $\partial V/\partial E = a = $ constant. With this, rearranging the above equation, making use of equation (5.9) and defining Λ as $-\partial\theta_c/\partial E$, yields

$$\Lambda = \theta_c \left[\frac{a}{V} + \frac{g}{R\theta_a'}\right]\left[\frac{1}{1 + (c/d)\theta_c} + \frac{gT}{R\theta_a'}\right] \quad (5.17a)$$

Note that any changes in the surface temperature θ_c are due to climatic and elevation changes. In other words, the oxygen isotope data are a direct measure of θ_c:

$$\theta_c = f\delta + f^* \qquad (5.17b)$$

Equation (5.17b) now needs to be applied both in Antarctica (Byrd Station) and in Greenland (Camp Century). Consider first Antarctica, and make use of an empirical relation discovered by Phillpot & Zillman (1970), who found that the strength of the Antarctic inversion was linearly dependent on surface temperatures. Thus in equation (5.17a),

$$\theta_a = m\theta_c + n \qquad (5.17c)$$

where, from Phillpot and Zillman data, $m = 0.437$ and $n = 137.8224$ K: both θ_a and θ_c are in K.

Equations (5.17a, b, c) now give Λ as a function of surface temperature alone, provided T is known. Here it will be assumed to be constant. Data from Phillpot & Zillman (1970) show a large variation in the thickness of the inversion (from 700–300 m), but for Byrd they claim a figure of between 400–500 m. Here, 450 m will be taken for T.

For Greenland, there seems to be no known relation such as equation (5.17c); in fact, good inversion data exist apparently only for Station Centrale (Putnins, 1970). To a crude first approximation, the inversion height, T, is about 350 m, and has a strength of about 6.2 °C. In the absence of other information, it will be assumed these values are constant and will hold at Camp Century. Thus equation (5.17c) for Greenland will have $m = 1$, and $n = 6.2$ K; again, θ_a and θ_c are in K.

Computational procedure

With Λ computable (equations (5.17)) and with a relation between gas content and elevation (equation (5.12)), the separation of the climatic from the elevation-induced temperature change is relatively simple.

(1) With the known values of V and δ at some depth in the core (that is, at the nth point in the core), and the known values of f, f^*, c and d, we find E_n from equation (5.12):

$$E_n = \frac{\theta_0}{\gamma_\alpha} (1 - [\{V/\theta_0(c + d/\theta_c)\}$$
$$\times \{1 - T\gamma_i/\theta_c\}^{g/\gamma_i R}]^{\gamma_a R/g}) - T \qquad (5.18)$$

where E_n is the elevation at the nth point in the core (that is, at time t). T can be estimated from data in Phillpot & Zillman (1970) or from Putnins (1970); γ_i will then follow as $\gamma_i = (\theta_c - \theta_a)/T$; γ_a can be taken to be the dry adiabatic lapse-rate ($\gamma_a = 0.01$ °C m^{-1}) (see, for example, List (1966)) so that $\theta_a' = \theta_0 - (E_n + T)\gamma_a$. From the analysis showing $V = aE + b$, it can be seen that the values of γ_a and γ_i are not critical since they cancel out in a first approximation. Even relatively poor estimates of these would not be expected to create much ultimate error in E_n. Note also that the term containing T (in both equations (5.17a) and (5.18)) is a magnitude smaller than its associated terms, so again, gross errors in the inversion thickness should not contribute greatly to the errors in E (or Λ).

(2) From the previously determined E and the gas constant at the last higher point in the core, approximate $\partial V/\partial E$. Now find V and δ between the two current points:

$$\bar{V}_n = (V_n + V_{n-1})/2 \qquad \bar{\delta}_n = (\delta_n + \delta_{n-1})/2$$

where n is the current point and $(n - 1)$ the adjacent higher datum point in the core ($n = 0$ at the top of the core). Replace V and δ in equations (5.17) by $\bar{V}$ and $\bar{\delta}$ and so find Λ_n midway between n and $(n - 1)$: this value of Λ_n is now assumed to hold over that depth.

(3) Since θ_c at any depth contains both climatic and elevation components, remove the latter which is estimated to be

$$\theta_{elev} = \theta_c^0 - \left(\int_0^t \Lambda \frac{\partial E}{\partial t} \, dt \right)$$
$$\approx \theta_c^0 - \bar{\Lambda}_n (E^t - E^0)$$

where t is the current time, θ_c^0 is the ice surface temperature at the time corresponding to the topmost point of the core, and $\bar{\Lambda}_n = (1/n)\Sigma\Lambda_n$.

Since time and depth are equivalent, E^t is identical to E_n and E^0 to E_0. Thus

$$\theta_{climate} = \theta_c^t + \bar{\Lambda}(E^t - E^0)$$

or

$$(\theta_{climate})_n = f\delta_n + f^* + \bar{\Lambda}_n(E_n - E_0) \qquad (5.19)$$

(For the results shown here, there has been some slight smoothing of the data: $X_i = (X_i + \frac{1}{2}(X_{i-1} + X_{i+1}))/2$ where the X_i are δ, D or V data.)

Data

Observations of V and δ and at Camp Century and Byrd stations were used, and are given graphically in Figures 5.14 and 5.15. These were supplied to the Cambridge Workshop by Lorius and Johnsen.

Results

Use of oxygen isotope data

Figure 5.16 shows the past climate and elevation at Byrd derived from the data and model presented in this section; Figure 5.17 shows those obtained at Camp Century. Points to note in interpreting these figures are as follows.

(1) The 'elevation of the past' is the height of the *source* point of the ice of the core above or below the elevation of the ice surface at the time when the ice now at 200 m depth was deposited.

(2) The 'climate of the past' is relative to the climate at the zero point in the core (200 m depth in these cases).

Figure 5.14. Data for ice core from Byrd Station, Antarctica showing δ values (dashed line) supplied by Johnsen for this study and total gas content V (full line), from Raynaud & Lorius (1973).

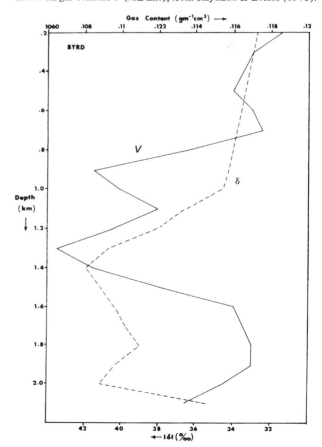

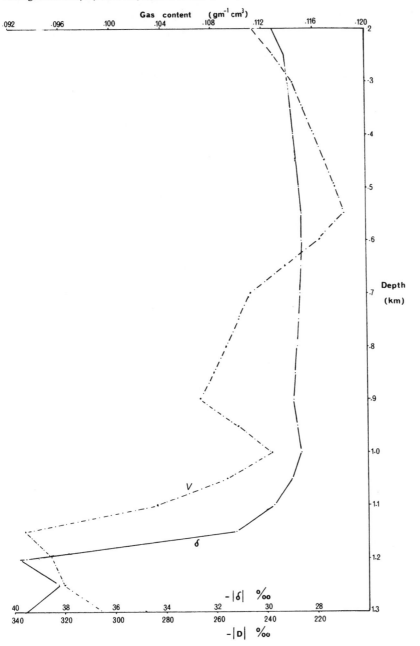

Figure 5.15. Data for ice core from Camp Century, Greenland
showing oxygen δ values supplied by Johnsen for this study and
total gas content, V, from Raynaud & Lorius (1973).

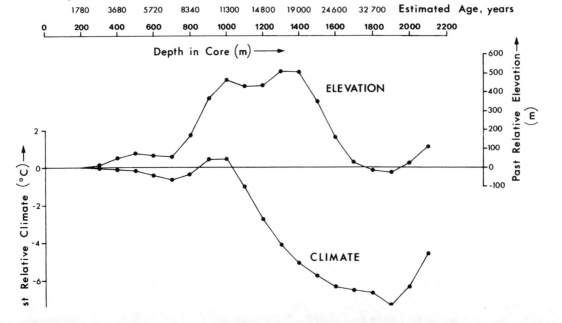

Figure 5.16. Computed past elevations and climate for Byrd
Station. $\theta_c = -|\delta| + 279.29$; $V = -1.66 \times 10^{-5} E + 0.138$;

$V_c = 7.4 \times 10^{-4} \theta_c - 0.057$; $T = 450\,\mathrm{m}$; $\theta_a = 0.437\theta_c + 137.8$;
$-\partial\theta_c/\partial E = \bar{\Lambda} = 0.0062\,^\circ\mathrm{C\,m^{-1}}$.

(3) Unlike earlier theory which used a constant close-off volume (V_c), the present analysis determines values of Λ for each step of the calculation by equation (5.17). In spite of smoothing of the input data, there are still considerable variations in the values of Λ because of the short intervals over which they are computed.

Results of computation over 20 intervals for smoothed data for the Byrd ice core give a mean value of $\Lambda = 0.0062 \pm 0.0022\,^{\circ}\mathrm{C\,m}^{-1}$, while the mean value over 23 intervals at Camp Century give $\Lambda = 0.0099 \pm 0.0042\,^{\circ}\mathrm{C\,m}^{-1}$. Although the standard deviations are large, the figures from the two boreholes are similar. Furthermore, they agree approximately with the value expected from field measurements of surface ice temperatures as a function of altitude, and with the adiabatic lapse results (moist and dry) in the atmosphere. This gives confidence in the validity of the model and some indication of the approximate accuracy of the derived climatic records.

Byrd Station

It is possible to date, tentatively, the ice in the core through computer modelling, as described in sections 5.4 and 5.5. The ages plotted above the abscissa of Figure 5.16 are those shown in Figure 5.50, but it must be stressed that these dates are tentative and a second estimate of dates is given in section 3.7. The differences are significant but not large and agree at around 11 000 a BP. With that caution in mind, the graphs are interpreted

as showing some warming since 7000 a BP, with earlier cooling, a warming before 11 000 a BP, with an ice age before 15 000 a BP or so. Of course, since the ice becomes more and more compacted in lower depths, the dating becomes correspondingly uncertain.

The past elevations at, or near, Byrd are also given in Figure 5.16. They appear to have fallen by less than 100 m over the last 5000 a (see also section 3.7) and changed little from 7000–5000 a BP. During the ice age the elevations near Byrd were some 400–500 m higher than at present, reaching a maximum between 19 000 and 17 000 a BP.

At the end of the ice age, the general warming trend appears to have lasted from about 25 000–11 000 a BP, while this was associated with some thickening of the ice to 19 000 a BP. From then until 11 000 a BP, while the warming continued, the ice thickness was fairly uniform with a small net thinning. From 11 000 a BP to the present the ice was generally thinning, rapidly at first.

This analysis raises the question of why the ice sheet remained so thick while temperatures showed a considerable rise from 28 000–11 000 a BP. First, the cold period need not coincide with the thickest ice. A slightly warmer global climate would mean increased moisture content of the atmosphere, and hence it is likely that ice sheets could expand, particularly near their outside boundary where the moisture can easily penetrate. Secondly, Figure 5.16 has been computed with no account being taken of the flow of the ice over

Figure 5.17. Computed past elevations and climate for Camp Century, Greenland. $\theta_c = 1.7143|\delta| + 301$;

$$V = -1.66 \times 10^{-5}E + 0.1355;\ V_c = 7.4 \times 10^{-4}\theta_c - 0.057;$$
$$T = 350\,\mathrm{m};\ \theta_a = \theta_c + 6.2;\ -\partial\theta_c/\partial E = \bar{\Lambda} = 0.0104\,^{\circ}\mathrm{C\,m}^{-1}.$$

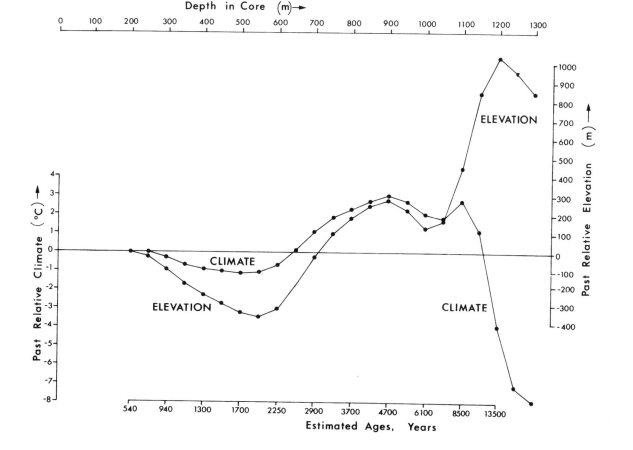

the past 30 000 a, but in fact, the ice now in the Byrd core will have come from some distance away and from a higher elevation, as discussed in sections 5.4 and 5.5. This could account for most of the thinning from 19 000–11 000 a BP. Without a good model of the ice flow (preferably, of course, three-dimensional) it is not possible to separate changes in ice thickness from 'advection of elevation'.

Nonetheless, the relatively low elevations, at time greater than 25 000 a BP, are still unexplained. That this feature is 'real' and not simply an inherent by-product of the model is given substantiation by other independent data, notably sea-level changes and changes in the Wisconsin ice-sheet front. Chappell (1978, Figure 5.2.2) gives graphs of these two quantities which clearly show low sea levels and a maximal Equatorward extent of ice-sheet at about 20 000 a BP. However, from 50 000 a BP to this time the ice boundaries in the northern hemisphere were much further north and the sea levels were much higher. It would seem clear, then, that the ice sheets of both hemispheres were at a maximum extent of thickness some 20 000 a ago, and were of substantially smaller extent in the 30 000 or so years preceding that time.

The reasons for these changes are still unclear, but one mechanism does suggest itself and is treated by Chappell; namely, changes in solar radiation due to variations in the orbital parameters of the earth. These changes were first discussed by Milankovitch (1938) and arise primarily from changes in the earth's orbit's ellipticity, in changes in the tilt of the earth's axis to the orbital plane and to the precession of the equinoxes. These changes and the resulting variation of insolation at the earth's surface are treated by Vernekar (1972) and Birchfield & Weertman (1978).

The process which suggests itself is a lowered radiation input which leads to a growth of ice sheets to such an extent that, when the radiation increases again, the ice sheets are sufficiently large that ablation does not destroy them. In fact, radiational losses through the increased albedo may cause continued growth which will slow, stop or reverse only when the radiation increases to the point where summer ablation exceeds winter accumulation. Thus there will be an out-of-phase relation between the solar radiation variations and the ice extent; that is, essentially, between the global climatic temperatures and ice cover, a relation greatly complicated by feedback effects (increased albedo, changes in moisture content of the atmosphere due to changing rates of evaporation), by changes in the general circulation of the atmosphere, by isostatic changes, and so on. To verify the importance of the Milankovitch radiation curves requires numerical modelling: preferably an atmospheric and cryospheric one, but at the very least a three-dimensional glaciological one, in which the variations of insolation are used as a forcing function to change accumulation/ablation patterns over the ice.

Camp Century
Figure 5.17 gives the previous climatic history as deduced for Camp Century in Greenland. The tempera-ture/δ relation is, for θ in $°K$ and δ in $°/_{oo}$,

$$\theta = -1.714|\delta| + 301.2 \quad K$$

while all other constants remain unchanged.

These values are not those used in equation (5.8), and come, as do the constants used here for Byrd, from Dansgaard *et al.* (1973); those of equation (5.8) were supplied, at the workshop, by Johnsen. However, the exact choice of values for these constants is not critical: a brief examination of equations (5.17*a* and *b*) will show that the difference in climatic temperatures deduced using both sets of values is of the order of 0.2 °C or less. The difference in deduced elevation is of the order of 3 m or less.

The age–depth relation was computed by Young (see Figure 5.22) and gives a relation similar to that of Dansgaard & Johnsen (1969*a*) and to Figure 1.9. Thus the warm period around 900 m depth occurs at about 5000 a BP and that at 1150 m is about 8500 a BP; whilst the transition from cold to warm climate (1150–1200 m) is from 11 000–13 500 a BP. The magnitudes of the fluctuations at Camp Century and Byrd are similar, although the former show greater variations in warming and cooling. The Greenland data show slightly colder temperatures than at present around 2000 a BP, with slight warming since that time.

The past elevations near Camp Century are in phase with the past temperatures throughout almost the entire depth of the core, the only out-of-phase portion being from 1100–1200 m. Indications again are that the thickest ice was just at (or even during) the transition from very cold to warm climatic conditions. This feature has already been discussed in relation to the Byrd core and it would seem that the end of the cold temperatures is not coincident with maximum ice thickness at these latitudes. As well, errors in the dating may be of increasing importance for older ice, and advective processes may also play a part. These are outlined in section 5.5.

5.4

Techniques for the analysis of temperature–depth profiles in ice sheets

W.F.BUDD & N.W.YOUNG

Introduction

The temperature distribution in large ice sheets depends on a large number of factors. These include such independent variables as the temperature at the surface, the snow accumulation rate, the ice thickness, the motion of the ice and the geothermal heat flux at the base. In addition, since these parameters can vary with time as well as position, it would be necessary to know them all as functions of time in order to derive the temperature profile correctly.

However, since we do not know the history of ice sheets sufficiently well to prescribe the time changes of all these parameters, a reasonable procedure is to prescribe as many of the features as are known and allow the others to vary systematically over plausible ranges in order to study the temperature profiles that result. By comparing such derived profiles with measured profiles it is possible to establish those values of the parameters and their past variations that give rise to temperature profiles which match the measured profiles as closely as possible. As more of the present values of the various parameters and their past history become known, the more narrow the ranges of the other variables become to give the required matching to the measured temperature depth profile.

Thus, by studying calculated temperature profiles in comparison with measured profiles, various features of the past history of the ice sheet can be deduced as new additional data become available.

This scheme appears sound in principle, but in practice there are generally still too many unknowns which require resolving before a unique solution can be established. In anticipation of the description of the model, the unknowns include the past history of the ice sheet, elevation, temperature, accumulation rate and the thermal and flow properties of the ice. With regard to the latter, the effect of large-scale crystal anisotropy appears to be an important effect that has still not been clearly identified.

Description of the models

The heat conduction equation

We consider the general heat conduction equation for the temperature θ at position x_i in a moving medium where the velocity is u_i at time t

$$\nabla(K\nabla\theta) + \rho c u_i \cdot \nabla\theta + Q = \rho c \frac{\partial\theta}{\partial t}$$

where K is the ice thermal conductivity, c is the ice thermal capacitance, ρ is the ice density, and Q is the rate of internal heat generation by the motion per unit volume. The parameters K, c and ρ can be functions of position and time through their dependence on temperature-volume.

Considerable simplification can be made when studying the temperature in a vertical column by choosing axes with x horizontal in the plane of flow, z vertically upwards and y across the line of flow.

Let u, v and w be the components of velocity in the direction of x, y, z. The character of the dimensions and symmetry of large ice sheets away from rock obstructions gives rise to greatest variations of temperature in the vertical, the next greatest in the line of flow and the least across the flow. This leads to the following simplifications

$$\frac{\partial^2\theta}{\partial x^2} \approx \frac{\partial^2\theta}{\partial y^2} \approx v\frac{\partial\theta}{\partial y} \approx 0 \qquad (5.21)$$

The heat conduction equation then reduces to

$$\frac{\partial}{\partial z}K\left(\frac{\partial\theta}{\partial z}\right) + \rho c w \frac{\partial\theta}{\partial z} + \rho c u \frac{\partial\theta}{\partial x} + Q = \rho c \frac{\partial\theta}{\partial t} \quad (5.22)$$

Internal heating

Stress. The primary generation of internal heating is from the dissipation of energy due to the horizontal shear stress τ_{xz} and strain rate $\dot{\epsilon}_{xz}$. For the most part other stresses can be neglected, although for flow over an irregular bed the longitudinal stress and strain rates may also become significant.

The large-scale average shear stress at depth z in the ice is the important stress governing the motion. This is given by

$$\tau_{xz} = \bar{\rho}g\alpha z \qquad (5.23)$$

where g is the gravitational acceleration, α is the large-scale smooth surface slope and $\bar{\rho}$ is the average density over the depth. The heat dissipation rate is given by

$$Q = 2\tau_{xz}\dot{\epsilon}_{xz} \qquad (5.24)$$

and $\dot{\epsilon}_{xz}$ is determined from τ_{xz} and the flow law for ice.

Flow law for ice. Although the flow law of ice depends on many parameters such as the stress magnitude (primarily the second invariant of the stress tensor), the stress configuration (which is important for non-linear flow laws) and temperature (relative to the pressure melting point), the crystal orientations, crystal size, density, impurity content and so on, only a few of these are important in the present context. The density in an ice sheet varies from about 0.3–0.4 Mg m^{-3} at the surface to over 0.9 Mg m^{-3} by about 100 m depth. Since at the surface the horizontal shear is zero, the variation of the density in the upper layers of thick ice sheets has little effect on the horizontal velocity profile.

Recent measurements, reported in Lile (1978) and Lile (unpublished), indicate that crystal size alone is not an important variable for the shear strain rate over the stress range relevant here. Crystal orientation, however, is very important because ice sheets tend to develop strong regional anisotropies, with polycrystalline ice having a predominance of basal planes horizontal throughout a major layer of the ice sheet and deforming several times more rapidly than randomly oriented ice. These last two facts indicate that the development of crystal anisotropy in ice sheets and its effect on the ice flow rates should be taken into account in the modelling. The descriptions given by Budd (1972) indicate in a general way how these developments may be characterised.

Near the base of the ice sheet the effect of morainic layers may also be important. However, since measurements by Hooke, Dahlin & Kauper (1972) show that this type of ice usually deforms less rapidly than pure ice, the problem is not expected to be severe.

This leaves as the most important parameters for the flow law, the stress magnitude τ and the temperature θ^* relative to the pressure melting point θ_m.

The pressure melting point is determined from the hydrostatic pressure at depth z by

$$\theta_m = -P\rho gz \qquad (5.25)$$

where $P \approx 0.0085$ °C bar^{-1}.

For computational work an empirical tabular and interpolation scheme can be used to represent the flow law to match measured values as closely as possible, for instance see Budd, Jenssen & Radok (1971a).

Alternatively, an expression can be used of the form

$$\dot\epsilon_0 = \dot\epsilon_c \sinh\left(\frac{\tau_0}{\tau_c}\right) e^{\nu\theta^*} \qquad (5.26)$$

where $\tau_0, \dot\epsilon_0$ are the octahedral shear stress and strain rates and $\dot\epsilon_c, \tau_c$ and ν are constants which can be chosen to fit the experimental data for randomly oriented polycrystalline ice reasonably closely. To take account of effects of anisotropy, the parameter $\dot\epsilon_c$ may be made a function of relative depth to give increased strain rates in relation to the degree to which the crystal basal planes approach horizontal.

Horizontal velocity

For the case of the only significant stress being the horizontal shear stress τ_{xz} the horizontal velocity-depth gradient is given by

$$\frac{du}{dz} = 2\dot\epsilon_{xz} = \sqrt6\,\dot\epsilon_c \sinh\left(\frac{\sqrt{\tfrac{2}{3}}\bar\rho g\alpha z}{\tau_c}\right) e^{\nu\theta^*} \qquad (5.27)$$

If there is no basal sliding the velocity at any depth may be obtained by integration from the base upwards. In particular the surface velocity is given by

$$u_s = \int_{z=0}^{Z} \sqrt6\,\dot\epsilon_c \sinh\left(\frac{\sqrt{\tfrac{2}{3}}\bar\rho g\alpha z}{\tau_c}\right) e^{\nu\theta^*}\,dz \qquad (5.28)$$

Thus if we know the temperature profile (θ^*) and the flow parameters ($\dot\epsilon_c, \tau_c, \nu$) then a velocity profile can be calculated. Alternatively, given the surface velocity or an average velocity and the temperature profile, then if the above form of flow law is assumed an effective flow law parameter $\dot\epsilon_c$ can be calculated.

The internal heating Q_z can be obtained from

$$Q_z = 2\tau_{xz}\dot\epsilon_{xz}$$

$$= \bar\rho g\alpha(Z-z)\sqrt6\,\dot\epsilon_c \sinh\left|\frac{\sqrt{\tfrac{2}{3}}\rho g\alpha(Z-z)}{\tau_c}\right| e^{\nu\theta^*} \qquad (5.29)$$

Since this involves the temperature (through θ^*) and the equation for temperature involves Q_z, this last equation has to be solved simultaneously with

$$\frac{\partial}{\partial z}\left(K\frac{\partial\theta}{\partial z}\right) + \rho cw\frac{\partial\theta}{\partial z} + \rho cu\frac{\partial\theta}{\partial x} + Q_z = \rho c\frac{\partial\theta}{\partial t} \qquad (5.30)$$

The horizontal velocity u is obtained from (5.27) but the vertical velocity w has yet to be determined.

Vertical velocity

Strain rate through the ice sheet. We first of all consider the body of the ice sheet, neglecting the firn layer, and then add a term for the effect of compaction at the surface. For an ice sheet in steady state, with an accumulation rate A m of ice per year, the vertical velocity near the surface is given by

$$w_s = A \qquad (5.31)$$

If the strain rate is uniform over the depth the vertical velocity at height z above bedrock is given by

$$w_z = \frac{Az}{Z} \qquad (5.32)$$

In general, however, if the base is not sliding we can expect the longitudinal strain rate to vary from its surface value to zero at the base in a similar way to the velocity profile, that is if the shape of the velocity-depth profile changes slowly along the line of flow

$$\left.\frac{du}{dx}\right)_z \propto u)_z \qquad (5.33)$$

For zero base velocity a similar result applies for $\partial v/\partial y$ because the transverse strain rate is proportional to the divergence of the flowlines and the forward velocity u.

Hence, since we have

$$\dot\epsilon_x + \dot\epsilon_y + \dot\epsilon_z = 0 \qquad (5.34)$$

the vertical strain rate may be expected to vary with depth in a similar way to the horizontal velocity, that is

$$\frac{\partial w}{\partial z} \propto u_z \qquad (5.35)$$

$$w \propto \int_0^z u_z\,dz$$

Therefore

$$w_z = A\frac{z\bar u_z}{Z\bar u_s} \qquad (5.36)$$

where $\bar u_z$ and $\bar u_s$ are the average velocities from the base to height z and the surface respectively.

Firn layer compaction. Above the level of constant density the vertical velocity is increased due to the compaction. This can be calculated from the density ρ as a function of depth z. This results in a higher vertical velocity at the surface due to compaction which decreases with depth as the density approaches that of ice ρ_i. At the surface, the annual layer thickness is $A\rho_i/\rho$, and similarly the velocity at depth z due to settling within the upper layer is $A\rho_i/\rho$, see Budd (1966a). Thus the effect of compaction gives a vertical velocity of

$$w_z' = w_z \frac{\rho_i}{\rho} \tag{5.37}$$

This expression can be used for studies concerning the upper firn zone, but for the bulk of the ice thickness w_z can be taken as the vertical velocity according to equation (5.36).

Non-steady state. For considerations of a non-steady-state ice sheet, the simplest procedure is to use the same expressions for the vertical velocity (which is then relative to the moving surface) using the prescribed accumulation rate and adopt a vertical coordinate scale proportional to the ice thickness; then specify the surface temperature boundary condition to include the effect of elevation and temperature changes, and transform the differential equation as required, see Budd *et al.* (1971a). The net result is that the ice thickness changes with time but the temperature profile can still be calculated as a function of relative depth.

System of equations and boundary conditions
Equations
Because of the complex feedback between velocity, temperature and time, the system of equations are only tractable to numerical solution. The following system defines a closed set of equations, boundary conditions and specified input parameters required for a solution.

Heat conduction:

$$\frac{\partial}{\partial z}\, K\, \frac{\partial \theta}{\partial z} + \rho c w\, \frac{\partial \theta}{\partial z} + \rho c u\, \frac{\partial \theta}{\partial x} + Q_z = \rho c\, \frac{\partial \theta}{\partial t} \tag{5.38}$$

Internal heating:

$$Q_z = \frac{du}{dz}\, \bar{\rho} g \alpha (Z - z) \tag{5.39}$$

Horizontal velocity:

$$u_z = \int_0^z \sqrt{6}\, \dot{\epsilon}_c \sinh\left(\frac{\sqrt{\tfrac{2}{3}} \rho g \alpha (Z - z)}{\tau_c}\right) e^{\nu\theta*}\, dz \tag{5.40}$$

Vertical velocity:

$$w_z = A\, \frac{z}{Z}\, \frac{\bar{u}_z}{u_s} \text{ (below the firn layer)} \tag{5.41}$$

Relative temperature:

$$\theta* = \theta - P\bar{\rho}gZ \tag{5.42}$$

Input ice thermal parameters:

$$K(\theta), c(\theta), \rho(z), P$$

Input ice flow parameters:

$$\dot{\epsilon}_c, \tau_c, \nu$$

Specified data: ice surface elevation $E(x, t)$

ice thickness $Z(x, t)$

accumulation rate $A(x, t)$

surface temperature $\theta_s(h, x, t)$

geothermal gradient γ_G

flowline divergence $\partial Y/\partial x$ or balance velocity $\bar{U}(x, t)$

Boundary conditions
At the surface the accumulation rate and the surface temperature can be specified. If preferred these parameters could be specified at some other level, for instance below the firn layer.

At the base, two quite distinct situations may exist. If pressure melting point is reached, the basal boundary condition becomes $\theta_b = \theta_m$ and the melt rate is given by

$$M = \frac{\gamma_G - \gamma_b}{K L} \tag{5.43}$$

where γ_b is the calculated basal gradient; and L is the latent heat of fusion for ice.

If pressure melting point is not reached, then the basal gradient at the ice rock interface (γ_G) could be specified. The problem is that this is generally not known and hence a range of possible values needs to be covered. In addition, since long time changes of temperature may also penetrate the bedrock, when no melt occurs the temperature gradient in the rock may differ substantially from the geothermal gradient. Hence in considering time changes, the geothermal gradient deep below the interface should be used as the basal boundary condition and the heat conduction in the rock considered as well.

Balance velocities
Since the flow rate of ice (that is primarily the parameter $\dot{\epsilon}_c$) is not accurately known for the temperature, stress, and crystal fabric conditions commonly found in ice sheets, it is sometimes preferable to supply as input the surface velocity (u_s) or average velocity through the column (U). In each case, if one of $\dot{\epsilon}_c$, u_s or U are supplied then the other can be calculated. If the velocities are unknown then an estimate of the average velocity can be obtained from a steady-state assumption to calculate a 'balance velocity' $\bar{U}$ from the accumulation rate A, the ice thickness Z and the flowline patterns by

$$\bar{U} = \frac{1}{Y\bar{Z}} \int_0^x A Y\, dx \tag{5.44}$$

where Y is the distance and $\bar{Z}$ the mean depth between a pair of neighbouring flowlines at distance x, while within the integral A and Y are functions of x.

Application of models
The single column model
The above system of equations can be applied to a vertical column of an ice sheet if the input as described can be specified, and in addition the horizontal temperature gradient ($\partial\theta/\partial x$) can be prescribed as a function of depth.

For an ice sheet approximating a uniform slab on an inclined plane with uniform conditions over large distances, the horizontal temperature gradient may be expected to be approximately uniform with depth if melting does not occur. In general, however, it is necessary to calculate profiles upstream and downstream to obtain estimates of this gradient and the estimates can be generally improved by iteration. This leads directly to the use of flowline models.

However the single column model is very useful to study the effects which variations of the input have on the resulting temperature profile. In particular, time changes of Z, θ_s, A, and U, can be studied directly by following the solutions of the above system with time which result from prescribed variations of the input over time. The main advantage of the single column model is that it is possible with a computer to scan over large ranges of many variables to study the temperature profiles sensitivity to them and thus find appropriate best-fit solutions.

Simplified models

For many purposes it is convenient to use simplified versions of the general system of equations (5.38)–(5.42) to study certain gross effects which are not significantly affected by the more complex refinements. Such simplified models have been discussed by Budd *et al.* (1971*a*) and Budd, Jenssen & Young (1973). Here we just list those which are made use of in later sections. In the applications to the various borehole profiles the effects of these various refinements are illustrated by numerical examples.

(a) Constant thermal parameters. Equation (5.38) becomes

$$k\frac{\partial^2\theta}{\partial z^2} + w\frac{\partial\theta}{\partial z} + u\frac{\partial\theta}{\partial x} + \frac{Q}{\rho c} = \frac{\partial\theta}{\partial t} \quad (5.45)$$

where k is a mean thermal diffusivity for the column. A closer approximation to (5.38) is obtained by specifying k in (5.45) as a function of θ but still neglecting the derivative of conductivity. It needs to be pointed out that this procedure is not satisfactory for the firn layer in which the thermal parameters vary greatly with density as well as temperature.

(b) Basal heating. In this case the internal heating Q is considered to be concentrated to the base of the ice. This allows (5.45) to be written as

$$k\frac{\partial^2\theta}{\partial z^2} + w\frac{\partial\theta}{\partial z} + u\frac{\partial\theta}{\partial x} = \frac{\partial\theta}{\partial t} \quad (5.46)$$

and the basal boundary condition $\gamma_b = \gamma_G$ is replaced by

$$\gamma_b = \gamma_G + \frac{\tau_b U}{K} \quad (5.47)$$

where τ_b is the base stress and U is the average column velocity.

(c) Uniform strain rate. If the strain rate is uniform with depth and non-zero this strictly implies that some

sliding must take place upstream or downstream. However it can give a close approximation where the bulk of the shear is near the base. The vertical velocity then becomes proportional to depth and equation (5.46) may be written

$$k\frac{\partial^2\theta}{\partial z^2} + A\frac{z}{Z}\frac{\partial\theta}{\partial z} + u\frac{\partial\theta}{\partial x} = \frac{\partial\theta}{\partial t} \quad (5.48)$$

It is important to note that this does not imply a uniform velocity over the depth such as in block flow as considered below.

(d) Uniform velocity. This is the case for block flow in which case the velocity at all levels (u) may be taken as equal to the average velocity through the column (U). Equation (5.48) may then be written

$$k\frac{\partial^2\theta}{\partial z^2} + A\frac{z}{Z}\frac{\partial\theta}{\partial z} + U\frac{\partial\theta}{\partial x} = \frac{\partial\theta}{\partial t} \quad (5.49)$$

(e) Constant advection. For the case of uniform velocity and constant longitudinal temperature gradient ($\partial\theta/\partial x$) over the depth (or some special combination in which their product is constant) the advection term may be replaced by a constant, say R, so that equation (5.49) becomes

$$k\frac{\partial^2\theta}{\partial z^2} + A\frac{z}{Z}\frac{\partial\theta}{\partial z} + R = \frac{\partial\theta}{\partial t} \quad (5.50)$$

(f) Steady-state column. For the case in which the rate of temperature change is a constant with time but a function of depth, say $\dot\theta(z)$, equation (5.50) becomes a steady state or a function of z only

$$k\frac{\partial^2\theta}{\partial z^2} + A\frac{z}{Z}\frac{\partial\theta}{\partial z} + R = \dot\theta(z) \quad (5.51)$$

(g) Constant warming rate. For the case in which $\dot\theta(z)$ is a constant over the depth, this can be combined with the advection term to give the (surface) warming rate, S, by

$$S = R - \dot\theta \quad (5.52)$$

and the differential equation reduces to the simple form

$$k\frac{\partial^2\theta}{\partial z^2} + A\frac{z}{Z}\frac{\partial\theta}{\partial z} + S = 0 \quad (5.53)$$

This equation has simple analytical solutions (see Budd *et al.*, 1971*a*) and includes the most important parameters for the temperature profile, namely A, Z and S and the boundary conditions, say θ_s and γ_b. It is very useful for any first gross analysis for a given situation.

(h) Combined models. Various combinations of the above models may be useful in certain situations depending on the data available. For example, the variable thermal parameters may be included in any of the subsequent models. Alternatively a steady-state equation may be used which still has varying strain rate and advection with depth, internal heating and varying thermal parameters, namely

$$\frac{\partial}{\partial z}\left(K\frac{\partial \theta}{\partial z}\right) + \rho c A \frac{z}{Z}\frac{\bar{u}_z}{\bar{u}_s}\frac{\partial \theta}{\partial z} + \rho c u \left(\frac{\partial \theta}{\partial x}\right)_z + Q = \rho c \dot{\theta}$$

(5.54)

where the temperature gradient $(\partial \theta/\partial x)_z$ can be a function of depth or treated as a constant. This particular equation is very useful for single column analyses and is also readily adaptable to flowline techniques.

The flowline model

To begin with, consider the model applied to a steady-state ice sheet. If the surface contours of the ice sheet are constant the flowlines are constant. Consider a vertical column near the centre of an ice sheet; the ice moves through that column at speed u. If a moving column is considered, for example, moving at the average column velocity $U(x)$ along the flowline, then the horizontal advection is given by

$$(u - U)\frac{\partial \theta}{\partial x}$$

(5.55)

The above model can then be used with this advection rate and the horizontal temperature gradient $\partial \theta /\partial x$ can be obtained from the last backward differences, provided that the time and space differences are sufficiently small.

For this model it is convenient to use a variable vertical scale, $\zeta_1 = z/Z$ as described by Budd *et al.* (1971*a*), to allow the column to grow or shrink with the ice thickness and preserve the surface and basal boundary conditions as specified.

For the steady-state model, then the data is prescribed simply as a function of distance along the flowline, namely $E(x), Z(x), A(x), U(x), \theta(x), \gamma_b(x)$. In addition to calculating the temperature and velocity distributions, the model can also be used to calculate particle paths and ages of the ice in the column as it moves along. The position (x_1, ζ_1) of an ice particle at time t_1 from its arrival on the surface is given by

$$x_1 = \int_0^{t_1} u \, dx \quad \text{and} \quad \zeta_1 = \int_0^{t_1} w \, dt$$

where u and w are determined, as described above, along the flowline, relative to the prescribed bedrock profile.

Non-steady-state models

The variety of possible non-steady-state models is largely due to the wide scope of possible variations of one or more of the input parameters, together or separately, with time and as a function of distance along the flowline.

A simple case would be to vary the surface temperature as a function of time and study the response of the temperature in the ice. In reality, however, as climate changes go on the ice sheet reacts to them, changing in thickness and velocity, and no doubt affecting also the temperature and accumulation rate through feedback with the environment.

If these changes are large, then it is also possible for the ice sheet to change substantially in shape and, as a result, the pattern of flowlines could be altered drastically. In order to study this, three-dimensional models of the type devised by Jenssen (1977) are required.

Use of borehole and ice core data

Record of time and space variations

The main question is whether borehole and ice core data can be used to uniquely determine the past history of the ice sheet and the climate. From the outline of the model it is apparent that if we know the past data on the following input,

surface temperature θ_s
surface elevation E
(or ice thickness Z)
accumulation rate A
the ice flow properties $\dot{\epsilon}_c$
(or the average horizontal velocity U),

then, provided the flow patterns have not changed too greatly, we should be able to use a flowline model to calculate the ice temperature–depth profile θ. If just one of the above input parameters is unknown then it should be possible by trial and error to find the value which matches a measured temperature profile.

All the above data are required as a function of time and space, or at least distance along a flowline. The ideal case would be to have a series of boreholes along a flowline, but it is first of all worthwhile to consider what can be deduced from a single borehole.

Stable isotopes and mean surface temperature

In order to use the stable isotopes as an indication of past surface temperature the following factors must be considered:

(1)   the current measured variations of mean isotopic and temperature values over periods of years, decades and centuries;

(2)   the variation of mean surface isotopic values with elevation and mean surface temperature;

(3)   the variation of mean isotopic ratios and temperature with changes in the ice-sheet elevation and time;

(4)   the variation of the isotopic ratios with temperature associated with changes in atmospheric circulation or structure which occur with climate and ice sheet changes.

If it is possible that real surface temperature changes can be separated from other extraneous effects on the isotopic ratios, then a single ice core could be used to give the first input data. However it must be realised that the scale of the time changes is dependent on the depth–age relation which must be calculated from some assumptions of the past behaviour of the ice sheet and that the *ice is derived from positions removed from the site according to the age and the past velocities.*

Gas volumes and elevations

If reliable total gas volumes per unit mass (at STP) in the core can be obtained without losses, then it seems possible that, combined with the surface temperature data derived from the isotopic ratios, it may be possible to deduce the past surface elevation changes. Again this applies to ice derived from further inland (see sections 5.3 and 5.5 for further details).

*Accumulation and strain rates from
seasonal variations of stable isotopes*

If the seasonal variations in the stable isotopic
ratios can be detected to great depths, then the past
accumulation rates can be estimated, provided the
effects of horizontal and vertical strain can be accounted
for. Again for a single profile the answer is not unique.
However for a series of boreholes along a flowline the
results are much less ambiguous.

Ice flow properties and past velocities

The ice flow properties can be studied in several
ways, such as by deformation rates on the core simulating
in situ conditions, by borehole inclination measurements
and by independent measurements of the surface velocity.
If these methods give reasonable agreement, the flow
parameter ϵ_c can be determined, and also an estimate can
be made of the state of balance. If it is assumed that the
flow parameter remains constant over the period considered, the past velocities can be calculated from the
model, provided that the elevation profile is known,
along with the temperature profile. The effects of the
development of recrystallisation on the flow properties
of ice will also have to be considered.

A further point here that needs consideration is
that to some extent the input data could be overspecified
by such a series of borehole data because the velocity,
elevation, and past balance are related. This provides
a further check on the calculated values.

Age of the ice from stable and unstable isotopes

By tracing the annual layers, where possible,
through the ice sheet, an age–depth relation can be
established, Figure 1.9. However, to obtain many points
through each year back to 10^5 a BP is a formidable
task, and hence a certain amount of sampling and
interpolation may be required in the depth–age
determination.

Independent means of dating by unstable isotopes,
such as use of ^{14}C and ^{32}Si by Paterson *et al.* (1977), is
most valuable since errors in the ages affect the calculated
temperature profiles.

From the above sources of data it appears that
a good case can be made for the possibility of determining
both the past history of the ice sheet and the climate
from a series of boreholes along a flowline provided
enough support data is determined and the location has
certain suitable features. In addition, it is also clear that
a considerable amount can be determined from a single
borehole by trying a variety of possibilities in the model.

Such studies are carried out, for some of the few
detailed measured boreholes obtained to date in section
5.5.

5.5

Application of modelling techniques to measured profiles of temperatures and isotopes

W.F.BUDD & N.W.YOUNG

5.5.1   Camp Century, Greenland
5.5.1.1   Background

(a) Large-scale data. To begin with, we examine
the large-scale features of the Greenland ice sheet in
order to place Camp Century in perspective. Figure
5.18 shows the ice sheet elevation contours and broad-
scale flowlines from Budd *et al.* (1982).

Camp Century is shown on a particular flowline
which, if the contours were valid and steady state

Figure 5.18. Elevation contours (in km) and flowlines for the
Greenland ice sheet. Camp Century is shown by C.

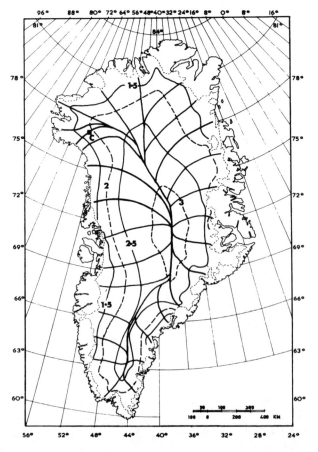

applied, could be traced to the centre of the ice cap.

From the data of ice surface elevation E, net accumulation rate A, and ice thickness Z, (or bedrock height b), steady-state 'balance velocities' $\bar{U}$ have been calculated according to the technique described by Budd *et al.* (1971*a*), and summarised in the previous sections using equation (5.44).

The net result, for the large scale, over the ice sheet, is shown in Figure 5.19. Sources for A are from Benson (1962) and Mock (1967). For ice thickness, the data from Holtzscherer (1954) has been used. From the surface temperature (θ_s) of Mock & Weeks (1966) and the elevations given above, smoothed temperature elevation gradients $\Delta\theta_s/\Delta E$ have been calculated as shown in Figure 5.20 for distances along flowlines.

(b) Particle paths and ages. Since the velocities of the ice sheet along the supposed flowline through Camp Century are unknown, some assumptions need to be made. The balance velocities suggest that an average speed of $10\,\mathrm{m\,a}^{-1}$ is reached by Camp Century. The borehole inclination measurements of B. Lyle Hansen (personal communication) first suggested that a surface velocity of about $10\,\mathrm{m\,a}^{-1}$ applied at the borehole site, but a later interpretation suggests $5.5 \pm 0.3\,\mathrm{m\,a}^{-1}$ (see section 4.2), while the survey measurements of S. J. Mock (personal communication) also favour a value of about $5\,\mathrm{m\,a}^{-1}$.

Irrespective of the present velocity, the velocity in earlier times could well have been faster or slower,

probably varying with the ice thickness. In addition, the possibility of rapid surges of the ice sheet should also be kept in mind.

Hence, for the time being, particle paths have been calculated for steady state, firstly for the calculated balance velocities of $10\,\mathrm{m\,a}^{-1}$ and secondly for the case of half this velocity which could be appropriate for a more rapidly diverging ice sheet and match the later borehole and survey reports. The resultant particle paths and ages are shown in Figures 5.21(*a*) and 5.21(*b*). The method uses a variable strain rate and velocity with depth as calculated from the temperature and velocity profiles of equations (5.38), (5.40) and (5.41). From these the ages of the ice at the various depths in the core at Camp Century are obtained for each case. These are shown in Figure 5.22, together with those obtained by Dansgaard & Johnsen (1969*a*) for a uniform strain rate and according to a specially prescribed two-layer strain rate close to that of equation (5.6*a*) and (5.6*b*).

In addition to the ages of the core, the particle paths give information on the source and position of the origin of the ice in the core at various depths for the prescribed velocities under the steady-state assumption. Although the flowline results for $5\,\mathrm{m\,a}^{-1}$ match that of Dansgaard and Johnsen's curve closely down to $1100\,\mathrm{m}$, the latter gives rise to much older ice in the basal layers. The $5\,\mathrm{m\,a}^{-1}$ and $10\,\mathrm{m\,a}^{-1}$ results become very similar near the base and are closer to the uniform strain rate results there than is the Dansgaard & Johnsen curve.

Figure 5.19. Balance velocities, $\bar{U}$ (m a^{-1}), for the Greenland ice sheet.

Figure 5.20. Temperature–elevation gradients ($-10^{-2}\,^\circ\mathrm{C\,m}^{-1}$) along flowlines for the Greenland ice sheet from Budd *et al.* (1982).

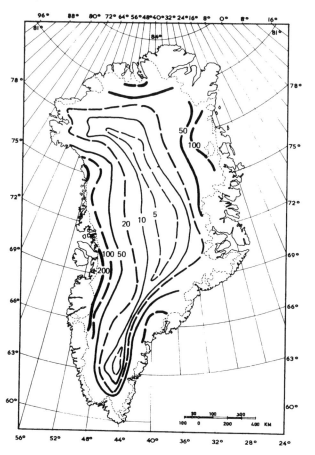

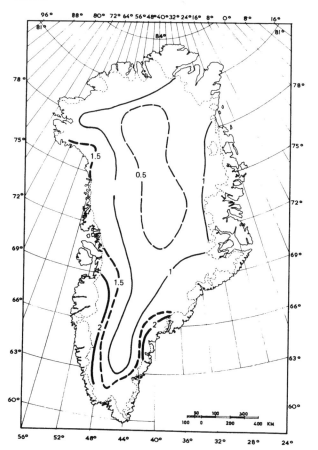

Since, as is shown later in Figure 5.42, the flowline velocity profiles used here are much closer to the measured profile, the resulting derived age–depth relations are also to be preferred here for self-consistency within the present treatment.

Finally, the resultant temperature profiles and isotope profiles are shown in Figures 5.23(a) and 5.23(b) together with the measured values from B. L. Hansen (personal communication) and Dansgaard, Johnsen, Møller & Langway (1970). Figure 5.23(b) includes a smooth approximation to the $\delta^{18}O$ values in the upper 1100 m used in the calculations.

Obviously the steady state of the present regime as prescribed in this calculation has not applied over the time period relevant to the temperature profile. The deviations of the isotope profile from the steady state suggest a cause of the discrepancy, namely that from above the 1000 m depth ($\approx$8000 a BP) there has been a tendency for cooling rather than warming ($\approx$1.5 °C) required by the present regime. To examine this further we turn to a study of the temperature profile in the column at Camp Century.

Figure 5.21. (a) Modelled flowline of the Greenland ice sheet passing through Camp Century showing calculated particle paths (full lines) and ages (broken lines marked in 10^3 a) for the case with a velocity at Camp Century of 10 m a^{-1}.

(b) Modelled flowline of the Greenland ice sheet passing through Camp Century showing calculated particle paths (full lines) and ages (broken lines marked in 10^3 a) for the case with a velocity at Camp Century of 5 m a^{-1}.

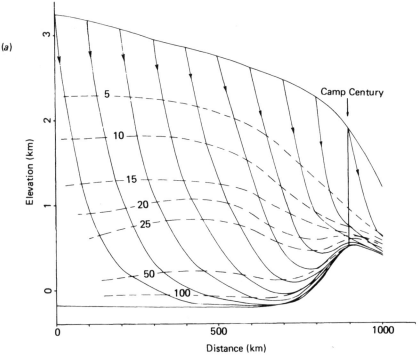

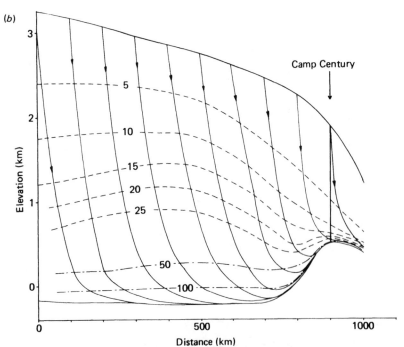

Figure 5.22. Camp Century age–depth relations as follows:
(1) uniform strain rate model calculation;
(2) Dansgaard & Johnsen (1969a) two-layer model;
(3) flowline model with velocity at Camp Century of 10 m a⁻¹;
(4) flowline model with velocity at Camp Century of 5 m a⁻¹.

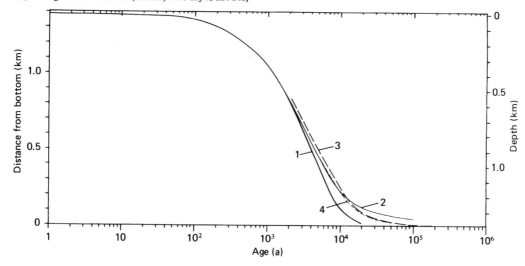

Figure 5.23. (a) Camp Century flowline steady-state temperature profiles for different velocities, and base gradients compared to the measured profile.
(1) Velocity at 10 m a⁻¹, base gradient at 0.018 °C m⁻¹.
(2) Velocity at 5 m a⁻¹, base gradient at 0.018 °C m⁻¹.
(3) Velocity at 5 m a⁻¹, base gradient at 0.022 °C m⁻¹.
  (b) (i) Measured Camp Century isotope–depth profile from

Dansgaard et al. (1970). (ii) Isotope–depth profile converted to surface temperature of origin profile using (i) with two conversion factors. A smooth curve has been used in calculations through the upper 1100 m. Also shown, is the steady-state profile from the flowline calculation with $U = 10$ m a⁻¹ at Camp Century.

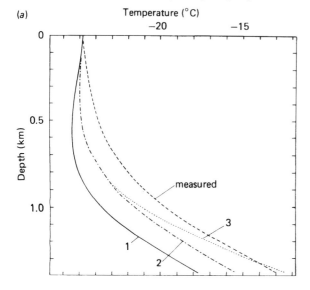

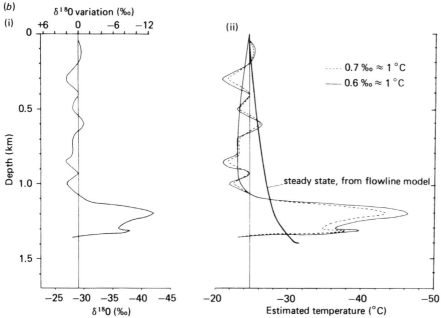

5.5.1.2 *Single column temperature profiles*

(a) Steady-state column. The single column model provides a very powerful technique for studying the gross effect of all the important parameters on the temperature profile.

To begin with we turn to the simplest equation, which includes the main parameters, and afterwards show what small differences are produced by the more complex refinements.

To begin with, for basal heating we take

$$k\frac{\partial^2\theta}{\partial z^2} + \frac{Az}{Z}\frac{\partial\theta}{\partial z} + S = 0$$

with $\theta = \theta_i$ at $z = z_i$

and $\left.\begin{array}{c}\theta = \theta_j \\ \text{or}\quad \gamma = \gamma_j\end{array}\right\}$ at $z = z_j$.

Thus the parameters to be considered are the diffusivity k, the accumulation rate A, the ice thickness Z, the advection or warming rate S, and two temperatures at different levels or a temperature and a temperature gradient. For *exact* fitting it does not matter which of these two alternative boundary conditions are used since the result is the same. For the present context the temperature at 200 m depth, well below the firn layer, and at the base are chosen from the measured profile, and the rest of the profile is used to derive standard deviations between the measured and calculated temperature profiles.

Of the above parameters, the ice thickness Z is the most accurately known so is set at 1386 m from Hansen & Langway (1966). The thermal diffusivity is a function of density and temperature, but below the firn the density dependence may be neglected, hence it is taken as a function of temperature only (the measured temperature as a function of depth) and the profile is only matched below the firn level. The layer of the firn is considerably more complex than the rest

of the ice since it involves varying density, compaction rates, and thermal parameters, and furthermore is grossly affected by short-term changes in surface temperature and accumulation rate. Thus the upper layer is best treated by studying it as a separate project using the deviations of the measured temperature profile from the calculated steady-state profile (see section 5.2). Here we concentrate on the temperatures below 100 m depth.

This leaves just A and S as unknowns. Hence solutions of the equation were scanned over a large area of the A–S domain and the standard deviations from the measured profile calculated. Contours can then be constructed over the domain to indicate the range of values of A and S which give a close fit.

This procedure can also be carried out with the other models to see if better fits can be obtained and which differences result in the best-fit parameters. Even the simple basal heating model gives a good fit with a minimum standard deviation (SD) of 0.04 °C (as shown in Figure 5.24, from Budd *et al.*, 1971*b*) with best-fit values of $A = 0.36$ m a^{-1} and $S = -0.37 \times 10^{-3}$ °C a^{-1}. The computed profile is so close to the measured profile that it is impossible to see the differences on the plotted scale so these differences are also shown on a five-fold expanded scale in Figure 5.24.

The layer heating model gives negligible improvement to the degree of best fit, but gives a slight shift in the values of the best-fit parameters, A and S. The contours of the SD of the differences between the measured and computed profiles over the A–S domain are shown in Figure 5.25. It is clear from this figure that the minimum-SD region is located on a long trough that represents close fits to the measured profile for values on a line through the minimum with

$$-\frac{A}{S} = \frac{0.01 \text{ m a}^{-1}}{0.03 \, 10^{-3} \,^\circ\text{C a}^{-1}}$$

for values below 200 m depth. Deviations are shown on a scale enlarged five-fold.

Figure 5.24. Single column model fit to measured temperature–depth profile at Camp Century (from Hansen & Langway, 1966)

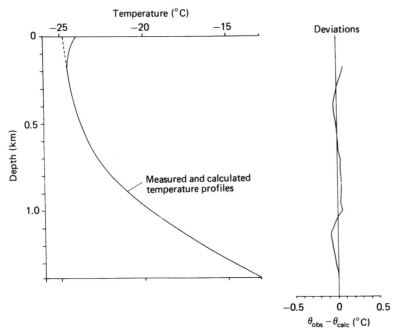

The nomogram of Figure 5.26 shows clearly how the various parameters Z, A, S, and γ_b affect the base temperature for small deviations from the best-fit values. For small deviations the changes are approximately linear, and a small variation in one or more of the parameters can be compensated for by a corresponding variation of another to preserve the close matching to the measured temperature profile. No significant improvement in the degree of best fit was found by adding the complexities of variable strain rate and advection and variable thermal parameters to the equation,

$$\frac{1}{\rho c}\frac{\partial}{\partial z}\left(k\frac{\partial\theta}{\partial z}\right)+\frac{Az}{Z}\frac{\bar{u}_z}{\bar{u}_s}\frac{\partial\theta}{\partial z}+u\left(\frac{\partial\theta}{\partial x}\right)_z+\frac{Q}{\rho c}=0$$

although the contours of best fit over the A–S domain are slightly shifted. Figure 5.27 shows the slight changes to the temperature profile made by variable strain rate and advection using the same values of A, S and the other parameters. The shifts in the best-fit troughs over the A–S domain are shown in Figure 5.28.

It is significant that the best-fit values give vertical velocities as close as can be expected to the measured surface accumulation rate which applies irrespective of steady state. The advection however, is a cooling compared to the warming of steady state of about $+0.15$ to $+0.3\times10^{-3}\,^{\circ}\mathrm{C}\,\mathrm{a}^{-1}$, shown in Figure 5.28 for velocities of 5 and 10 m a^{-1}, with the present surface slope of about 3×10^{-3} obtained from Mock (1965). This agrees also with the isotopic data as shown in Figure 5.23b.

The difference in warming rates between steady state and that deduced from the isotopes and the temperature profile could be due to either a climatic cooling of about 0.2–$0.3\times10^{-3}\,^{\circ}\mathrm{C}\,\mathrm{a}^{-1}$ or a rising of the surface of about 20–30×10^{-3} m a^{-1}, or to some contributions from each. However, for non-steady state many other possibilities may also exist, as shown below.

Accumulation rates: Table 3.3 on p. 69 shows measurements of snow accumulation around Camp Century from 0.34–0.40 m a^{-1} of ice equivalent. The best fit to the temperature profile for this range would be compatible with a range of cooling rates from about 0.2–$0.4\times10^{-3}\,^{\circ}\mathrm{C}\,\mathrm{a}^{-1}$. In order to consider possible past variations of accumulation rate we examine the annual layers measured in the isotopic profiles of Johnsen *et al.* (1972) shown in Figure 1.7. Here it is quite clear that the annual layers thin greatly with depth, as would be expected from vertical strain rates. In addition, since the accumulation rate decreases upstream this may also contribute to the reduction in the thickness of the annual layers. Figure 5.29 compares calculated decreases in the annual layer thicknesses using various strain rates with the measured values from the ice core, and also with the surface accumulation rates (A_0) at the points of origin of the steady-state particle paths upstream. The measured values of annual layer thickness agree most closely with those calculated from strain rates decreasing with depth similarly to the average horizontal velocity. The decrease in the accumulation rate upstream only makes a slight difference in the upper 1000 m. Although errors are proportionally large in the lower layers, the results suggest that the accumulation rate has been reasonably constant since the ice age change, except for the value at 800 m which is a little high. Many more such determinations are desirable to avoid effects of short-term variations (see figure 1.9).

Warming rates, temperature and elevation changes: Now, to interpret the derived cooling rate we may turn to the

Figure 5.25. Single column fit to Camp Century temperature profile. Contours of standard deviations over the surface vertical velocity, A, and advection, S, domain.

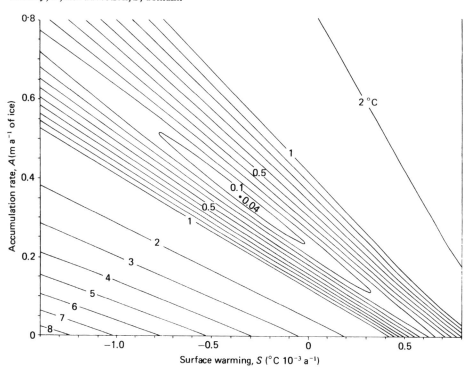

measured profile of isotopic ratios through the core of Dansgaard *et al.*, (1970) drawn on a depth scale in Figure 5.23(*b*). Two curves are shown against a temperature scale by using the two conversion factors of 0.7 $^o/_{oo}$ $\delta^{18}O = 1\,^\circ C$ and 0.6 $^o/_{oo}$ $\delta^{18}O = 1\,^\circ C$.

Assuming such a temperature interpretation holds for the present context, it is apparent that a cooling has taken place since the warm ice around 900 m depth was deposited. From the depth–age graph of Figure 5.22 this ice may be estimated to be about 5000 a old. From the smooth curve drawn through the isotopic values this net cooling can be gauged at about 1.2 $^\circ$C/5000 a or about $0.24 \times 10^{-3}\,^\circ C\,a^{-1}$. This agrees very closely with the best-fit steady-state value for the model with variable strain rate and advection, as shown in Figure 5.28, and provides a very good fit to the temperature profile with an accumulation rate of about 0.38 m a^{-1} of ice.

Turning now to the gas volumes measured by Raynaud & Lorius (1973) we construct an elevation scale depending on the pressure and temperature variation with elevation in a similar way to that described in section 5.3. The net result is shown in Figure 5.30. Here we construct a smooth curve through the measured

points as shown to relate to the smoothed isotope profiles in order to concentrate on the major anomalies over many thousand years.

In addition, the steady-state flowline elevations calculated from the particle paths of Figure 5.21 are shown by the dashed curve to represent plausible elevations of origin, for steady state, for the ice at various depths in the core. At around 700 m depth, the smoothed curve of elevation derived from gas volumes indicates that the surface level was about 200 m below the present surface. This corresponds to about 3000 a BP, and suggests that an average rise of surface level of about 200 m per 3000 a or 0.067 m a^{-1} is quite plausible. As a result of this, the surface temperature could fall at 0.7–1.0 $^\circ$C per 100 m of increase in elevation. Thus a temperature fall of some 0.4–$0.7 \times 10^{-3}\,^\circ C\,a^{-1}$ over the last 3000 a is reasonable. The average value over the last 5000 a could be somewhat less and thus comparable to the effective net cooling obtained from both the temperature and isotopic profiles. This rate of rising is of the order of 67 mm a^{-1} which is less than 20% of the current accumulation rate and would correspond to only modest deviations from balance or steady state.

Figure 5.26. Nomogram showing the sensitivity of the computed Camp Century temperature profile to variation in the key parameters. The changes in base temperature are shown for

variations in each of A, S, Z and γ_b about values which give a close fit to the measured profile.

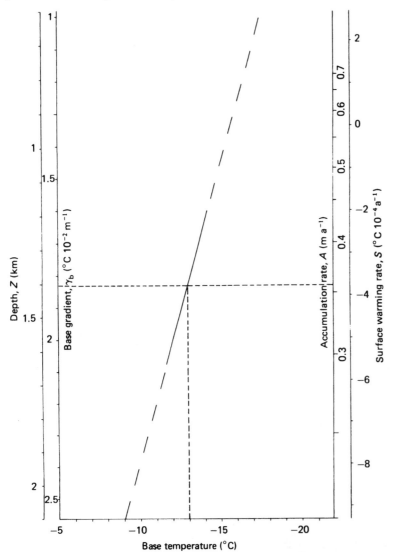

However the data on gas volumes is very coarse and sparse at this stage, and so it must be emphasised that the numerical results could well be in error by a factor of two. The main point to be made here is that the independent information from profiles of temperature, isotopes and gas volumes provides a powerful method for isolating the different effects of climate and elevation change. In fact it can already be inferred that any long-term deviation from balance is positive and less than 10%, and the average temperature change

Figure 5.27. Effects of refinements to the single column model on the calculated temperature depth profile at Camp Century.
(1) Original profile with uniform strain rate and advection.
(2) Variable advection rate only.
(3) Variable strain rate only.
(4) Both variable advection and variable strain rate.

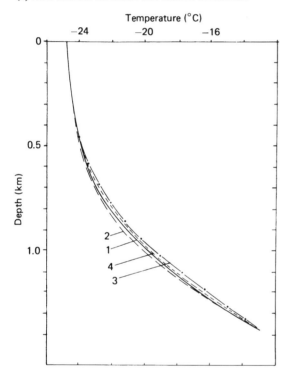

Figure 5.28. Shifts in the lines of best fit to the Camp Century temperature profile over the (A–S) domain for various refinements to the single column model.

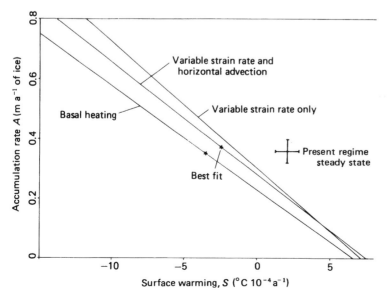

with time is not significantly more than would result from a rising surface due to such an imbalance.

So far only the steady state has been considered. In order to clarify the period over which the steady state is relevant we consider the passage of the effect of a time-varying temperature at the surface through the ice sheet at Camp Century.

(b) Non-steady-state column. The propagation of periodic temperature variation into ice sheets of various depths and accumulation rates has been discussed by Budd (1969) and Robin (1970). If the accumulation rate A is high compared to the speed of propagation of the temperature wave $v = \sqrt{(2\omega k)}$, where ω is the frequency and k the diffusivity, then the penetration of the variation is dominated by vertical advection. If v is large compared to A, then the conductive propagation dominates. For Camp Century, with $A \approx 36\ \mathrm{cm\ a^{-1}}$, periods of the range 10^3–10^4 a have propagation rates comparable to A.

In practice the vertical velocity decreases to zero towards the base, so the most straightforward way of studying the effects of time variations of temperatures at the surface, on the temperature profile, is to prescribe θ_s as a function of time t as a boundary condition for the simple equation

$$\frac{\partial \theta}{\partial t} = k \frac{\partial^2 \theta}{\partial z^2} + \frac{Az}{Z} \frac{\partial \theta}{\partial z}$$

and treat this as an anomaly on the steady-state profile. Provided the amplitude of the variation is not too large, this gives a good approximation since other terms only depend slightly on θ (k for instance and Q).

The resultant anomaly depends on the thickness of the layer concerned and the boundary condition at the lower surface. For the present context we wish to examine the effect of running an anomaly comparable to that represented by the $\delta^{18}O$ profile through Camp Century. An arbitrary generalised cold-wave anomaly was constructed as shown in Figure 5.31(*a*), starting at 60 000 a BP, rising to its maximum cold of 9 °C at 50 000 a BP, continuing to about 12 000 a BP then warming to the present at 10 000 a BP, rising to 2.5 °C warmer at 7000 a BP and then smoothly approaching zero at the present. Since the high frequency components of an anomaly die out rapidly with time, only the general outline of the anomaly is relevant. This is illustrated by the passage of a pulse of 2000 a duration through the ice sheet at Camp Century as shown in Figure 5.32. Here the vertical velocity varies with depth by equation (5.32) and it is quite clear that the propagation rate of the pulse is governed by this. In addition, the amplitude has dropped to 15% of the surface value by the bottom and the pulse has diffused out to many times its original duration in time. For this calculation the basal condition was taken at 2.5 km below the ice–rock interface. It is quite clear that this is ample depth for that time period. However, the much longer duration of the 'ice age' period changes, as des-

cribed above, requires a much deeper domain. In this case the temperature variation was studied for each of the locations where we have deep temperature data, namely Camp Century, Byrd and Vostok. For each of these a layer about 5 km thick was studied, so that in the case of Camp Century the bottom boundary reached about 3.6 km below the ice–rock interface. The boundary condition at that level in the rock was taken as the base gradient prescribed constant.

Although the very deep interface is slightly affected by the variation it is not important here because this makes negligible difference to the ice layer. In fact the variations in the rock would be greatly dependent on the temperature conditions much prior to 60 000 a ago which are not well known.

Hence we turn to the penetration of the long wave

Figure 5.29. Variation in thickness of annual layers with depth for the Camp Century core for various models compared to the measurements from δ ^{18}O variation by Johnsen *et al.* (1972). Circles, from δ ^{18}O measurements by Johnsen *et al.* (1972); dashed line, uniform strain rate, $A_s z/Z$ with $A_s = 0.36$ m a^{-1}; crosses, accumulation rate of origin from particle paths A_0; dots, uniform strain rate with accumulation of origin $A_0 z/Z$; squares, variable strain rate from $A_s z/Z$. $\bar{u}_z/\bar{u}_s$.

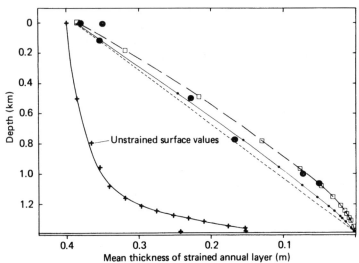

Figure 5.30. Steady-state elevation of origin for the Camp Century core from the flowline calculations (dashed line) compared to those derived from the total gas content in the core from the measurements of Raynaud & Lorius (1973), shown in (a) and as a fine full line in (b). The heavy full line in (b) represents the smoothed curve used in the subsequent calculations.

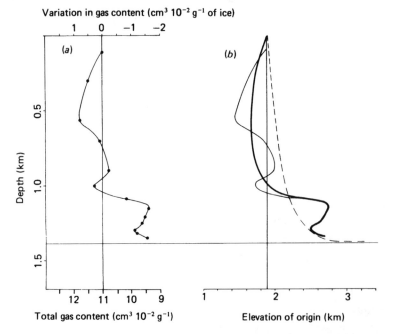

at the Camp Century site as shown in Figure 5.31(b).
Here it is apparent that the majority of the wave has
passed through the ice leaving less than a few tenths of
a degree Centigrade remaining above the 1000 m depth.
The most important feature is that the base gradient
over the bottom 500 m is decreased by about $0.004\,^{\circ}\mathrm{C\,m^{-1}}$
by the anomaly. Since the measured base gradient at Camp
Century is $0.018\,^{\circ}\mathrm{C\,m^{-1}}$ we can infer that the undisturbed
gradient would be about $0.022\,^{\circ}\mathrm{C\,m^{-1}}$ which is more
appropriate to a mean continental shield value, especially
when a component of 0.002–$0.005\,^{\circ}\mathrm{C\,m^{-1}}$ due to
frictional heating is included.

The basal temperature would be about $2\,^{\circ}\mathrm{C}$ colder
due to such an ice-age cold wave.

We now turn to Figure 5.26 again, which shows that
small deviations of the various parameters in the heat
conduction equation and its solution are approximately
linearly related. Thus it is easy to see that a close match
to the measured profile is obtained by increasing the
base gradient to $0.022\,^{\circ}\mathrm{C\,m^{-1}}$ (which would give about
$2\,^{\circ}\mathrm{C}$ shift in the base temperature) and at the same time
changing the other parameters to keep the net change in
the base temperature zero. For example, an accumulation
rate of $0.4\,\mathrm{m\,a^{-1}}$ and a surface cooling of $0.002\,^{\circ}\mathrm{C\,a^{-1}}$
may give a closer agreement with the current regime and
isotope data.

It is now becoming quite apparent that there are
many other interpretations still possible which could
explain the measured temperature profile. This is due
to our lack of knowledge of some essential data, for

instance the conversion relation between δ value and
mean temperature, the velocity of the Camp Century
site and the past history of the flowline. All that can be
said at this stage is that the measured profile can be
closely matched by plausible inferences from the
available data.

Before we turn to the flowline model analysis
of the temperature field, to clarify some of the past
history we note that the flowline studies so far have
been used primarily to calculate steady-state effects.
For this the basal boundary condition has been set as
either

(1)   a prescribed base gradient equal to an estimated
geothermal flux, with layer heating, or plus
a frictional heating term for the basal heating
model, or

(2)   a pressure melting point base temperature if
pressure melting is reached.

For thick ice and long-term temperature changes
over the last few ten thousand years these boundary
conditions at the ice-rock interface are satisfactory
even for non-steady-state changes. From the above,
however, it is clear that for thin ice, such as at Camp
Century, if melting at the base does not hold, then for
non-steady-state studies it is necessary to consider
conduction in the rock as well. The inclusion of deep
layers of bedrock in flowline calculations requires
considerably more computer time, and although such
calculations may be practical with faster machines in

Figure 5.31. Penetration of non-steady-state temperature
variations at Camp Century into the bedrock, with the base
gradient constant at 2.5 km below rock–ice interface.

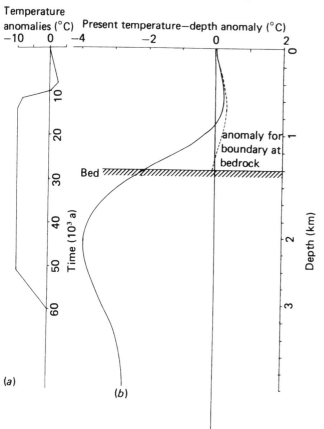

(a)

(b)

Figure 5.32. Passage of a 2000 a temperature pulse through the
ice-sheet at Camp Century with the gradient constant at 2.5 km
below ice–rock interface.

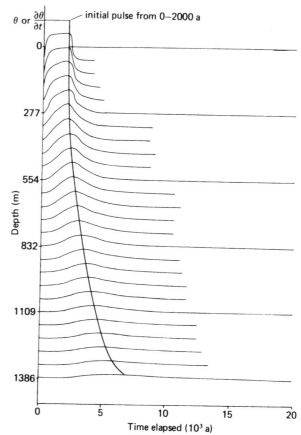

the future, for the present prescribed base gradients have been used.

Therefore it is necessary to examine the difference between the temperature in the ice obtained from a prescribed base gradient at the interface and one prescribed deep in the rock. It is clear from Figure 5.31 that the part of the past temperature record estimated from the isotope profile which is relevant to the temperature anomaly in the ice is the rapid change of temperature of 10–20 °C over a few thousand years about 10 000 a BP.

Thus we examine the propagation of a temperature rise over a period of 2000 a during the following 20 000 a through the ice at Camp Century, with the present being expected to correspond to about the 10 000 a time. The amplitude is arbitrary, but in this circumstance we can think in terms of a shift of 10 or 20 °C. Figure 5.33(b) shows the results of the propagation of such a change through the Camp Century ice with the basal gradient held constant. Figure 5.33(a) shows the corresponding propagation for the condition that the geothermal gradient at the depth of 2500 m below the ice–rock interface is kept constant.

We see that 10 000 a after the rise commenced, with the deep boundary conditions, the resultant temperature at the base had risen to only 0.43 of its final (steady-state) value compared to 0.76 of its final value if the base gradient remains constant at the ice–bedrock interface. In other words, the basal temperature anomaly due to the remaining effect of the last ice age is more than doubled when account is taken of conduction into bedrock.

Furthermore, with the deep boundary conditions a significant change in base gradient is shown to occur. After 20 000 a, the difference of basal temperature between Figures 5.33(a) and (b) remains similar, with the basal temperature still lagging by 0.34 temperature units for the deep boundary conditions. The difference between the two cases indicates the discrepancy to be expected from flowline calculations that prescribe the geothermal flux at the ice–rock interface rather than deep in the rock when the ice thickness is 1.4 km.

The actual magnitudes depend on the isotope-temperature conversion factor. Since it is the effect of the total anomaly of the surface temperature, due to both elevation and temperature changes, the range suggested by Figure 5.23(b) is about 10–20 °C. The actual period over which the change takes place is not quite so relevant because of the rapid diffusion, that is whether it is 2000 or 4000 a for the change of 10–20 °C does not matter so much for the anomaly at the base 10 000 a after it occurs. From Figure 5.33(b) showing the difference between the anomalies from the two conditions, it is apparent that the discrepancy for a 10–20 °C shift could be a 3.5–7 °C shift in the base temperature and a 0.01–0.02 °C m^{-1} shift in the base gradient. This would imply that the observed base gradient of 0.018 °C m^{-1} would be 0.028–0.038 °C m^{-1} without the ice-age anomaly. For the case of the 10 m a^{-1} velocity, the contribution from frictional heating is about 0.005 °C m^{-1}. This means that the geothermal gradient could be expected to be 0.023–

Figure 5.33. (a) Passage of a 2000 a warming pulse through Camp Century with a gradient 2.5 km below the bed constant. (b) Passage of a 2000 a warming pulse through Camp Century with a base gradient constant at the ice–bedrock interface. On the right is shown the temperature difference between the 10 000 a curve of (a) and (b).

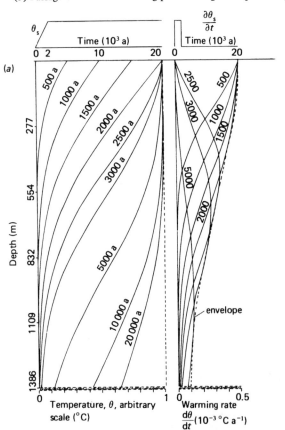

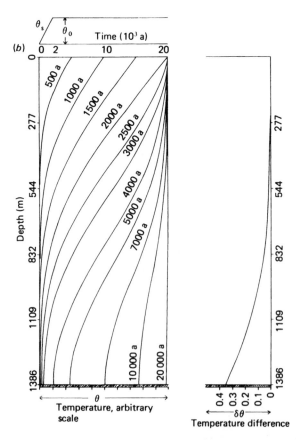

0.033 °C m⁻¹ rather than the anomalously low value of 0.013 °C m⁻¹ implied by steady state.

These results differ somewhat from the calculations carried out for the cold anomaly starting 60 000 a BP with the residual anomaly at the base of the ice being less for the latter. This will be due to the differing boundary conditions, both over the past 14 000 a (especially the 3 °C warming of the past 8000 a in Figure 5.31) and to the remnant effect of the long warm period prior to the cooling from 60 000–50 000 a BP. Although our knowledge of past boundary conditions is limited, especially regarding the date of the beginning of the cold period and the duration of the preceding warmer period, we consider that Figure 5.31 gives a more realistic estimate of the temperature at bedrock than can be derived from Figure 5.33. Hence the anomaly in the ice is also shown by a dashed curve in Figure 5.31. For our purpose this last anomaly is negligible, and thus we can consider the anomaly produced with the deep boundary condition as representative of the effective difference between our flowline calculations with prescribed base gradients and the actual measured profile at Camp Century.

Thus, with an idea of the effect of heat conduction in the rock, we turn to non-steady-state flowlines with prescribed base gradients and will then examine possible effects of bedrock conduction on the results.

5.5.1.3 Flowline temperature studies

(a) Steady-state flowline. To begin with we return to the steady-state flowline. The relevant input data required is illustrated in Figure 5.34 which shows the parameters surface elevation E, bedrock elevation b, accumulation rate A, balance velocity U, and surface temperature θ_s as functions of distance along the flowline and time prior to reaching Camp Century.

The particle paths calculated from these data provide the steady-state values to be expected in the borehole profile at Camp Century for such variables as elevation of origin, surface temperature of origin, accumulation of origin and so on. By comparing these steady-state values with the measured values, as shown in Figures 5.23, 5.29 and 5.30, estimates can be made on how the surface values at the points of origin differed from the steady-state input data.

Figure 5.35 examines the ice core data for isotopes and gas volumes and interprets them in terms of elevation and temperature change. It is apparent that the temperature (isotope) changes are greater than would be expected from the elevation (gas volume) changes alone, which could hardly exceed 0.01 °C m⁻¹. By subtracting the elevation temperature effect from the total temperature changes a residual climatic temperature change may be deduced. Such a residual climatic temperature variation is shown in comparison with the total inferred temperature in Figure 5.36. It should be noted that the net climatic change is only about half of the total temperature change. However, it must be pointed out that at this stage the conversion factors between isotopes, temperature, gas volumes and elevation are still not well known and so such differences may be

considerably in error. Here the emphasis is on discussing techniques; precise inferences will need to await more accurate data and information on such conversion factors.

In the meantime we take the described temperature and elevation values obtained from the isotopic and gas volume data as given and use them to prescribe input values along the flowlines as functions of distance and time instead of the corresponding steady-state values. It is noted here that it is the temperature and elevation differences from steady state that need to be added to the steady state to give the required input data. Also it is the total temperature change, and not just the climatic change, that is relevant to the resultant temperature profile.

(b) Non-steady-state flowline. The resultant modified inputs for a non-steady-state calculation are shown in Figures 5.37 and 5.38. The inferred elevation changes are compatible with an ice sheet some 600 m thicker near Camp Century, and some 200 km more extensive than the present ice sheet, about 15 000 a BP. In addition the temperature would have been about 8 °C lower than required for the elevation change, that is a residual climatic cooling for that time of about 8 °C.

It needs to be noted that these modifications to the input data have been made at the positions of the calculated steady-state particle paths. Of course, new particle paths could be calculated and these used to obtain new positions of origin and the process repeated.

Figure 5.34. Steady-state flowline input data for the Camp Century flowline.

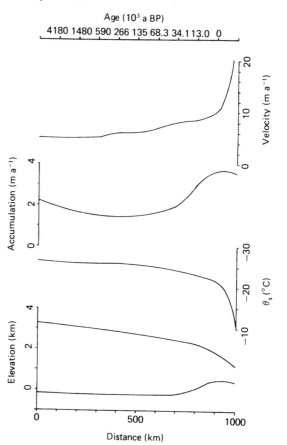

However at this stage these discrepancies may be considered as second-order effects as the past velocities and ice surface shape are still undetermined.

Hence we carry out the new flowline calculations with the data as shown in Figures 5.37 and 5.38 as a first approximation. The resultant temperature profiles are shown in comparison with the original steady-state profiles and the measured profile at Camp Century in Figure 5.39. Temperature profiles are shown, firstly for the effect of the surface temperature alone, changing as

prescribed, and secondly with both the temperature and ice thickness changing as prescribed. It is clear that the effect of the temperature variations are far more important than the ice thickness changes on the resultant temperature profile.

The basal gradient for this flowline study, for both the original steady state and the current non-steady state, was obtained from the measured base gradient at Camp Century adjusted for the effects of varying frictional heating along the route. It is apparent from Figure 5.39 that at Camp Century the upper 400 m of the measured and non-steady-state temperature profiles agree, whereas the steady-state profile is quite different.

Figure 5.35. Estimation of past elevations and temperature from gas volume (full curves) and isotope profiles (dashed curves). (*a*) smoothed data; (*b*) original data.

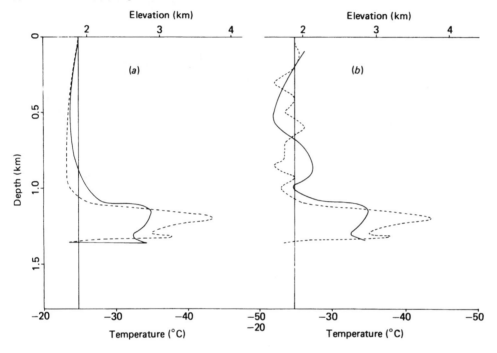

Figure 5.36. Comparison of residual climatic temperature change (*a*), with estimated total temperature change (*b*), for the Camp Century core.

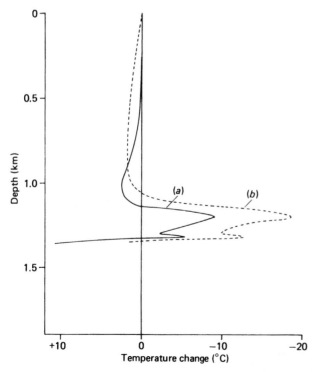

Figure 5.37. Non-steady state flowline temperatures (dashed line) superimposed on present temperatures (full line) for Camp Century flowline.

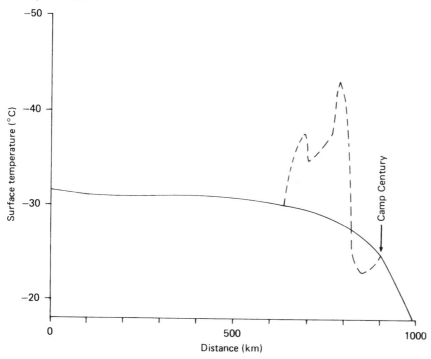

Figure 5.38. Non-steady-state flowline particle paths (arrowed lines) and ages (dashed lines with figures for 10^3 a). The three surface elevation profiles shown are for estimated maximum extent (short dashes), present ice sheet (continuous line) and that prescribed for the Camp Century moving column flowline (longer dashes).

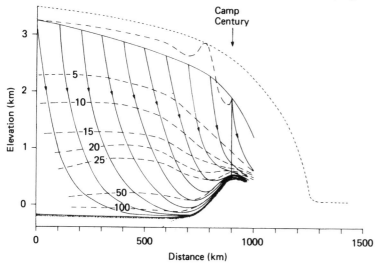

Figure 5.39. Camp Century flowline temperature profiles at marked distances from the origin are compared with the steady-state profiles, all using the present base gradient.

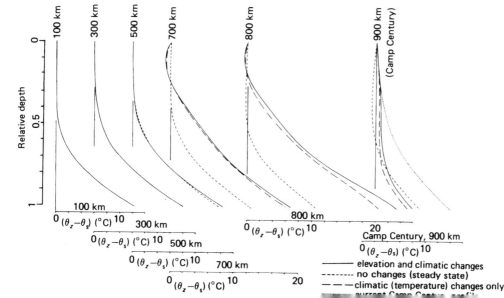

The bottom parts of the profiles are quite different, and so to interpret this we consider the effects of heat conduction in the bed. First of all we note from Figure 5.31 that a temperature change of about 20 °C over a few thousand years as shown in Figure 5.37 could cause a discrepancy in base gradient of 0.01 °C m^{-1}. Also about a 4.7 °C lowering of the base temperature could result from conduction in the rock not represented by the condition of a prescribed gradient at the ice–rock interface. In order to clarify this effect two further flowline calculations were carried out with the basal gradient increased by 0.0105 and 0.0165 °C m^{-1} respectively.

The resultant temperature profiles at Camp Century are shown in Figure 5.40. The first, with an increase of 0.0105 °C m^{-1} in γ_b, brings the base temperature up to the measured one but leaves a discrepancy higher up. A third calculation, with $\gamma_G = 0.028$ °C m^{-1}, gives closer agreement in the profiles than with $\gamma_G = 0.026$ °C m^{-1} down to about 850 m with thereafter an increasing discrepancy in temperature and base gradient towards the bedrock. This discrepancy agrees in form and magnitude with that derived from the effect of heat conduction in bedrock and ice for a temperature variation of the magnitude of that of Figure 5.37 10^4 a after the event.

Thus we may conclude that the measured Camp Century temperature profile can be matched closely using the isotope-derived temperatures and the flowline characteristics with a larger base gradient, provided conduction into the bed is considered. Hence the undisturbed geothermal gradient is considerably higher than the gradient measured in the borehole. However it must still be remembered that so far no account has been taken of possible different velocities in the past which may also have contributed to the effect on the basal gradient through frictional heating.

(c) Alternative matchings of the Camp Century temperature profile using the isotope–depth profile. The two most important unknowns involved in the relation between the isotope and temperature profiles are the conversion factor from isotope variations to temperature and the depth–age scale.

The knowledge of past velocities is very important because it affects not only the temperature and isotope profiles but also the time scale. All the time scales shown in Figure 5.22 are obtained from steady-state assumptions. If velocities have been greatly different in the past then the time scales could be grossly in error. In particular if the ice cap has risen and lowered to the extent shown in Figure 5.38 then considerable variations in velocity must be expected. Furthermore, if the rapid change in isotope values about 10 000 a BP was associated with rapid surge behaviour, this could have involved velocities of one or two orders of magnitude greater than at present. Such a velocity change, together with the associated large strain rates, could have grossly affected the time scale in that region. Thus other possibilities must also be kept open at this stage which give rise to the correct isotope and temperature profile but allow the velocity and time scale to be different from that which applies for steady state under the present regime.

Although the previous section derives a past history and set of conditions which give rise to a close matching to the measured temperature profile and is compatible with the other data, under the assumptions made concerning the velocities and time scale it is not difficult to derive other close fits to the measured temperature profiles from alternative histories. Some examples are discussed below. The question to be answered is, can other histories which give matching of the temperature profile be compatible with the other data, such as the isotope profile and be substantially different from the history derived in the previous section?

The main feature of the isotope profile which affects the temperature profile at Camp Century is the rapid warming about 10 000 a BP and the subsequent

Figure 5.40. Camp Century flowline non-steady-state temperature profiles with increased base gradient, compared to measured profile.

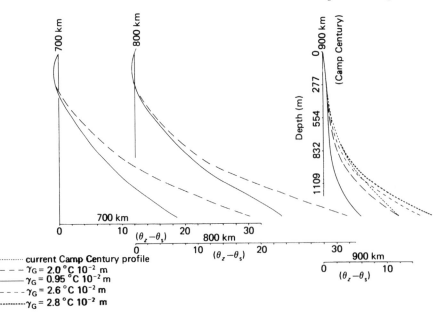

slight cooling. Hence a variety of magnitudes of such a warming and cooling, over a variety of time scales around the one used previously, were examined to study what kind of changes would be needed to give a close match to the measured profile. Some combinations which give very close matching are shown in Figure 5.41.

Although these calculated profiles give close matching, even for those with quite small amplitude warming (equivalent to a high $\delta^{18}O$–temperature difference conversion factor) the amount of subsequent cooling needs to be too high to give reasonable agreement with the isotope profile. This is seen to apply irrespective of the time scale because the lower the rate of cooling the longer it must last to give matching.

Since it is clear from Figure 5.39 that this result is not greatly affected by ice thickness change, it is concluded that the previous analysis is substantially unique. This shows that the matching of both the isotope and temperature profiles requires the consideration of the conduction in the bedrock and a higher geothermal gradient ($\gamma_G \approx 0.026$–$0.028\,^\circ C\,m^{-1}$) than implied by the present base gradient.

It still remains to examine the effects of changes in velocity during the large changes of ice thickness which this analysis suggests have taken place since the maximum of the glaciation.

5.5.1.4   Velocity considerations

In calculating the surface velocity from the stress, temperature and flow properties of the ice integrated through the column according to equation (5.28), large errors can accrue in the result due to errors from the surface slope, ice thickness, temperature profile and flow properties of the ice. Such large errors could cause instabilities in calculations of the type described in the previous section if the flow of the ice were governed by such a velocity calculated from a flow law. This type of calculated velocity is called a 'dynamics velocity' here, to distinguish it from the 'balance velocity' which is used as input and is calculated from an integral of the mass influx over the surface and the steady-state assumption.

Nevertheless, whatever the input velocity used, the program calculates a velocity–depth profile using a flow law such as that given by Budd *et al.* (1971*a*), either by a formula or look-up table. The shape of the dynamics velocity profile is used in the calculation of the internal heating distribution. The velocity profile is also used to calculate the particle paths by normalising its surface value to the input value.

Such a use of separate input velocity and calculated velocities is called a 'decoupled velocity' model because the calculated velocity does not prescribe the flow. Decoupled models are necessary when we do not know the present state of the ice cap as regards the balance, temperature distribution, past history, or flow properties of the ice, well enough to prescribe the correct input. Overspecification could lead to lack of assimilation of incompatible data.

To run a 'coupled velocity' model it is necessary to run the model as time-dependent so that the initial data is assimilated and all parameters and data are compatible. If such a system is run with data of the present (or estimated past) regime then there is no guarantee that the present surface profile (for example) would result, unless there was some special procedure adopted by adjusting some unknown parameters to supply the match.

Thus at this stage we wish to examine the following questions.

(1)   How well do ice flow properties derived from laboratory studies give a matching to measured velocity profiles?

Figure 5.41. Alternative matching of the Camp Century temperature profile by various warming and cooling sequences. (1) measured profile; (2) steady-state flowline. (*a*) Warming of 5 °C/2000 a from 13 000–11 000 a BP followed by (3) cooling of 0.0005 °C a^{-1} from 11 000; (4) cooling of 0.0008 °C a^{-1} from 11 000; (9) no cooling. (*b*) Warming of 10 °C/2000 a from 15 000–13 000 a BP followed by (5) no cooling; (7) cooling of 0.0008 °C a^{-1} from 13 000 a BP. (*c*) Warming of 5 °C/2000 a from 15 000–13 000 a BP followed by (6) no cooling; (8) cooling of 0.0008 °C a^{-1} from 13 000 a BP.

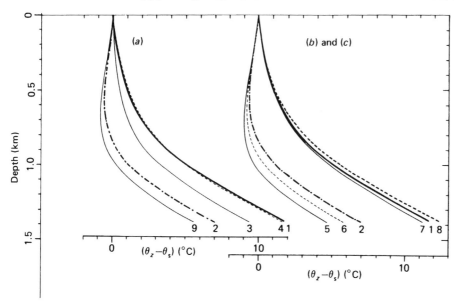

(2) What is the present velocity distribution in the ice sheet and its present state of balance (assuming we know well enough the distribution of net accumulation, ice surface and bedrock elevation)?

(3) If the present 'dynamics velocities' (and therefore flow parameters) are adequate, what can we say about the velocities of the ice sheet in the past when it had a greater extent and different thermal regime?

To answer the first question we examine the measured velocity profile at Camp Century by B. L. Hansen (personal communication and Figure 4.7) as shown in Figure 5.42 in comparison with those calculated from both the simple analytical formula and the look-up table of Budd *et al.* (1971*a*).

The agreement is quite close, whereas that calculated from higher power laws and more severe temperature dependence formulae give rise to velocity profiles which are too much like block flow compared to the measured profile.

Because of the low shear stresses in the Camp Century flowline profile further inland, this effect becomes even more noticeable in the interior. This is illustrated by Figure 5.43 which shows the progression of the computed velocity along the flowline.

Figure 5.44 shows normalised velocities calculated for the same average velocity as the input velocity. The magnitude of the dynamics velocity obtained at Camp Century was about 6 m a^{-1}. Since the velocity at the site according to the latest interpretation of borehole data is 5.5 ± 0.3 m a^{-1} (Figure 4.7), the agreement is close, although perhaps fortuitous due to our lack of knowledge of the flow law.

To answer the second question we consider the magnitudes of the dynamics velocities further upstream along the flowline. In general they have a tendency to be somewhat lower than the input balance velocities. This suggests that the interior could be building up both in thickness and velocity. However, since the temperatures and flow properties are not well known, this is speculation, but it does agree with conclusions from other evidence in section 2.5.

Finally, as to the question regarding the velocities in former times, as for example during the period of maximum extent, we consider the results of a flowline run with this former state as input.

The results suggest that the velocities were somewhat greater along the flowline inland of Camp Century but not much more proportionally than the ice thickness. The reason for this is that, although the thickness increases, the slope decreases and hence the base stress only slightly increases. Secondly, the increase in thickness does increase the base temperature, in spite of the

Figure 5.43. Calculated velocity profiles along the Camp Century flowline from the generalised flow law of Budd *et al.* (1971*a*) and the derived temperature distribution.

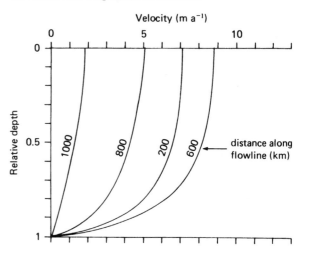

Figure 5.42. Various calculated velocity profiles for Camp Century compared with the profile calculated from the measurements of B. Lyle Hansen (personal communication). All curves have been normalised to the same surface velocity.

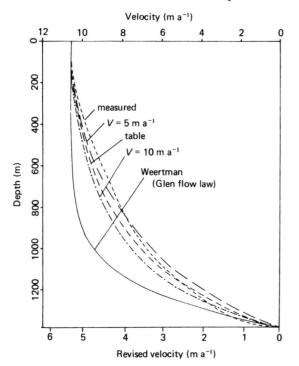

Figure 5.44. Normalised velocity–depth profiles calculated for the Camp Century flowline to show the change in shape of the profiles with distance towards the coast.

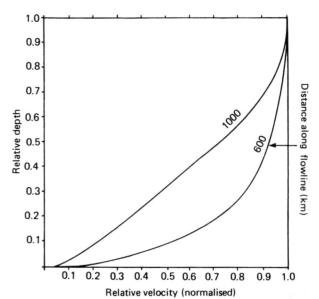

lower surface temperature from higher elevation. Thus, if the temperatures are generally as low as indicated by the analysis of the Century profile, then this gives a compensating effect on the velocities by the time the colder temperatures penetrate through the ice.

Consequently, unless there has been a substantial change such as a rapid surge, it appears that the present analysis is not greatly affected by otherwise expected variations in past velocities.

The possibility of a surge, or series of surges, of the Greenland ice sheet should certainly not be over-

Figure 5.45. Balance flux rates (10^{-2} km^3 km^{-1} a^{-1}) over Greenland from Budd *et al.* (1982) showing that on the west coast the high fluxes are favourable for high surging potential according to the method of Budd (1975).

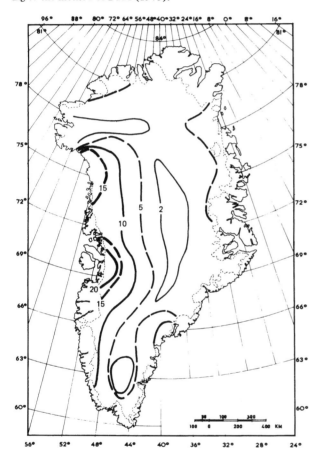

Figure 5.46. Estimated former extent of the Camp Century flowline using the borehole data, the edge of the continental shelf and the base stress distribution.

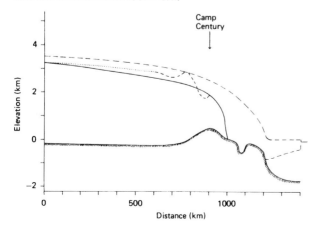

looked. From the type of model of ice mass surges discussed by Budd & McInnes (1974) and Budd (1975) it appears that surging could be possible when the flux rate becomes high enough. In Figure 5.45, taken from Budd *et al.* (1982), showing computed present 'balance flux rates' over Greenland, there appear to exist, at present, zones on the western side which could perhaps develop surge potential. At present this high flux seems to be catered for by the fast sliding Greenland outlet glaciers. However, during the maximum extent the rock in between was probably also covered, with the ice extending out to near the edge of the continental shelf before floating (compare Figure 5.46). Under such circumstances possible surging cannot be ruled out, at least until some further modelling calculations have been undertaken.

5.5.2 Byrd, West Antarctica
5.5.2.1 Background

Because the technique of analysis is essentially the same as that described for Camp Century in Greenland, the discussion is kept brief here with emphasis on the results.

Firstly we consider the map of the large-scale area with surface elevation contours from Bentley & Chang (1971). Lines are drawn orthogonal to the contours to represent the general regional flow patterns, as shown in Figure 5.47. The location of the ice movement net from Byrd to the ice divide, as described by Whillans (1973) and in section 3.7 is shown on the map. This line seems to be close to the general trend of flowline direction, although some crossing over occurs in places. Whillans' report gives the most accurate data, so this has been used for the basis of a flowline study (see Figure 5.48).

The resultant particle paths and ages are shown in Figure 5.49. From this, the age–depth relation for the Byrd core, as shown in Figure 5.50, is obtained.

Figure 5.47. Byrd basin in West Antarctica. 100 m elevation contours from Bentley & Chang (1971) are shown and flowlines drawn orthogonal to contours. The Byrd strain net is located between the divide, D, and Byrd, B.

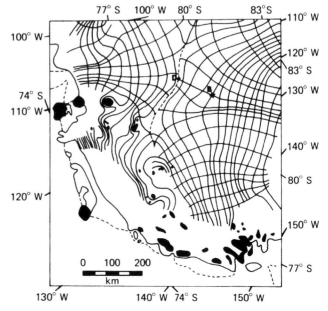

Figure 5.48. Byrd flowline steady-state input data.

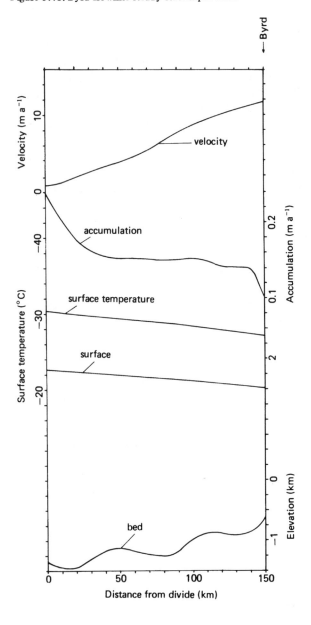

5.5.2.2   *Single column analysis*

We now turn to a study of the analysis of the temperature profile. Previous studies have been carried out by Radok *et al.* (1970) and Budd, Jenssen & Young (1973). Figure 5.51 shows some comparisons of steady-state profiles compared with the measured profile for a variety of models with different degrees of sophistication. Although slight improvements to the goodness of fit are obtained, these are somewhat academic since it is apparent from the deviations that there is a small but significant pattern of departure from the measured profile which suggests some definite non-steady-state conditions have applied.

For the various models, patterns of the standard deviations, over the accumulation rate (A) and warming rate (S) domain can be calculated. A typical example is shown in Figure 5.52 for layer heating. This indicates that a good fit is found for the present accumulation rate but the warming rate of about $0.5 \times 10^{-3}\,°\mathrm{C\,a^{-1}}$ appears to be substantially higher than the present steady-state value of about $0.35 \times 10^{-3}\,°\mathrm{C\,a^{-1}}$ obtained from a surface slope $\alpha \approx 2.9 \times 10^{-3}$, and a velocity $U \approx 12\,\mathrm{m\,a^{-1}}$.

We now turn to the measured isotope profile from Epstein, Sharp & Gow (1970). This shows ice of substantially colder origin prior to about 10 000 a BP (see Figure 5.53).

In order to study the effects of such a temperature variation at the surface on the deep temperature profile we return to the analysis of the long-term temperature anomaly of arbitrary amplitude described in 5.5.1.2(*b*). For Byrd, two cases were considered. For the first case the temperature at the base was assumed not to reach pressure melting point; for the second case the base temperature was assumed to remain at pressure melting. The latter is considered to be more likely. It can be seen from Figure 5.54(*c*)(ii) (with the base held at melting point) that for a 10 $°$C perturbation the anomaly would be about 1.8 $°$C maximum at about the 1500 m level. Since the magnitude of the temperature

Figure 5.49. Byrd flowline steady-state particle paths (full line) and ages (10^3 a) (dashed line).

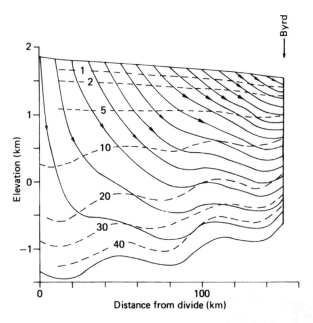

Figure 5.50. Byrd borehole steady-state flowline age–depth relation.

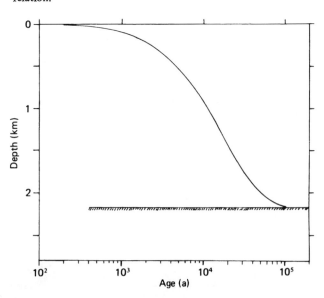

scale in Figure 5.54 is arbitrary, an 8 °C perturbation similar to that which might be expected from Byrd would result in the anomaly maximum being about 1.4 °C at 1500 m depth. This magnitude is compatible with the deviations from steady state if a lower warming rate is used. In fact the best-fit steady-state warming rate represents a weighted mean over the depth and time scale appropriate to the heat conduction and advection at that location. Figure 5.54(c) suggests this time period is from about the centre of the anomaly which represents about the last 30 000 – 40 000 a.

5.5.2.3 *Flowline calculations*

We now turn to the flowline calculations because over this time period the ice has come from a considerable distance up the flowline. If, instead of the normal decrease of temperature with elevation upstream, as shown in Figure 5.48, the temperatures are set at values derived from the isotope profiles (using an assumed conversion factor of $1\,^0/_{00} \approx 1\,^{\circ}C$) and the surface elevations of Figure 5.55, then a new non-steady-state temperature profile may be calculated as shown in Figure 5.56 together with the corresponding steady-state flowline temperature and the measured profile. A similar calculation has also been carried out in which the surface elevation deduced from total gas content as well as the temperatures are changed. It is clear from Figure 5.56 that the non-steady-state temperatures as input give the better fit to the measured profile. Furthermore, the change in thickness does not affect the result greatly but could improve the fit perhaps if correct magnitudes

Figure 5.51. Measured (*a*) and calculated (*b*) temperature profiles through the Antarctic ice sheet at Byrd. Deviations between the measured and different best-fit profiles are shown for models of increasing sophistication with (1) basal heating, (2) layer heating, (3) variable advection and (4) variable strain and variable advection.

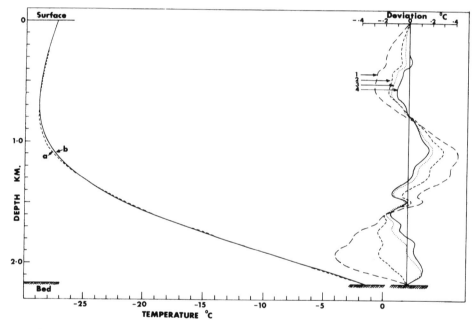

Figure 5.52. Byrd single column fit to measured temperature profile. Standard deviation (°C) contours of the differences over the A–S domain.

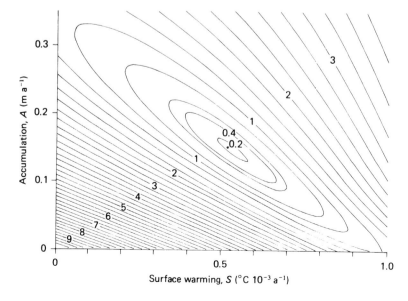

Figure 5.53. Oxygen isotopic ratios for Byrd Station (after Epstein *et al.*, 1970) compared with curve calculated using flowline model.

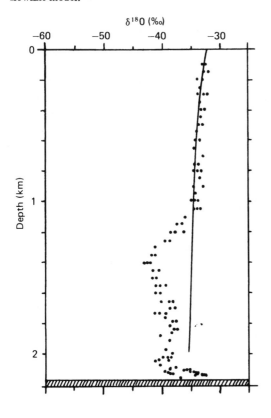

were used. The greatest discrepancies are in the lower layers, but it is expected that the layer heating model together with flow enhancement due to the measured crystal anisotropy could explain such small differences.

There are still too many unknowns at this stage, however, to allow confident improvement to the matching. These include lack of knowledge on the temperature–isotope conversion, the past elevations and velocities, past accumulation rates and the geo-thermal flux along the line. Hence we conclude that the measured temperature and isotope profiles may be considered compatible as far as we can tell at this stage.

5.5.3 Vostok, East Antarctica
5.5.3.1 Background

To begin with we examine the map of the large-scale area around Vostok showing the surface and bedrock elevation contours obtained from the SPRI map described by Drewry (1975*b*) as shown in Figure 5.57. The general flow patterns of the region are indicated by the orthogonals to the contours as shown in the figure. This general flow pattern is confirmed by the velocity deduced for Vostok from astronomic measurements reported by Liebert (1973) which gives a speed of $3.7 \, \mathrm{m \, a^{-1}}$ to the south east, that is towards the Ross Ice Shelf.

From the flowline drawn through Vostok the input data for a steady-state flowline calculation was obtained as shown in Figure 5.58. For the present context the balance velocity has been taken to increase approximately linearly from the divide. The net accumu-lation rate has been taken uniform at 25 mm of ice per year over the distance.

Figure 5.54. Penetration of a non-steady-state ice-age type temperature variation (*d*), through various locations of the Greenland and Antarctic ice sheets. (*a*) Vostok; (*b*) Camp Century; (*c*) Byrd (i) base gradient at 5 km depth constant; (ii) base temperature constant at ice–rock interface.

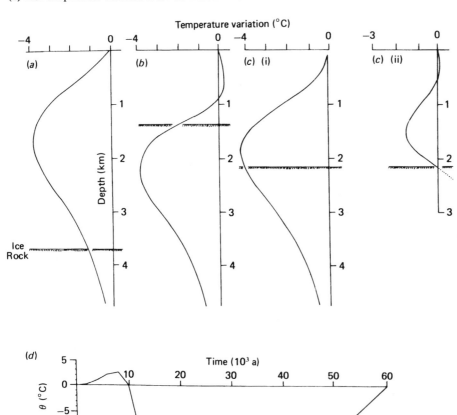

Figure 5.55. Byrd flowline non-steady-state input data.

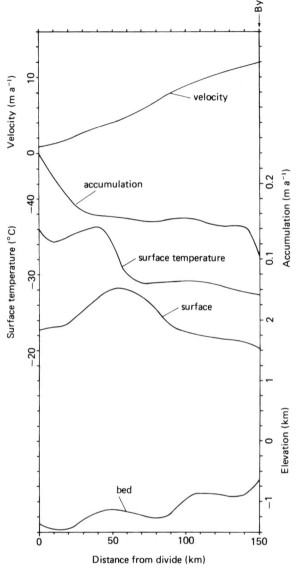

For this data the particle paths and ages result as shown in Figure 5.59. From these results the age-depth curve of Figure 5.60 and the positions of origin of the ice in the core at Vostok can be estimated.

5.5.3.2 Steady-state column

An analysis of the matching of the temperatures of the top 500 m reported by Barkov & Uvarov (1973) with temperatures calculated from steady state was presented by Budd, Jenssen & Young (1975) (Figure 5.61). From this analysis projections of the temperature profile right through to the bed for steady state were also made. Since then additional temperature data has been presented by Barkov, Vostretsov & Putikov (1975) to the depth of 780 m. This new data is also found to match the calculated temperature profile.

The deviations from the steady state are small (about ±0.01 °C) except in the upper firn layer. Again no attempt is made here to match the temperatures in the firn layer; this deserves the attention of a special project because of the intricacies already referred to in 5.5.1.2(*a*). We note here also that the density at Vostok is substantially lower in the upper layers than at Byrd or Site 2 in Greenland as shown in Figure 5.62.

It was pointed out by Budd *et al.* (1975) that even the simplest model gave a close fit at Vostok, and the inclusion of any of the complex refinements of the more sophisticated models makes negligible difference over the depth being matched. Even the layer-heating model gives only a slight difference from the basal heating, which is noticeable only in the profiles near the bed.

The standard deviation contours over the *A–S* domain are primarily straight lines in the region of interest. The pattern compared to those of the Camp Century and Byrd profiles are shown in Figure 5.63. The minimum line passes through the points (*S* = 0,

Figure 5.56. Byrd flowline non-steady-state temperature profiles compared to steady-state flowline model and measured profiles. (1) Temperature change only; (2) temperature change plus elevation change; (3) temperature change plus reduced elevation change; (4) reduced temperature change plus elevation change.

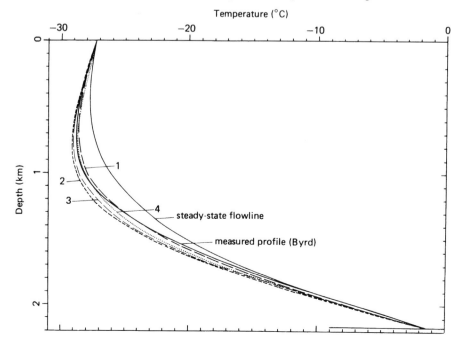

Figure 5.57. Plan maps of the ice sheet in the region of Vostok (V) from the SPRI map (Drewry 1975*b*) showing the surface (full line) and bedrock (dashed line) elevation contours and the estimated general flowlines from dome B.

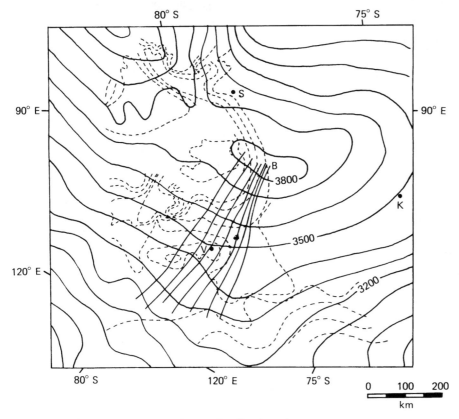

Figure 5.58. Input data for the steady-state flowline through Vostok. Accumulation is constant ($0.025\,\mathrm{m\,a^{-1}}$) along the flowline.

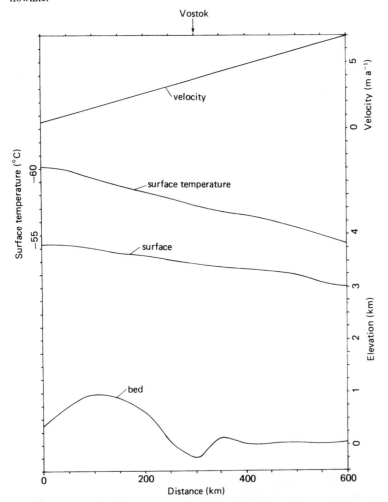

$A = 25 \text{ mm a}^{-1}$) and ($S = 0.05 \times 10^{-3}\,^\circ\text{C a}^{-1}$, $A = 20 \text{ mm a}^{-1}$). With a warming rate equal to the steady-state down-slope advection, namely $S = 0.024 \times 10^{-3}\,^\circ\text{C a}^{-1}$, the best-fit accumulation rate amounts to 23 mm a^{-1}. This is close to the measured values around Vostok given by Barkov et al. (1975) and other recent reports. It is important to note that small proportional changes in the accumulation and advection rates can cause corresponding changes in the surface temperature gradient according to the formula given by Budd (1969),

$$\gamma_s = \gamma_b\,e^{-y^2} + S/A\ 2y E(y)$$

where $y = \sqrt{(AZ/2k)}$, and $E(y)$ is the integral of Dawson's integral. Since the temperature gradient of the measured profile varies from about $0.007\,^\circ\text{C m}^{-1}$ near the surface to $0.01\,^\circ\text{C m}^{-1}$ at 700 m depth, a good match to the

Figure 5.59. Particle paths (arrowed lines) and ages (10^3 a) (dashed lines) for the flowline through Vostok (V).

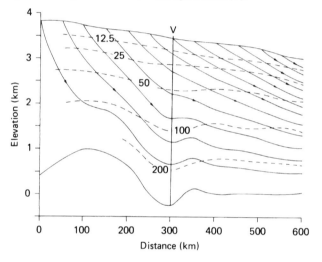

Figure 5.60. Vostok age–depth profile from flowline calculations.

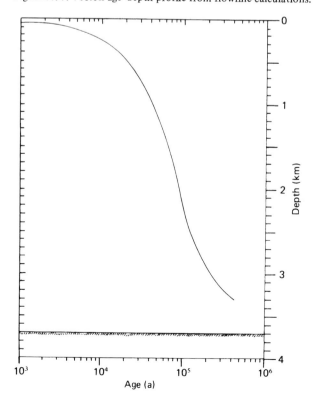

measured profile can be obtained by matching the surface gradient. The three parameters relevant for this matching are A, S and γ_b. A nomogram showing the variation of the surface gradient with these variables is shown in Figure 5.64.

With this background we now turn to the case of non-steady state, as revealed by the ice core isotope profile.

5.5.3.3 Non-steady-state changes

The measured isotope profile down to about 500 m from Barkov, Gordiyenko, Korotkevich & Kotlyakov (1974), and to 950 m from Barkov, Gordiyenko, Korotkevich & Kotlyakov (1976, 1977) is shown in Figure 5.65.

From the time scale of Figure 5.60 it is apparent that prior to 10 000 a ago the ice had a considerably colder origin by about 4 ‰ δ^{18}O, or around 4 °C, depending on the conversion factor. By comparison the estimated steady-state flow warming of about $0.03 \times 10^{-3}\,^\circ\text{C a}^{-1}$ gives only 0.6 °C change back as far as 20 000 a. Thus in order to see what sort of anomaly results from a temperature variation at the surface of the type suggested by the isotope profile we return to Figure 5.54 which shows the passage of the large temperature perturbation of about 50 000 a duration through the ice sheet at Vostok. For such a study it is clear that the duration of the perturbation is very important for the temperature anomaly at Vostok. However since the Vostok isotope data does not go back that far the duration has to be estimated. For this the corresponding isotope profiles for Byrd and Camp Century were used as a guide.

The result illustrated in Figure 5.54 shows that most of the anomaly from such a perturbation would still be contained in the thick ice at Vostok. For a 4 °C perturbation the anomaly would be about 1.7 °C. As far as the top 780 m is concerned, over which we have data for temperatures, the main effect of the perturbation was to add a negative gradient of magnitude about $0.002\,^\circ\text{C m}^{-1}$, to the existing steady state.

From the above equation for the surface gradient, and Figure 5.64, it is apparent that if a close match to the measured profile is to be obtained using the isotope data converted to temperature as the input then a lower accumulation rate, lower ice thickness or high base gradient is required. In addition this would result in a warmer base temperature. Since the least well known of these parameters is the base gradient, the measured values of the other variables were taken as before for the flowline, and calculations for different basal gradients were carried out.

First of all a flowline calculation was run with the surface temperature determined from the borehole isotopic data with a conversion factor of about $1.0\,‰\,\delta^{18}$O $^\circ\text{C}^{-1}$.

The modified flowline input data is shown in Figure 5.66. The resultant temperature profile is shown in Figure 5.67 as curve (c) along with the measured temperatures (a) and the profile (b) obtained for steady state with the present measured surface temperatures along the flowlines as shown in Figure 5.58.

Figure 5.61. Measured (1) and computed steady-state (2) temperature profiles to 480 m depth at Vostok. The deviations between the profiles are shown on a ten-times expanded scale. Zero depth is taken at 8.16 m below snow surface.

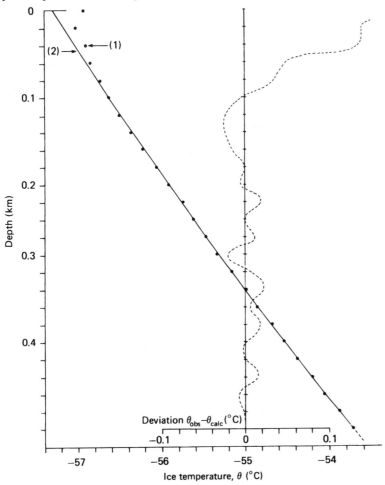

Figure 5.62. Density–depth profiles for various locations.

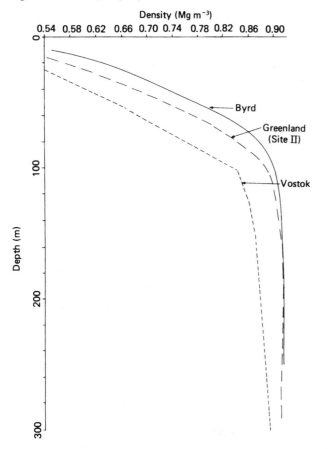

Figure 5.63. Standard deviations (°C) over the A–S domain for
the best-fit steady-state column models for Camp Century,
Byrd and Vostok.

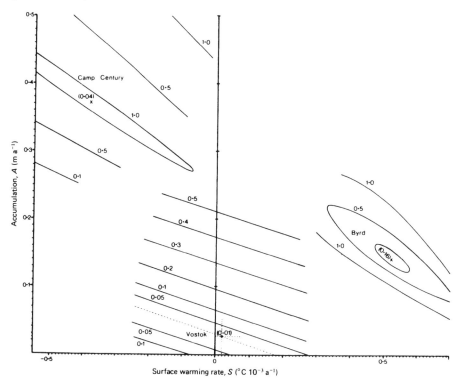

Figure 5.64. Nomogram for estimating change in the Vostok
surface temperature–depth gradient γ_s for changes in γ_b, A, S
and Z. Central values:
$A = 0.029\,\mathrm{m\,a^{-1}}$

$S = 0.02 \times 10^{-3}\,°\mathrm{C\,a^{-1}}$
$\gamma_b = 0.024\,°\mathrm{C\,m^{-1}}$
$Z = 3700\,\mathrm{m}$
$K = $ (variable thermal parameters)

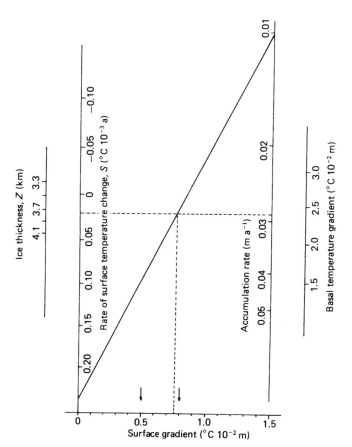

Since there is a gap in the isotope record in the top 50 m it was necessary to make some assumption about the recent surface values. Two extremes were considered.

(1) Firstly 20 sample running means (approximately 100 m averages as shown in Figure 5.65) from the $\delta^{18}O/^{16}O$ measurements to 950 m depth were interpreted as deviations from a present-day value of $-54.6\,^o/_{oo}$ and converted to temperature by

$$1\,^o/_{oo} = 1\,^\circ C$$

The result due to this input is given by curve (c).

(2) Secondly, because the scatter in the upper layers is so high, a constant temperature since 9000 a BP was also tried. The resultant temperature profile is shown in Figure 5.67 by curve (d).

Finally a second steady state with a lower base gradient $\gamma_b = 0.02\,^\circ C\,m^{-1}$ was also calculated with the other input the same. The result is shown by curve (e).

Figure 5.65. (a) The measured isotope–depth profile for Vostok from Barkov *et al.* (1974) to 500 m, and as provided by Kotlyakov to 950 m.
   (b) Smoothed curve from 20 sample means of (a) and the present surface values.
   (c) Calculated steady-state values from the present regime flowline model which results in the attached age–depth scale.

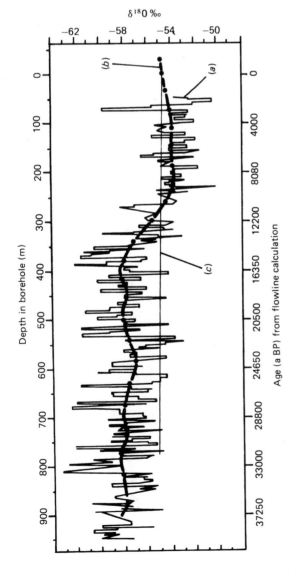

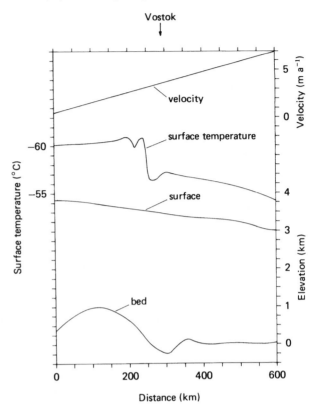

Figure 5.66. Vostok non-steady-state flowline data. Accumulation is constant (0.025 m a^{-1}) along the flowline.

Figure 5.67. The measured Vostok temperature–depth profile (a) (to 780 m depth) compared with temperature profiles calculated using the flowline model with accumulation = 0.025 m a^{-1}, $\gamma_G = 0.025\,^\circ C\,m^{-1}$; (b) present-day surface temperatures; (c) 20 sample ($\approx$100 m) running means from $^{18}O/^{16}O$ measurements to 950 m depth, interpreted as deviations from present-day value of $-54.6\,^o/_{oo}$ and converted to surface temperature change by $1 \times \delta^{18}O\,^o/_{oo} = 1 \times \delta\theta\,^\circ C$; (d) as for (c), except constant temperature from 9000 a (BP); (e) present-day temperatures, but $\gamma_G = 0.02\,^\circ C\,m^{-1}$.

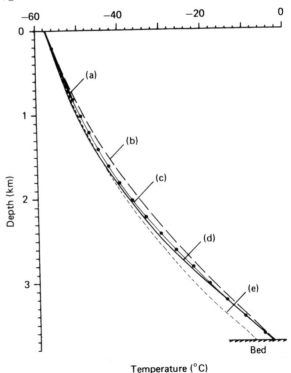

It is quite clear that the deviation caused by the isotope temperatures (c) compared to steady state (b) is as expected with a maximum at about 2 km depth of the order of 2 °C. The change in the surface gradient is small and much less than would result from a plausible lowering of the base gradient as shown by curve (e).

Nevertheless it has been shown that the measured temperature profile can be closely matched by the isotope-derived temperatures and the present-day ice-cap regime. It is also clear that deviations in the ice-cap parameters from the values obtained here for a close fit are limited and would need to be compensating as shown in Figure 5.54 to preserve the fit.

The results suggest that a base gradient of 0.025 °C m^{-1} is more appropriate than 0.02 °C m^{-1} and that as a result pressure melting at the base may be expected to occur, as is suggested by the presence of a sub-ice lake at Vostok (Robin, Drewry & Meldrum, 1977).

A lower conversion factor from isotopes to temperature would need a compensating variation in some other variable to give a close match.

The better fit of curve (c) of Figure 5.67 compared to (d) suggests that the cooling since about 8000 a BP is real and significant. This amounts to about 0.02×10^{-3} °C a^{-1} and again could be climatic or due to surface rising.

Since the deviations are small in the top 500 m, it will be most valuable to study the temperature profile at deeper levels, when available, to clarify the interpretation further.

5.6

Summary and conclusions

W.F.BUDD

The Camp Century results
Steady-state profiles

Firstly, when considering the degree to which the present Camp Century temperature profile approaches that of a steady-state regime, it is necessary to consider the steady state with horizontal advection, depending on the horizontal velocity, as shown in Figure 5.23(a), rather than with no temperature change as shown in Figure 5.2. It is clear there are quite substantial differences from the measured temperature profile and similarly from the measured isotope profile. The difference from that which could be expected from steady state is illustrated in Figure 5.23(b).

The inclusion of improvements to the simpler model, such as variable density and thermal parameters or non-columnar flow, only make relatively minor differences which cannot account for the major discrepancy which is the size of the difference between the maximum and minimum temperature of the profile. The depth Z* below which strain rates decrease to zero at bedrock used in section 5.2 gave most improvement when raised to reach the 800 m depth, in which case it was closer to the smooth velocity profiles used in section 5.4 and also the measured profile.

Close matching of a steady 'shape' profile (constant warming rate) with the measured profile could be obtained by using a variety of combinations of a lower accumulation rate and surface *cooling* advection rather than the warming required for steady state. The isotopic evidence from annual layering discounts the lower accumulation rate required, and the mean isotopic changes show a cooling of a magnitude similar to that required for the steady-state fit back to about 3000 a, but a lower average cooling would apply back to 5000 a ago. Beyond that, warming occurred with a very large change prior to about 10 000 a BP. The transient calculations show that these deviations from steady state have an important effect on the temperatures in the lower part of the profile.

Transient calculations

Firn layer. It is clear from section 5.2 that close matching was not achieved in the firn layer. This region is worthy of a separate study, in which case extreme care needs to be taken to avoid disturbances to the temperatures by the drilling itself, and several cores should be obtained to guarantee a reliable representative isotope profile. The fluctuations in the 5 a mean isotope data are too large to represent realistic temperature changes (see section 6.2). Smoother data should be used for this sensitive upper region for which smoothing by diffusion in the heat conduction equation is much less.

Because of the high variability of temperatures in the upper firn layer, and the arbitrariness of the zero of the temperature scale for the calculations, it is simpler to compare measured and computed profiles by reference to some deeper temperature such as the minimum in the measured profile between 150–200 m depth. The gross matching can then be judged in terms of the magnitude of the temperature difference between this level and the base.

Upper ice zone. The use of the isotope data as input to the transient temperature calculations provides a close matching for the measured temperature profile in the range 200–300 m depth relative to a fixed temperature in this layer. This result follows from both the column analyses of section 5.2 and the flowline analyses of Figure 5.39. This is particularly striking when compared to the profile to be expected from steady state. The single column analysis of section 5.2 gives agreement to greater depths. This seems to be primarily due to the time for the rapid warming being taken at about 12 500 a BP compared to that in section 5.5 being taken at about 9000 a, plus the fact that both analyses prescribed constant basal gradients at the ice–rock interface. The difference between the two can be understood in terms of Figure 5.33(*b*) concerning conduction in the bed as discussed below.

Overall profile and lower ice zone. For each case studied in section 5.2, and for the flowline studies which use the measured Camp Century base gradient in section 5.5, the difference between the minimum and base temperatures calculated is smaller in magnitude than that of the measured profile. That is to say, although matching in the upper ice layer is obtained, the calculated base temperatures are too cold. The effect of changing ice thickness, as derived from the gas volume and isotopic data, was also incorporated into both of the models, and although some slight improvement in the matching at Camp Century was obtained, the calculated base temperatures were still too cold (relative to the minimum temperature as fixed).

Although improved matching could be obtained by using higher warming rates, which could result from a larger $\delta^{18}O$ isotope to temperature ratio or from a lower accumulation rate, there are no grounds for the latter and the former is not necessary. It is shown in section 5.5 that the discrepancy can be adequately accounted for by heat conduction in the bedrock, together with a higher geothermal flux.

Bedrock heat conduction. For the Camp Century situation it is shown in section 5.5 that prescribing a constant temperature gradient at the base is inappropriate for the above transient-type calculations. In fact, if the boundary is fixed deep in the bed, below the level to which the changes considered here penetrate, then the resultant temperature near the base is several degrees colder than occurs with a constant base gradient, and the resulting base gradient is also substantially reduced. It was noted above, however, that the observed base temperature was higher (relative to the minimum) than that calculated with a constant base gradient. It is shown in Figure 5.40, with reference to Figure 5.33(*b*), that this can be accounted for simply by using a larger geothermal flux in the calculations. It is thus concluded that the measured basal gradient at Camp Century has been reduced below the geothermal gradient by the effect of conduction of the end of the ice-age 'cold wave' into the bed.

The resultant calculated temperature profile then agrees reasonably with the measured profile. The deduced geothermal gradient depends on the Camp Century velocity, but may be expected to be between 0.022–$0.026\,^{\circ}C\,m^{-1}$, which seems more plausible than the low value of 0.014–$0.016\,^{\circ}C\,m^{-1}$ which would result from the measured gradient of about $0.018\,^{\circ}C\,m^{-1}$.

Since it is shown in section 5.5 that this matching history is substantially unique, it is concluded that the use of the isotopic data for past surface temperatures is confirmed to a reasonable accuracy.

Interpretation of deduced temperature and elevation changes

The analyses of sections 5.2 and 5.5 show that the measured temperature profile confirms the total temperature changes implied by the δ values using the current conversion factors. This total change includes effects of both a climate change as well as an elevation change. Both section 5.3 and 5.5 find that, by using the gas volumes to determine past elevations, the deep ice, of apparently cold origin, also originated from considerably greater elevations than the present Camp Century site. In terms of approximate magnitudes, for a total change of about 18 °C around 10 °C was due to elevation change and some 8 °C due to net climate change in this region.

Furthermore, it is shown by the flowline model in section 5.5 that the deduced elevation change of about 1200 m consists of a component of downslope movement of about 400–600 m and an ice thickness change of 600–800 m.

At this stage these figures should only be taken as an approximate guide. Nevertheless, these analyses suggest that a series of boreholes along a flowline could furnish data to give a reasonably precise picture of the past history of both climate and ice-sheet changes by the techniques described here.

The Byrd results

It was shown by Radok, Jenssen & Budd (1970) that a close match to the measured temperature profile at Byrd could be obtained from a steady-state model with plausible values of accumulation and ice velocity as understood at that time. Budd, Jenssen & Young (1973) showed that this matching could be improved slightly by a more sophisticated model, and that a best-fit horizontal advection rate could be deduced along with an accumulation rate which was in agreement with the present observed value.

Since then Whillans (1973) has obtained accurate velocities and accumulation rates (A) at Byrd and upstream. It is then quite clear that the advection warming rate (S) required for best fit with steady-state model is too high by about a factor of 2. This is shown in Figure 5.52 by the steady-state analysis scanning over the complete field of A and S relevant for Byrd.

The isotope profile, compared to that expected from steady-state flow, as shown in Figure 5.53 from the flowline calculations, clearly shows evidence of a large, cold 'ice age wave'. Such a cold wave causes an anomaly to the steady-state temperature profile of the type shown in Figure 5.54(c).

This is more clearly brought out by the results of the flowline model analyses for steady state, with the present regime using Whillans' data, compared with that using the isotope profile as the source of past surface temperatures. The use of the isotope data at Byrd is somewhat complicated due to the lack of a well-established isotope–temperature or elevation relationship in that region. Nevertheless the use of a similar direct temperature–isotope conversion factor is supported by the agreement between the measured temperature profile and that computed from the isotopically derived temperatures as shown in Figures 5.56. In this case the present ice velocities at Byrd and upstream are well known, as are also the accumulation rates, and therefore the steady-state flowline temperature profile differs clearly and substantially from the measured profile. By contrast, that calculated from the isotopic temperatures gives quite a close fit to the measured profile.

The effect of past elevation changes as deduced from the gas volumes described in section 5.3 does not substantially affect the matching. A relatively small discrepancy still persists in the lower third of the ice thickness. It is expected that enhanced internal deformation heating due to the strong crystal anisotropy of the ice in this region can account for this discrepancy.

In summary, it appears that a net climate change of about 5 °C has occurred since the coldest period. A total elevation change of perhaps 500 m is inferred, of which about one third could be due to down-slope flow and the remainder to a decrease in ice thickness.

The Vostok results

The earlier work by Budd, Jenssen & Young (1975), performed before the isotope data were available, showed that the measured temperatures in the upper 500 m, below the firn layer, could be matched very closely by the present regime at Vostok for which the velocity, accumulation rate and ice thickness are well known. A basal gradient and basal temperature for best fit were deduced, as were possible combinations of accumulation rate A and surface advection rate S, given by a consistent A/S ratio shown in Figure 5.63.

Since then the temperature data has been extended to 780 m depth (see Barkov, Vostretsov & Putikov, 1975) and an isotope profile to 950 m has become available (Barkov *et al.*, 1976). Again a cold anomaly is indicated, of smaller magnitude than that at Byrd but similar timing for its ending some 10 000 a BP.

In the case of the thick ice at Vostok, with its very low accumulation rate, such a cold wave would produce a greatly diffused anomaly to the steady-state temperature profile as shown in Figure 5.54(a), with the peak about 2 km depth. The main effect on upper layers is to cause a slightly reduced positive temperature–depth gradient. Because of the very low accumulation rate and velocity at Vostok, the basal gradient, due largely to the geothermal flux, influences the surface gradient strongly. Since the basal temperature which was derived for steady state was below the pressure melting point, a higher base gradient can be used, until melting sets in, to compensate for the effects of the temperature anomaly associated with the isotopic cold wave.

The Vostok flowline analysis clearly shows that isotopic data, with a conversion factor of $1^0/_{00}/1$ °C, give a close fit to the measured temperature profile using the features of the present regime (Figure 5.67). The basal gradient and minimum geothermal flux required for this are deduced. In this case melting is found to occur at the base. Thus the geothermal flux and basal melt rate could be higher without influencing the gradient in the ice.

Since a gas volume profile is not available for Vostok, no calculations of previous elevations can be made. Hence, although we can infer a total temperature change from the isotope profile which is compatible with the observed temperature profile, it is still not definite how much of this change is due to climate change or ice thickness change. Furthermore, several unknown factors cause the conclusions to be less definite for the Vostok profiles. Firstly, as in the case of Byrd, the isotope variation with elevation upstream is not well known. Secondly, because of the extensive diffusion the temperature regime prior to 5×10^4 a BP (that is beyond the present isotope profile) is still relevant to the upper temperature gradient. Thirdly, the shift in the upper gradient caused by the anomaly is small, with the greatest effect being still deeper down. Finally, the observed temperatures are strongly affected by temperature changes over the last 2000 a for which the isotope data in the top 60 m is relevant but not yet available.

Thus in the case of Vostok it appears that a much clearer picture of the past changes can be determined when further data are collected. Even so it can already be said that a climatic warming from a regime about 5 °C colder some 10 000 a BP, as indicated by the isotopic profile, is also compatible with the present temperature–depth profile.

THE CLIMATIC RECORD FROM ICE CORES

G. de Q. ROBIN

6.1

The δ value–temperature relationship

Introduction

The verification in chapter 5 of a close connection between isotopic composition and mean temperature at any location on an ice sheet is impressive, and serves to advance our knowledge of the long-term variations of the climate, size and flow of the ice sheets of Greenland and Antarctica. On a shorter time scale, the link between temperature and isotopic composition is evident both in the seasonal cycle and at periods of 50 a or longer. At intermediate periods the situation needs clarification. In particular, it would be useful to know if mean annual δ values of polar snow can provide a record of former mean annual temperatures that will match the very effective record of mean annual snowfall preserved in annual layering.

The analyses in chapter 5 indicate that a component in the fluctuation of mean annual δ values which is not directly related to temperature is filtered out when data are smoothed over longer periods. We shall refer to this component as 'isotopic–temperature noise' ($\delta-\theta$ noise), whatever its cause or significance may be. However, processes contributing to this $\delta-\theta$ noise may well contain valuable information that would contribute to the other studies. If the $\delta-\theta$ noise component of annual δ values can be treated as random noise with a standard deviation $\sigma_{\delta\theta}$, it will cause an error in mean temperatures derived from mean δ values over a period of T years proportional to $\sigma_{\delta\theta}/\sqrt{T}$. This provides a simple hypothesis that appears consistent with the evidence. By noting the amplitude and duration of past fluctuations of δ values, we can then assess their significance as a past record of climate. We have already used these criteria when presenting isotopic profiles in chapter 4. In this chapter we present further statistical evidence in support of the hypothesis before discussing the physical factors governing $\delta-\theta$ noise. We then use the criteria as a basis for comparison of isotopic profiles from different ice cores and for comparison between the climatic record from polar ice cores and other climatic records.

Data correlation

A direct comparison of mean annual temperatures with corresponding mean annual δ values at sites on cold polar ice sheets over a long period would provide the best test of the hypothesis in the preceding paragraph. Unfortunately, there are no fully satisfactory records from one site on an ice sheet, but some temperature and isotopic records from nearby localities provide a limited test of the hypothesis.

Table 6.1 presents values of the standard deviations of mean annual temperatures (σ_θ) and of mean annual isotopic values (σ_δ) from a number of sites in Antarctica

and Greenland. We note a general tendency for similar numerical values of the standard deviations, which should be expected for a conversion factor of $d\delta^{18}O/d\theta \approx 1\,^{0}/_{00}\,^{\circ}C^{-1}$ if isotopic–temperature noise were absent or small. However, this is misleading as we show later, and such a correspondence may well have led to over-optimistic expectations of the short-term climatic significance of δ values. In Antarctica, values of σ_θ are similar to those of Greenland, while values of σ_δ over 15 a at Casey-Wilkes show similar numerical values to σ_θ at borehole P at 30 km distance.

At Antarctic inland stations, values of σ_θ are

Table 6.1. *Standard deviations of mean annual temperature (σ_θ) and of mean annual isotopic ($\delta^{18}O$) value (σ_δ) for selected polar stations*

	Latitude	Longitude	Elevation (m)	Period	Number of years	σ_θ (°C)	σ_δ ($^0/_{00}$)	Data source of δ values
Greenland Coastal stations								
Upernavik	72° 47′N	56° 07′W	35	1934–63	30	1.08		
				1875–1963	89	1.52		
Jakobshavn	69° 13′N	51° 03′W	31	1934–63	30	1.19		
				1874–1963	90	1.67		
Angmagssalik	65° 36′N	37° 33′W	29	1935–64	30	0.61		
				1895–1964	70	1.02		
Myggbukta	73° 29′N	21° 34′W	2	1922–58	36	0.90		
Inland stations								
Dye 3, Site C	65° 11′N	43° 50′W	2479	1938–70	32		0.99	Reeh *et al.* (1977)
Camp Century	77° 11′N	61° 08′W	1885	1770–1965	195		1.59	Estimated from 5 a mean values
Crête	71° 07′N	37° 19′W	3172	1670–1970	300		1.22	Estimated from 6 a mean values
Milcent	70° 18′N	44° 33′W	2450	1874–1963	90		1.28	From annual δ layering
Devon Island Ice Cap Station	75° 20′N	82° 30′W	1800	≈1915–50 / ≈1827–1950	≈35 / ≈124		1.16 } / 1.22 }	From samples of thickness estimated – one year
Antarctica Coastal stations								
*Argentine Islands	65° 15′S	64° 15′W	11	1948–75	28	1.62		
Esperanza	62° 24′S	56° 59′W	7	1954–78	25	1.08		From data of Aristarian (1980)
James Ross Is. (D)	64° 10′S	57° 45′W	1603	1954–78	25		1.23	From data of Aristarian (1980)
*South Orkneys								
Orcadas	60° 44′S	44° 44′W	4	1905–74	70	1.18		
Signy	60° 43′S	45° 36′W	24	1948–74	27	1.07		
*Belgrano	77° 46′S	38° 11′W	≈50	1955–75	21	1.25		
*Halley Bay	75° 31′S	26° 51′W	30	1957–75	19	1.05		
*Mawson	67° 36′S	62° 53′E	8	1954–72	19	0.75		
*Casey/Wilkes	66° 17′S	110° 32′E	(12) Wilkes	1957–71	15	1.04		
Borehole P, Law Dome	66° 09′S	111° 00′W	375	1957–71	15		1.11	From data of Budd & Morgan (1977)
*Dumont D'Urville	66° 40′S	140° 00′E	41	1957–72	16	0.61		
*McMurdo	77° 51′S	166° 40′E	24	1957–75	19	0.81		
Inland stations								
South Pole	90° 00′S	–	2912	1957–77	21	0.34	1.88	From data of Merlivat (private communication)
				1960–77	17	0.34	1.38	
*Vostok	78° 28′S	106° 48′E	3500	1958–73	15 only	0.74		
*Byrd	80° 01′S	119° 31′W	1515	1957–69	13	0.90		
Dome C	74° 39′S	124° 10′E	3240	1955–79	15		2.90	Benoist *et al.* (1982)

* Supplied by D. W. Limbert, British Antarctic Survey.

similar to or lower than those at other stations, the value of $\sigma_\theta = 0.34\,^\circ\text{C}$ at the South Pole being particularly low. In contrast to this, values of σ_δ at the South Pole and Dome C are unusually high. This appears to be due to an increasing amount of deposition noise when accumulation rates are low as will be discussed in section 6.2. In any case, the difference between σ_δ and σ_θ at the South Pole and the high value of σ_δ at Dome C indicates that isotopic–temperature noise must be high at these sites.

Table 6.2 surveys the relationship between $\delta^{18}\text{O}$ ratios and temperatures from various studies using the coefficients a and b in the relationship $\delta^{18}\text{O} = a\theta + b$ for the comparison. Part (1) of the table presents results of linear regression analyses of time series of mean isotopic and mean temperature values including the value of the correlation coefficient r. The value r indicates the proportion of the variance (σ_δ^2) that constitutes the temperature related signal in δ values. The remainder $(1 - r)$ indicates the proportion of noise (Reeh & Fisher, in prep.) causes of which are discussed later. The value of r should exceed its 95 per cent confidence limit shown alongside in brackets for the coefficients a and b to be considered satisfactory.

The shortest distance between a station at which a temperature record is available and an ice-coring site for which useful mean annual δ values are available is found on Law Dome where borehole P is 30 km from Casey Station. Figure 6.1, from Budd & Morgan (1977), shows the data used for the linear regression analysis listed in Table 6.2. The accumulation rate of 21 cm ice

Table 6.2. *Summary of isotopic – temperature relationships from various studies to determine coefficients* a *and* b *in* $\delta^{18}\text{O} = a\theta + b$

Study	a	b		Comment or reference
1 Time series, mean annual values				
(a) James Ross Is. Site D (δ) and mean Antarctic Peninsula (θ)	0.44	−12.5*	†$r = 0.57$ (0.40)†	Aristarian (1980)
(b) (1) Casey (θ) and ice core P Law Dome (δ)	0.61	−9.1*	$r = 0.57$ (0.51)	Author, see text
(2) Greenland Jacobshavn (θ) and Milcent (δ)	0.57	−9.5*	$r = 0.56$ (0.36)	Author, see text
(c) Seasonal (mean monthly) series				
(1) South Pole - surface θ	0.35	−33.9	$r = 0.69$ (0.57)	δ values from Aldaz & Deutsch (1967).
(2) South Pole - free air temp. above inversion	0.98	−15.6	$r = 0.66$ (0.57)	Temperatures from US Weather Bureau, analysis by author
(3) Syowa Station surface θ	0.69	−3.7	$r = 0.68$ (0.63)	Data from Kato (1978), analysis by author
(d) Trend line. Casey (θ) and ice core (P) Law Dome	0.90	−9.1*		Budd & Morgan (1977)
Trend line. James Ross Is. (δ) and Antarctic Peninsula (θ)	0.48			Aristarian (1980)
2 Areal analysis				
(a) East Antarctic ice sheet	0.75	−7.6		Lorius, Merlivat & Hageman (1969)
East Antarctic ice sheet	0.84	−7.5		Gordiyenko & Barkov (1973), section 3.3
East Antarctic ice sheet	0.99	0.4		From fig. 2 of Dansgaard *et al.* (1973)
(b) West Antarctic ice sheet	0.99	−6.5		From fig. 2 of Dansgaard *et al.* (1973)
(c) Antarctic Peninsula	0.9	−6.4		From Peel & Clausen (1982)
(d) Greenland ice sheet	0.63	−15.4		From fig. 2 of Dansgaard *et al.* (1973)
(e) Queen Elizabeth Islands, Arctic Canada	No effective relationship			Koerner (1979)
(f) Ross Ice Shelf	No effective relationship			See Figure 3.12
3 Cloud temperatures and collected falling snow				
(a) Roi Baudouin, Antarctica	0.9	−6.4		Picciotto, de Maere & Friedman (1960)
(b) Syowa Station, Antarctica	0.83	−6.6	$r = 0.72$ (0.32)	Data from Kato (1978), analysis by author
(c) South Pole (1)	1.4	4.0		Aldaz & Deutsch (1967)
South Pole (2)	1.21	−5.7	$r = 0.66$ (0.30)	Data from Aldaz & Deutsch (1967), reanalysis by author
4 Values used in sections 5.2 and 5.5 for temperature profile calculations				
(1) Station calculations, Camp Century, section 5.2	0.61			Increase of a may have improved fit of calculated to observed profile
(2) Flowline calculations, Camp Century, section 5.5	0.7			See text
(3) All calculations, Antarctica for Byrd and Vostok Stations, sections 5.2 and 5.5	1.0			See text

* Value of b adjusted to allow for difference between ice core site and meteorological station.

† Correlation coefficient r. Figures in brackets show 95 per cent confidence limits of r.

per year at this site is large enough to preserve annual layering of upper layers in spite of the effects of vapour diffusion (section 3.4), meltwater penetration and rough sastrugi. However, if allowance were made for transfer of material between annual layers by diffusion and meltwater penetration, the coefficient of $d\delta/d\theta$ could be up to 50 per cent greater than the $0.61^{0}/_{00}$ $°C^{-1}$ shown in Table 6.2. The trend-line study of the same data listed in Table 6.2 supports this suggestion.

A more detailed record of mean annual values of deuterium (δD) over a 25 a period has been presented in Aristarian (1980) for Dome D at 1550 m elevation on James Ross Island in Antarctica. The correlation with the temperature record at Esperanza station around 100 km to the north is not significant. However, the weather at Esperanza is frequently affected by the barrier winds studied by Schwerdtfeger (1975) which are unlikely to be of similar magnitude at 1550 m elevation on James Ross Island. Limbert (1974) has shown that climatic variations in the Antarctic peninsula region are similar over a wide region ranging from the South Orkneys, some 800 km east of James Ross Island, to the Argentine Islands, some 300 km to the south west. A linear regression analysis between mean annual regional temperature variations and mean annual values of δD on James Ross Island over 25 a gives a correlation coefficient of $r = 0.57$ (95 per cent confidence level of 0.40). The resultant coefficient a (converted to $d\delta^{18}O/d\theta$ equivalent) is lower than in other analyses, and a similar value is found by use of trend lines. This may be due to elimination of the local temperature 'signal' through using regional mean temperature variations. The limited evidence supports

the concept that the value of a (the gradient $d\delta^{18}O/d\theta$) decreases as the separation increases between the site of the isotopic profile and the temperature record with which it is correlated.

In Greenland, where temperature records for the past century are available at several coastal stations, we have compared temperature data from Jacobshavn with mean annual δ values from Milcent on the ice sheet some 280 km away in the hope of finding a significant relationship. Analysis of the full data over 90 a gives a relationship that falls below the 95 per cent level of confidence, but on dividing the data into three 30 a periods, results for the 1904–33 are well above the 95 per cent level of significance as shown in Table 6.2. Possibly changes of atmospheric circulation dominated variations at both sites during 1904–33, whereas other changes, such as the oceanic circulation, may have been dominant at the coastal station at Jacobshavn but not at Milcent during the other periods.

Analysis of data from the South Pole over a period of 14 a in Jouzel, Merlivat, Pourchet & Lorius (1979), and from L. Merlivat (private communication), produced coefficients which were very sensitive to small changes in selecting the summer peak in the isotopic record, so they have not been listed in Table 6.2. Furthermore, with an annual accumulation of about 8 cm a^{-1} ice, diffusion between layers smooths out annual fluctuations increasingly with the age of the layer. This affects the δ–θ relationship and also makes layers more difficult to identify. Merlivat considers her dating to be accurate to within 5 a back to 1887, but for our purpose an error of only one year in identification will make the comparison unsatisfactory.

The seasonal studies of time series in Table 6.2 come from mean monthly temperatures and δ values at the South Pole and Syowa Stations during single years. The effect of the surface inversion on data from the South Pole is obvious. The free air temperatures give a relationship that is similar to other studies while the low value of a when using surface mean temperatures indicates that atmospheric cooling in the inversion layer is an additional process that does not appreciably modify the δ values of the firn layers.

The second group in Table 6.2 covers areal studies. Over inland ice sheets and over the Antarctic Peninsula, where precipitation is dominated by orographic processes, areal analyses give values of a between 0.6 and 1.0 that are similar to results from most time series. In areas where precipitation is not dominated by orographic processes over gradually rising ice sheets, δ values appear to be dominated by the distance from the source of water vapour (see section 3.2) and the δ–θ relationship is not significant.

The studies of δ values in falling snow (part (3) in Table 6.2) in relation to cloud temperatures has been discussed in chapter 1 and in section 3.2.

Evidence from Byrd and Camp Century (part (4) of Table 6.2) indicates that the coefficient a has been of similar magnitude to the present over past millenia. This is evident both from the fit of calculated temperature-depth profiles based on isotopically derived surface

Figure 6.1. Law Dome, Antarctica. Isotopic ($\delta^{18}O$) profiles for borehole P showing both measured data and smoothed data using a 1 m running mean. Annual mean temperatures for Wilkes/Casey stations are plotted for the period 1957–71 at equivalent depths. (From Budd & Morgan, 1977, Fig. 2.)

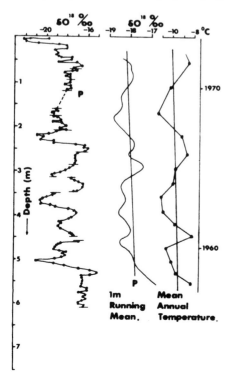

temperatures and from derived values of the former surface temperature lapse rate with elevation (Λ) in section 5.3.

Conclusion

Our understanding of the processes involved together with the statistical evidence in Table 6.2 leads to the following conclusions.

(1)   General climatic changes

We expect these to be recorded in isotopic profiles at any location on ice sheets or ice shelves irrespective of local isotopic gradients, since climatic effects on δ values are caused mainly by changes of distance to the source of water vapour. A value of $a = 0.7 \pm 0.3 \, ^0/_{00} \, ^\circ C^{-1}$ covers the magnitude of temperature changes over Antarctica and Greenland provided δ values are averaged over a sufficiently long period.

The coefficient b should be chosen to match the present mean δ value of surface layers at each ice coring site under study. This will average about $-9^0/_{00}$ at many sites, but values from 0 to $-16^0/_{00}$ have been noted in Antarctica and Greenland.

(2)   Correction for flowline dynamics

The best correction to δ values and temperatures for flow from higher elevations further inland is obtained from a knowledge of the present surface δ values and temperature profiles on the flowline leading to the ice coring site. This may vary widely with location, and the flowlines may have varied considerably in the past.

As a general guide in the absence of other data, coefficients of a and b obtained from areal studies listed in Table 6.2 for the appropriate areas should be used. In general, the coefficients of a lie between 0.6 and $1.0^0/_{00} \, ^\circ C^{-1}$.

(3)   Effect of elevation changes

These are shown most effectively by Figure 1.5. During condensation, as an air mass rises over an ice sheet, the effect of depletion of water vapour with removal of condensate (Rayleigh process) leads to the increased gradient of $d\delta^{18}O/d\theta$ with elevation shown in part (3) of Table 6.2 for cloud precipitation. Any increase of surface elevation due to thickening of the ice sheet should cause the same gradient of $d\delta^{18}O/d\theta$ with elevation. However, this will be compensated by an increased temperature inversion in surface layers, making the δ value–elevation (temperature) change similar to that shown by areal studies.

(4)   Absence of information

In the absence of specific information on the preceding points, use of a coefficient a of $0.8^0/_{00} \, ^\circ C^{-1}$ for general interpretation of Antarctic data should give derived temperature changes within about 30 per cent, while in Greenland a coefficient perhaps 25 per cent lower is suggested by the limited evidence available.

6.2

Isotopic – temperature (δ–θ) noise

The data in Tables 6.1 and 6.2 along with further studies of noise levels described in this section make it possible to estimate the error in mean temperature derived from mean δ values over a period of T years. This is given by $\sigma_{\delta\theta}/\sqrt{T}$ provided the distribution of δ–θ noise follows a normal distribution. The variance $\sigma^2_{\delta\theta}$ as we define it, is given by $(1-r)\sigma^2_\delta$ according to Reeh & Fisher (in prep.). Studies of deposition noise described later give the value of a major component of δ–θ noise which becomes the dominant component when accumulation rates are very low.

We first need to check the assumption that δ–θ noise follows a normal distribution. A study of 844 annual δ values in the top part of two ice cores from Devon Island gives a value of χ^2 well within that of a normal distribution, although this analysis covers both δ–θ noise and the δ signal related to temperature. At Dome C in Antarctica, Benoist *et al.* (1982) found that the distribution of δ values down to 105 m approximated to a Gaussian distribution. In this case the variance of δ values is dominated almost completely by the δ noise component of δ–θ noise. However, so many factors govern the δ–θ noise level that the assumption of random distribution is at best an empirical approximation. These factors are now discussed in more detail to show the complexity of the problem.

Processes governing the isotopic composition of surface snow are listed in columns (1) and (2) of Table 6.3, while column (3) lists parameters that are often compared with δ values in precipitation and ice cores. Discussions in chapters 1 and 3 cover most of the points listed in Table 6.3. In Table 6.4 we distinguish between factors producing δ–θ noise as we have defined it and noise in δ profiles only. By the latter we mean that component of the δ value variance which is not repeated in a second ice core from the same locality. It will be noted that two additional factors are listed as causing the δ–θ noise in addition to the factors that cause δ noise. We therefore expect δ–θ noise to be greater than δ noise. We shall discuss the causes of noise in Table 6.4.

Deposition noise

In the Antarctic, the only extensive study of δ noise by comparison of δ values between boreholes is that reported from Dome C by Benoist *et al.* (1982). They determined the spatial variability between 20 ice cores, each representing from 13–15 a of accumulation dated by β radioactivity. Assuming a normal distribution with time, the standard deviation of δD of $6^0/_{00}$ between these cores is equivalent to a mean annual variability of deposition noise of $\sigma_{N\delta(D)} = 23^0/_{00}$ equivalent to $\sigma_{N\delta}(^{18}O) = 2.9^0/_{00}$. This figure is much larger than the values of deposition noise of Reeh & Fisher in Greenland. At the South Pole, the value of $\sigma_\delta = 1.88$ shown in Table 6.1 is so much higher than the value of $\sigma_\theta = 0.34$ that the value of σ_δ must be dominated by δ–θ noise. Evidence from Dome C suggests that this will be almost entirely due to deposition noise ($\sigma_{N\delta}$), so we take $\sigma_{N\delta} = 1.80^0/_{00}$. Other estimates of δ–θ can be made for Law Dome and James Ross Island by using the values of σ_δ in Table 6.1 and of r in Table 6.2. The δ–θ noise variance given by $(1-r)\sigma_\delta^2$ will be the maximum possible value of the deposition noise. However, in our later discussion of comparable data from Greenland we estimate that there is also a contribution to δ–θ noise from rate of deposition noise. Applying this figure to data from Law Dome and James Ross Island, we therefore put the variance of deposition noise at these sites at two thirds of that of δ–θ noise, which gives us $\sigma_{N\delta} = 0.59^0/_{00}$ and $0.53^0/_{00}$ respectively for the two sites.

These admittedly rough estimates are plotted against the inverse of accumulation rate ($m^{-1}a$) in Figure 6.2 along with values of $\sigma_{N\delta}$ for the Greenland

region in Reeh & Fisher (in prep.). A line corresponding to $\sigma_{N\delta} = 0.125 A^{-1}$ gives a rough fit to the Antarctic points ($\sigma_{N\delta}$ in $^0/_{00}$ and A in $m\,a^{-1}$). This type of relation would be expected if the deposition noise is mainly a function of surface roughness as described by the theory of Reeh & Fisher, provided surface roughness is the same at all locations in spite of variations of accumulation rate and wind, etc. Although these assumptions are not very realistic, they provide a crude model to explain the Antarctic data. Because data from Greenland shown in Figure 6.2 cover a smaller range of accumulation rates, they do not help to clarify the problem in that area, which is discussed more fully in Reeh & Fisher (in prep.).

The broad conclusions on variations of deposition noise are consistent with correlations of δ values over longer periods between adjacent boreholes at the same site. On Devon Island Paterson *et al.* (1977) found a correlation coefficient of 0.965 between 50 a mean δ values from the surface to 13 m above bedrock at 300 m. In contrast, Benoist *et al.* (1982) found no obvious correlation between δ value profiles from two boreholes at Dome C when using a 170 a filter, but found some agreement between the two isotopic records with a 512 a filter.

Meltwater noise

Meltwater noise is difficult to estimate in terms of the process of surface melting and the depth of penetration of water before refreezing. Studies of temperature gradients in boreholes on Devon Island by Paterson & Clarke (1978) mention that at this site meltwater pene-

Table 6.3. *Factors governing isotopic composition of surface snow*

(1) Primary factors	(2) Secondary factors	(3) Related parameters
Isotopic composition of sea water	Type of precipitation:	Distances from sources of evaporation
Mean evaporation temperature	(a) frontal	to precipitation
Condensation processes, i.e.	(b) orographic	Latitude
immediate removal of condensate	(c) surface cooling	Altitude
or establishment of whole or partial	(i) rime	Mean surface temperature
equilibrium between vapour and ice,	(ii) inversion temperature	Mean temperature of air above surface
evaporation from snow-flakes in		inversion
lower atmosphere		
Mean condensation temperature		

Table 6.4. *Type and causes of noise affecting studies of isotopic (δ value) and temperature (θ) profiles in ice cores*

δ noise (comparison of 'δ profiles' over short distance)	δ–θ noise (comparison of mean annual δ values with mean annual temperatures)
Deposition variations	Deposition variations
Meltwater percolation	Meltwater percolation
	Deposition rate (i.e. summer/winter ratio)
	Temperature variations not related to deposition (inversion strength etc.)
Sampling errors	Sampling errors

tration seldom exceeds 1 m (Koerner, personal communication) but the effect may differ at other sites. If meltwater penetration allows a substantial fraction of meltwater from one annual layer to refreeze in a lower layer, it will affect the δ noise level. Measurements of δ noise between adjacent ice cores will not distinguish between that due to meltwater and that due to sastrugi, although absence of ice lenses will indicate that meltwater noise is not significant. Results in Fisher & Reeh (in prep.) show a higher level of δ noise at Crête, where meltwater penetration will be smaller than at Dye 2 in southern Greenland where ice lensing in upper layers is considerable and where they describe the δ noise as white (that is, random). This indicates that meltwater noise will generally be much lower than deposition noise.

Sampling noise

If annual layers cannot be identified accurately, as is often the case, δ noise will be introduced through inclusion of ice from adjacent years or omission of ice from the sampling year when measuring the mean annual δ levels. Even with clear annual cycles of δ values, as at Milcent, one cannot determine the precise limits of snow deposited over an individual twelve-month cycle. Where annual layering is not clear, as in studies on Devon Island (Paterson *et al.* 1977), samples cut with a length equal to the known mean annual accumulation at the site will introduce sampling noise. To determine errors due to adding fractional parts of annual layering to a given annual layer we will consider the case of δ values that vary regularly with depth z according to

$$\delta = A_0 + A_\delta \sin \frac{2\pi}{\lambda}(z + c)$$

where A_δ is the amplitude of δ fluctuations, A_0 their mean value, λ the annual layer thickness and c a constant. Then if we determine the mean isotopic composition $\bar{\delta}$ of a sample from a depth z to $z + (n + d)\lambda$, where n is an

integer and d a fraction, we get

$$\bar{\delta} = A_0 + \frac{A_\delta}{2\pi(n + d)}$$
$$\times \left[\cos \frac{2\pi}{\lambda}(d\lambda + z + c) - \cos \frac{2\pi}{\lambda}(z + c)\right] \quad (6.1)$$

Thus, not only is the effect of including an unknown fraction of an annual layer in our sample decreased in proportion to the total number of annual layers sampled, but it also depends on the phase angle $2\pi/\lambda(z + c) = 0$. Table 6.5 shows numerically how this sampling error falls with an increasing number of layers.

Since A_δ is typically around $5-10^0/_{00}$, and the measurement accuracy is typically $\pm 0.1 - \pm 0.2^0/_{00}$, there is no gain in accuracy by sampling more than 10 annual layers. The errors or 'noise' shown in Table 6.5 are an example of the 'blue' noise affecting measured δ values.

Equation (6.1) also serves to explain a pseudo-cyclic effect seen in some data plots. If successive samples of the length defined above are measured, the second will run from a depth $\{z + (n + d)\lambda\}$ to $\{z + (2n + 2d)\lambda\}$ while a third will run to a depth $\{z + (3n + 3d)\lambda\}$. The phase shift between the fractional component d/λ, $2d/\lambda$, $3d/\lambda$ will introduce an additional variation of δ that will run through a full cycle over $1/d$ layers. This effect, known as aliasing in computing studies, provides an explanation of the quasi-periodical variation seen in records such as Figure 4.11 which shows a cycle covering about three samples with an amplitude of the order of $0.5^0/_{00}$. With samples covering longer periods, random variations in annual layer thickness at most sites are likely to eliminate effects of aliasing. Again this points to the advisability of using mean δ values covering a period of at least 10 a for climatological purposes.

While we have presented our discussion in terms of regular sinusoidally varying layer properties, Table 6.5 also gives an indication of the magnitude of errors produced by the other factors mentioned earlier that lead to fractional sampling of layers.

Unwanted smoothing

We have shown in section 3.4 how diffusion in firn layers, associated with an average molecular movement (diffusion length L_0) of about 80 mm, effectively smooths out short-scale fluctuations of δ value and helps to produce a seasonal signal on which annual layering is

Figure 6.2. Deposition noise plotted against the inverse of the accumulation rate for Antarctic sites (filled circles) and for sites in Greenland (crosses).

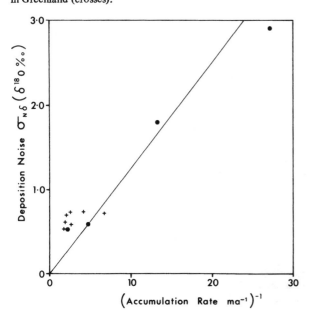

Table 6.5. *Sampling errors as a proportion of* A_δ *due to inclusion of whole plus a fraction (d) of annual layers with layering defined by equation (6.1)*

n	Maximum (d = +0.5)	Mean	Minimum
0	0.32	0.23	0
1	0.11	0.08	0
2	0.06	0.05	0
3	0.05	0.03	0
5	0.03	0.02	0
10	0.02	0.01	0

clearly seen, as shown in Figure 4.12 for Milcent. At the same time, the original amplitude of the seasonal fluctuations is decreased, and to regain the original amplitude, deconvolution methods are necessary (Figures 3.15 and 3.16). This process also affects the mean annual δ value due to exchange of water molecules by diffusion across the boundaries of annual layers. If the annual layer thickness is large compared to the normal diffusion length in firn (0.08 m) the effect on the mean annual δ value will be small, but for low accumulation rates the effect is considerable. Figure 6.3, which shows the filter factor in relation to depth, gives an appreciation of the smoothing effect in terms of distance from the sampling point. For an annual ice layer 0.6 m thick, only nine per cent of its volume is exchanged across its boundaries when diffusion is complete, while for a layer 0.2 m thick 30 per cent of the original ice will be lost after diffusion. For a layer of 0.04 m in thickness Figure 6.3 indicates that when diffusion is complete, the measured value (δ_{0m}) will be related to the original mean δ values of that layer (δ_0), and of adjacent layers $\delta_{+1}, \delta_{-1}, \delta_{+2}, \delta_{+3}$ etc., where the suffixes refer to mean δ values of annual layers above and below the reference layer, by

$$\delta_{0m} \approx 0.33\delta_0 + 0.30(\delta_{+1} + \delta_{-1}) + 0.20(\delta_{+2} + \delta_{-2})$$
$$+ 0.11(\delta_{+3} + \delta_{-3}) + 0.04(\delta_{+4} + \delta_{-4})$$
$$+ 0.01(\delta_{+5} + \delta_{-5}) \qquad (6.2)$$

Thus the measured value of a layer of 0.04 m of ice is weighted towards a running mean δ value over 5 a with further contributions to around 10 a. This smoothing effect is clearly seen in the studies of Dansgaard, Barkov & Spletstoesser (1977) on ice cores from Vostok, where annual accumulation is about 0.03 m a^{-1} of ice and even variations between annual layers are thoroughly smoothed out. In such a case diffusion prevents determination of climatic changes of δ value of periods much shorter than 20 a even if other isotopic–temperature noise were absent. In addition to effects of diffusion, meltwater

Figure 6.3. Form of filter function in relation to depth that smooths isotopic profiles of ice cores due to the diffusion length in firn of 80 mm.

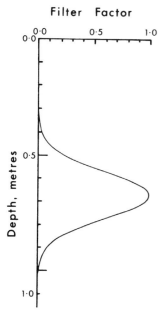

penetration discussed previously can also cause unwanted smoothing, but it is difficult to suggest its magnitude in areas where it is likely to be significant.

Causes of δ–θ noise in addition to those of δ noise
Rate of deposition noise

Variations in the proportion of snow deposited at different seasons, especially between summer and winter, will clearly produce fluctuations in mean δ values from year to year. These may even be related to the inverse of mean annual temperatures as we show in the next section. Such a negative correlation will tend to cancel out the normal correlation. Although 'rate of deposition' noise does not fall within our definition of δ noise, since its effects are expected to be the same for all ice cores in the same local area, it nevertheless forms part of the δ variance (σ_δ^2) that contributes to our δ–θ noise. If clearly defined boundaries to annual layering are used, then 'rate of deposition' noise may make an obvious contribution to the total δ–θ noise. A qualitative idea of the significance can be gained from the δ value profile at Milcent shown in Figure 4.12. In the absence of clearly defined annual layering, where annual samples are cut to a standard length, as in the work of Paterson *et al.* (1977) on Devon Island, sampling noise cannot be distinguished from noise produced from variations in rate of deposition.

Temperature noise

Formation of an inversion layer of cold air up to several hundred metres thick over polar ice sheets is primarily due to loss of heat from the surface layers by radiation. It forms under a clear sky except in summer months, but even in winter it is destroyed rapidly if thick cloud moves over a site. Variations of cloud cover therefore have a marked effect on mean surface temperatures of inland ice sheets, as will other variations of atmospheric circulation. Such factors produce noise in the temperature record that may be independent of the δ value recorded in precipitation. If snow falls more frequently and heavily during an individual winter season, a combination of rate of deposition noise and inversion temperature noise can occur. Abnormally heavy winter precipitation will lower the mean annual δ value of precipitation, while at the same time higher cloud cover will cause a decrease of surface cooling by radiation and result in warmer mean surface temperatures. The effect is shown on monthly data in Figure 6.4 which plots all the measured $\delta^{18}O$ values of Aldaz & Deutsch (1967) at the South Pole during 1964–5, along with mean monthly temperatures for the same period (horizontal bars) and the 20 a mean monthly temperatures (continuous curve). The month of June is seen to have been relatively warm, and this corresponds with frequent snowfall and cloud cover, and hence with a tendency to lower the mean δ value that month. To check whether the above effect can dominate the δ–θ relationship when using mean annual values, we show in Figure 6.5 a plot of mean annual accumulation rate (A) at a site 10 km from Dye 3 near the crest of the Greenland ice sheet against a plot of mean annual δ values at the same site. We can see that a year with high accumulation can have a lower (colder) mean δ value and the opposite with low accumulation, but the converse is also true. A linear regression fit shows no effective relationship between the two parameters, from which we conclude that at this site the tendency of a warmer air to carry more moisture and produce more precipitation during winter can be counterbalanced by other effects. It would be of interest to make similar comparisons at other sites with long records of annual δ values and accumulation rates.

Relative importance of different sources of δ–θ noise

Whereas the value of σ_δ increases rapidly with decreasing accumulation rate on the Antarctic plateau, values of σ_θ in Table 6.2 for Vostok and the South Pole show no such increase over values for coastal stations. Since the coefficient a in Table 6.2 shows no great change with location, we conclude that on the Antarctic plateau deposition noise is the dominant source of noise. We can therefore use the value of σ_δ at Dome C to estimate errors in interpretation of δ values as a record of past temperature. This gives a 95 per cent confidence level for estimating the climatic significance of mean δ values over periods of 25, 100 and 400 a of ± 1.2, ± 0.6 and $\pm 0.3^0/_{00}$ respectively.

For Antarctic coastal stations and Greenland, the similar numerical values of σ_θ and σ_δ shown in Table 6.1,

along with the significant values of r in Table 6.2, indicate that the δ–θ noise component given by $(1-r)\sigma_\theta^2$ and $(1-r)\sigma_\delta^2$, must also be of similar magnitude. We use the value $(1-r)=0.44$ for Jacobshavn–Milcent in Greenland, and values of σ_δ in Table 6.1, to give an approximate indication of δ–θ noise at various sites around Greenland for comparison with the very detailed studies of deposition noise of Reeh & Fisher (in prep.). Their estimates are based on comparisons between σ_δ profiles at adjacent localities in Greenland, so they cover deposition, meltwater and sampling noise shown in column 1 of Table 6.4, but do not cover rate of deposition noise which should be largely absent from adjacent boreholes. Although rate of deposition noise contributes to our δ–θ noise, it forms part of the variance of σ_δ^2 and not of σ_θ^2. The analyses of Reeh & Fisher cover at least two boreholes from each site, while our figures were based on a single borehole. Although some of the raw data used was the same as in Table 6.1, the analyses were independent. Reeh & Fisher (in prep.) tabulate only their noise component $\sigma_{N\delta}^2$ and the signal to noise ratio $r/(1-r)$. From this we have calculated the variance of the δ signal $\sigma_{S\delta}^2$ and the total variance $\sigma_\delta^2 = \sigma_{S\delta}^2 + \sigma_{N\delta}^2$. This figure is compared in Table 6.6 with values of σ_δ^2 from the present analysis. We also compare values of $\sigma_{N\delta}^2$ from comparisons between boreholes of Reeh & Fisher (in prep.) with estimates of $\sigma_{N\delta}^2 = 0.44\sigma_\delta^2$ suggested by these studies of δ–θ noise.

Figure 6.4. Isotopic δ ^{18}O values of individual snowfalls (dots) and of rime (triangles) at the South Pole from November 1964 to October 1965. Also shown are mean monthly temperatures

for the same period (horizontal bars) while the continuous line shows the average of the mean monthly temperatures over a twenty year period.

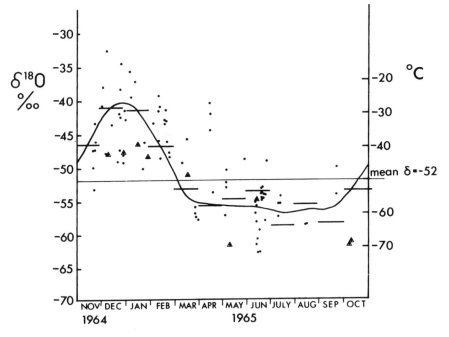

Figure 6.5. Mean annual values of δ ^{18}O$^0/_{00}$ (thin bars) and of accumulation A in cm a^{-1} (continuous line) for site 10 km from Dye C on the Greenland ice sheet plotted for the period 1937–70 (based on Reeh *et al.*, 1977).

Although there is considerable variation, Table 6.6 shows rough agreement between the two studies. The variance of δ noise between adjacent boreholes $\sigma_{N\delta}^2$ in column (1) is seen to vary from approximately the same as our estimate of $0.44\sigma_\delta^2$ when we use the value of σ_δ derived from Reeh & Fisher's figures, to one-half $0.44\sigma_\delta^2$ when σ_δ values of column (6) are used. The differences represent the rate of deposition noise plus any differences or errors in the analysis of data. They suggest that on average, the rate of deposition noise may contribute up to one third of the total variance σ_δ^2, and up to two-thirds of the variance can be due to deposition, meltwater and sampling noise.

We have little evidence for assessing different types of temperature noise, although the value of $(1-r)\sigma_\theta^2$ from Table 6.1 suggest that the noise variance of temperature contributing to δ–θ noise must be around $0.5\,(^\circ\mathrm{C})^2$ for stations in Antarctic coastal regions and Greenland. At James Ross Island, the lack of correlation with the record from Esperanza, 100 km away, compared to a correlation coefficient of 0.72 with regional mean temperatures indicates that the local temperature noise component in the Esperanza region that is unrelated to the regional temperature–isotopic component must be of the order of $(1-r)\sigma_\theta^2$ or around $0.8\,(^\circ\mathrm{C})^2$. We have already suggested that low values of σ_θ at the South Pole indicate that temperature noise does not increase inland from the coast. Greater variability of climate in some coastal regions suggests that changes in atmospheric and oceanic circulation have a greater effect there than on the inland plateau.

Conclusion

The concept of δ–θ noise is itself an attempt to summarise evidence on the δ–θ relationship over a wide range of time scales. We can conclude our survey by reviewing how our model works in practice.

It is notable that very little has been published on the relationship between mean annual values of temperature at an individual station and isotopic ratios. In *Stable Isotope Glaciology*, Dansgaard, Johnsen, Clausen & Gundestrup (1973) showed some correspondence between ten-year running means of temperature at coastal stations around Greenland and similar means of δ values from stations on the ice sheet, but this was little better than qualitative. However, correspondence between isotopic and temperature records using 50 or 60 a smoothing of data have been shown in Dansgaard *et al.* (1975) and Hammer (1980). In these studies temperature variations are of the order of $\pm 1\,^\circ\mathrm{C}$. Since this is some three times the expected noise level suggested by our studies, the general clarity of these correlations is consistent with our estimated level of δ–θ noise.

We conclude that our estimates of the level of δ–θ noise form a useful guide for interpretation of δ value records as climatic indicators.

Table 6.6. *δ noise estimates $(\delta^{18}\mathrm{O}\,^0/_{00})^2$*

	Reeh & Fisher					This study	
	(1) $\sigma_{N\delta}^2$	(2) $r/(1-r)$	(3) $\sigma_{S\delta}^2$	(4) σ_δ^2	(5) $0.44\sigma_\delta^2$	(6) σ_δ^2	(7) $0.44\sigma_\delta^2$
Dye 3	0.28	2.4	0.67	0.95	0.41	0.98	0.43
Milcent	0.37	1.1	0.41	0.78	0.34	1.64	0.71
Crête	0.49	1.4	0.69	1.18	0.51	1.49	0.65
Camp Century	0.54	2.0	1.08	1.62	0.70	2.53	1.10
Devon Island	0.55	1.3	0.72	1.27	0.55	1.49	0.98

6.3

Regional and global trends: the climatic record

If climatic changes affect large areas at the same time, climatic indicators such as δ values should vary simultaneously at different sites. In this section we study records from Byrd Station and Dome C in Antarctica and from stations in Greenland to examine the areal extent of climatic fluctuations recorded in isotopic profiles.

There are two requirements for comparison of isotopic records. First, the time scales should be known with sufficient accuracy for effective comparisons. Second, the data should be averaged over a sufficiently long period to be confident that isotopic δ–θ noise is small compared to the temperature related component of the mean δ values. From the previous section we see that 100 a mean values should produce a satisfactory signal to noise level for this purpose at most sites except on the high Antarctic plateau.

A comparison between 100 a mean δ values over some 900 a at Byrd Station and Dome C in Antarctica was presented in Robin (1981). This used 95 per cent confidence limits derived from the Law Dome and West Greenland, which are shown in Table 6.2, which provided the only available estimates of δ–θ noise when the comparison was made in 1979. The rather short run of nine 100 a mean values used for comparison suggested that the error estimates were realistic, but the later results of Benoist *et al.* (1982) do not support this conclusion.

For comparison of records from Dome C and Byrd over a long time period, in Figure 6.6(*a*), we plot 100 a running mean δ values for the Byrd ice core using the age scale shown in Figure 5.50, and 4 m mean values for Dome C spanning about 100 a of Lorius, Merlivat, Jouzel & Pourchet (1979). Figure 6.6(*b*) shows the same data using approximately 300 a mean values for both stations.

Although relative changes of surface elevation between Byrd and Dome C could affect the comparison, it would take 2500 a of the present annual accumulation at Dome C to increase the surface elevation by 100 m, and hence decrease δ values by 1 °/oo in the absence of

any outflow of ice. A similar change at Byrd would take about one-third the time, but when allowance is made for outflow of ice it is clear that events of duration less than 1000 a must be due to climatic changes affecting δ values.

We show 95 per cent confidence limits on Figure 6.6(*a*) and (*b*) to indicate the reliability with which we can interpret isotopic changes on each record. On Figure 6.6(*a*) these limits refer to a 100 a period and on Figure 6.6(*b*) to a 300 a period. We must also remember that the curve for Dome C on Figure 6.6(*a*) gives mean δ values over successive 4 m (about 100 a) intervals, rather than a running mean. Many of the sharp peaks on this record are of similar magnitude to the 95 per cent confidence level. Some of these peaks do match similar peaks on the Byrd record, especially two peaks at 10 000 a BP (Byrd time scale) and another group from 13 000–14 000 a BP, and hence give a strong indication that both sites were affected by the same climatic fluctuation that lasted around one century.

The 300 a smoothing on Figure 6.6(*b*) cuts out further noise, especially with smoothing of the Dome C record. This makes it easier to follow the similar form of the two records over the periods from 10 000 a BP to 16 300 a BP and 25 000 a BP to 32 000 a BP (Byrd time scale). However, the sharper peaks with 100 a smoothing on Figure 6.6(*a*) define the links between the two records more precisely. The apparent difference of dates between the two records is caused by a slow drift in relative errors of dating of the two profiles. It is clear that there is sufficient correspondence between the two sections of these records to establish the links between events shown on Figure 6.6(*a*).

The links indicate that dates on the two records correspond closely to 2600 a BP, drift apart by about 800 a at 10 000 a BP (Byrd dates) and then to 1300 a apart at 30 000 a BP. This suggests that the accumulation rate used in dating the Dome C ice core was 10 per cent high from about 2000–10 000 a BP in relation to Byrd Station, then fell to 2.5 per cent too high on average from 10 000–30 000 a BP.

Also shown on Figure 6.6(*a*) are 100 a running mean δ values for Devon Island and Camp Century ice cores over the period for which confidence in dating applies (see section 4.2). Some tentative links between smaller events common to both records are suggested since the 95 per cent confidence limits are smaller than in the Antarctic, and the stations are only 500 km apart.

A clear link between events at Camp Century and Byrd Station occurs from 10 000–11 500 a BP and the events are seen less clearly at Dome C on Figure 6.6(*b*). The decrease in amplitude of these events in the Antarctic to one-third that in Greenland is in agreement with predictions from global climatic models (for instance Manabe & Stouffer, 1980) of the response of the two polar regions to disturbances of global scope, such as changing CO_2 content of the atmosphere or changes of solar radiation. The times of the events correspond closely to those known as the Allerød sequences in the northern hemisphere. According to Flohn (1978) these lasted less than 2000 a. Several glacial readvances took place apparently simultaneously, and were accompanied by climatic variations all over the globe. Flohn puts the Younger Dryas readvance as occurring after 10 900 a BP, compared with the cold isotopic dip in Figure 6.6(a) of 10 600 a BP (Byrd) and 10 400 a BP (Camp Century). The small differences of date are unlikely to be significant as ice core dates are based on glaciological evidence and other dates on ^{14}C analysis. We consider that this evidence confirms the date of the Byrd core at this time. However, dating of events on the Dome C core have been linked to corresponding events in marine cores that have been dated by ^{14}C methods, so the Dome C dates are preferable prior

to 15 000 a BP. Lorius *et al.* (1979) used an accumulation rate that decreased to around 75 per cent of the present by 14 000 a BP. While this has brought the date scales of of the two cores closer, it is surprising that it did not produce a greater effect.

Curves of the past climate of Camp Century and Byrd derived in section 5.3 are also shown on Figure 6.6. These take into account evidence from total gas content as well as isotopic data. The slow drift apart of the isotopic record from the derived climate curve at Byrd from about 10 000 a BP to the present is presumably due to thinning of the ice sheet and/or downslope motion as

Figure 6.6. (a) Variation of δ [18]O values with time measured on ice cores from Arctic and Antarctic locations. Running means of about 100 a are shown for the Byrd, Camp Century and Devon Island stations, while that for Dome C shows mean values over successive intervals of about 100 a. The time scales for different cores are discussed in the text and in Chapter 4. The curves of derived climate are those shown in Figures 5.16 and 5.17. They

are plotted using the gradient $d\delta\,^{18}O/d\theta = 1\,^o/_{oo}$ and an arbitrary base level. 95 per cent confidence limits for isotopic–temperature conversion are shown on the left. Dotted lines link similar isotopic events on two profiles, and hence indicate the relative errors in dating the two profiles.

(b) Data in (a) smoothed by use of 300 a running means of the isotopic profiles for Byrd and Dome C.

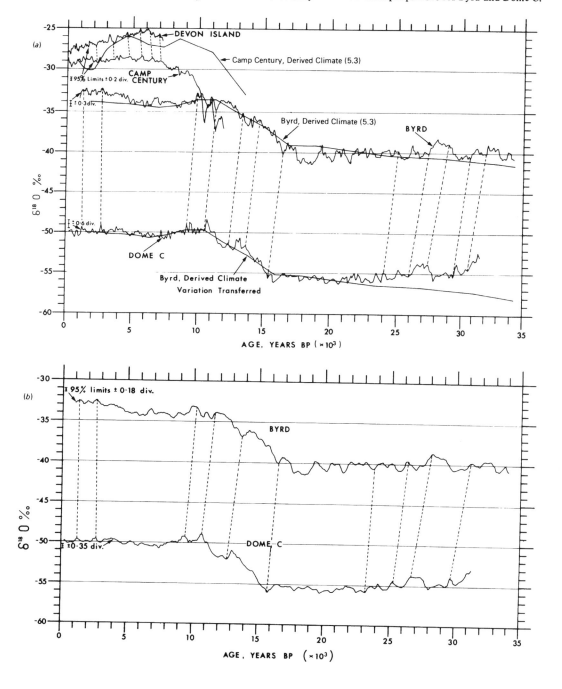

discussed in sections 3.7 and 5.5. We have transferred the derived climate curve from Byrd to the Dome C record by use of corresponding events, to allow for differences of dating. In this case the derived climate curve matches the isotopic record quite closely back to 20 000 a BP, which suggests that there has been little change of elevation around Dome C over this period.

Prior to 20 000 a BP, if the derived climate curve is correct, it appears that Dome C, as well as Byrd Station, was slowly increasing in elevation. Isotopic values suggest that the increase from about 30 000 a BP to about 21 000 a BP was greater by perhaps 150 m at Dome C, but from 21 000 a BP to 18 000 a BP the ice in the Byrd core appears to have a relative gain of about 200 m elevation in relation to Dome C. Furthermore, the changes of elevation indicated by comparison with the derived climate curve appear to have taken place around 1000 a earlier at Byrd than at Dome C.

Elevation changes also have a large effect on δ values at Camp Century, while the plateau situation of

the Devon Island ice cap limits its possible elevation changes. Fisher (1979) shows the maximum effect of these changes on δ values as $\pm 1.5^0/_{00}$, but over limited periods they are likely to be less. We should therefore expect the derived climate record for Camp Century in Figure 6.6 to be similar to the δ value record of Devon Island. However, it is only from about 3000–7000 a BP that there is any similarity of trends. The inverse problem of using the difference of δ values between the two records as a measure of the changing elevation of ice at Camp Century has been presented in Fisher (1979). However, the dating of basal ice of both cores is unsatisfactory, although useful links are provided by microparticle concentrations at certain levels. Fisher shows a maximum elevation difference of about 1600 m at 24 000 a BP and a more or less steady fall from that date to the present. Only around 13 500 a BP do these elevations agree with the shorter record for Camp Century derived in section 5.3.

From the global viewpoint, we note that the

Figure 6.7. Variation of 10 a mean δ values with time measured in ice cores from the region of Greenland. Dating of ice cores from Dye 3, Milcent and Crête is determined by annual isotopic layering and is within 3 a at all levels. The age of the ice core

from Camp Century should be accurate to about 3 per cent while dating of the ice core from Devon Island discussed in section 4.2, should approach this accuracy.

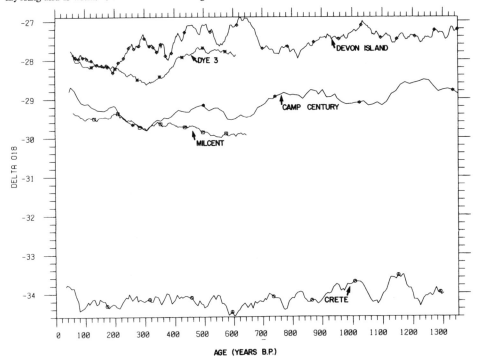

Figure 6.8. Temperature differences in degree Centigrade between 10 a mean values for the periods 1931/40 and 1941/50 for stations around the north Atlantic and Greenland. All

figures refer to mean values from instrumental records. From Kirch (1966, Fig. 29(c)).

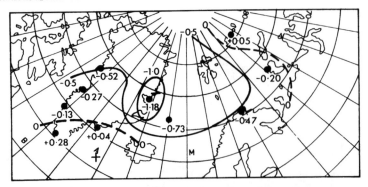

warmest climate in Antarctica during the Holocene
occurred around 11 000–10 000 a BP, at the end of
the Allerød period, compared with maxima around
6000 a BP at Devon Island and at 4500 and 8500 a BP
on the Camp Century derived climate curve. The relative
dates are unlikely to be in error by more than 500 a. They
show quite clearly that the warmest period at high southern latitudes did not coincide with the warmest Holocene
period in northern latitudes.

Greenland region

Comparisons of δ records from different sites
around Greenland are shown in Figure 6.7 using 10 a
mean values. Clearly the noise that gives a 95 per cent
confidence level of $\pm 0.9^0/_{00}$ of δ value for 10 a means
is too great to expect much correlation over individual
decades. Variations with a cycle time of around 100 a
on the records from Devon Island are not reproduced on
the Camp Century records. The latter corresponds better
with that of Dye 3 from southern Greenland, rather than
Crête in central Greenland. The warm spell at the start
of the record in the middle of the twentieth century is
almost the only feature common to all records. A similar
difference in trends at the different sites around Greenland is seen when comparing 10 a mean temperature
changes from one decade to the next. Figure 6.8, from
Kirch (1966) shows the shift between 1931–40 and
1941–50 of mean temperatures around the North
Atlantic and Greenland. We see that, over these decades,
while southern Greenland became warmer central
Greenland became cooler. The evidence from Figures
6.7 and 6.8 suggests that the lack of similar isotopic
trends between sites in the Greenland region may well
be due to lack of similar climatic trends between the
same sites. On physical grounds, especially the limited
agreement between isotopic and temperature variations
over limited regions, it appears that the isotopic variations still indicate climatic changes at individual sites.

In spite of the variability shown on Figure 6.7,
Dansgaard *et al.* (1975) found a good correlation
between the isotopic record at Crête and temperature
records for Iceland and England. They used a 60 a low
pass filter smoothing on all data, which is reproduced
in Figure 6.9. The correspondence between the Crête
record in Figure 6.7 and the filtered record in Figure 6.9
is clear. The agreement of the Crête isotopic record and
the later (instrumental) sections of the records from
Iceland and England is striking. It is also clear from
Figure 6.7 that similar correlations would not be found
between records from other sites in Greenland and the
records from Iceland and England.

Comparison with other climatic records

Since variations in the climate of polar regions
form part of the global pattern, some comparison with
variations elsewhere is appropriate. This is of most value
when using evidence similar to that shown in Figure 6.6.
This is well dated to 30 000 a BP and the effects of δ–θ
noise and changed dimensions of the ice sheet are known.
However, several published comparisons cover evidence
from the complete ice cores to bedrock from Byrd and

Figure 6.9. Comparison between δ ^{18}O concentrations in the ice
core from Crête (left) and temperature from England and
Iceland. Curves are smoothed by a 60 a digital low pass filter
except for the dashed record for England. Dashed curves are
based on indirect evidence and the dash–dot section of the
Icelandic record on systematic ice observations. From
Dansgaard *et al.* (1975, Fig. 2).

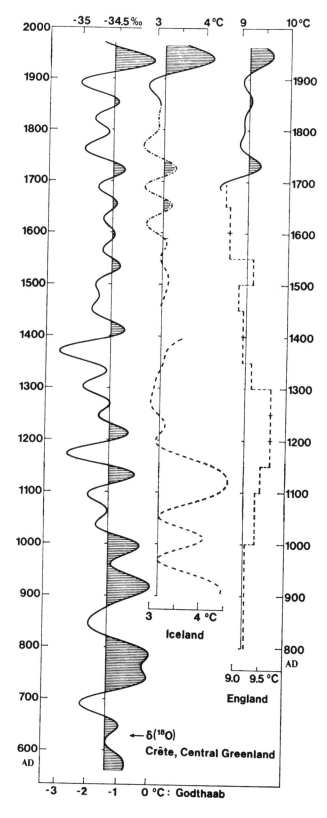

Camp Century, so we shall review such evidence briefly.

For a general comparison with worldwide climatic trends, we use the climatic index of Kukla, Berger, Lotti & Brown (1981). Figure 6.10(a) shows variations of their index (heavy line) over the past 130 000 a. The index presents an average of data from three sources: pollen data from France and Greece, sea-level data from some 13 sites around the world and isotopic data from four deep sea cores similar to the record in Figure 3.3. Details of how the mean index was obtained are given in their paper. In addition to the heavy line in Figure 6.10(a), thin lines show upper and lower limits of the index. The upper thin line clearly presents sea-level variations while the lower thin line is presumably dominated by $\delta^{18}O$ measurements on benthic fauna and possibly by some pollen measurements.

Figures 6.10(b) and (c) show the δ value profiles plotted on the same time scale for Byrd and Camp Century. Data for Byrd was plotted as 1000 a mean values and the time scale is that of Figure 5.50 which agrees

closely with northern hemisphere data at 11 000 a BP. The dating of the Camp Century plot is that shown in Figure 10c of Dansgaard *et al.* (1973) which differs only slightly from that shown in curve 2 of Figure 5.22.

Continuous radio-echo layering to about 1800 m at Byrd indicates that continuity of the record is satisfactory to around 30 000 a BP, and Figure 6.6 suggests that dating errors may not exceed ±3000 a at this level. Lack of sudden changes of δ values to 2050 m in Figure 4.23 suggests that continuity may continue to that level (about 60 000 a BP), but the time scale below 1800 m may be variable due to less regular deformation of the basal layers. From 2050–2164 m sudden jumps of δ values in Figure 4.23 suggest lack of continuity, but the trend of 1000 a mean δ values in Figure 6.10(b) between 100 000 a BP and 60 000 a BP is consistent with the gradual development of an ice age over this period.

Dating of the Camp Century ice core is within about 300 a to around 9000 a BP, and continuity of the

Figure 6.10. Comparison of δ profiles to bedrock at Byrd Station and Camp Century with the climatic index of Kukla *et al.* (1981).

(a) The climatic index is shown by the heavy line while the light lines show the upper and lower limits of index derived from the data sets used in compilation. Dating accuracy is shown by horizontal bars.

(b) δ profile to bedrock at Byrd Station using the age scale of Figure 5.50. The dashed line from 0 to about 13 000 a BP shows the derived climate curve of Figure 5.16. Also shown by the longer dashed line is the surface elevation curve from the same figure.

(c) δ profile to bedrock at Camp Century using the time scale of Dansgaard, Johnsen, Clausen & Gundestrup (1973).

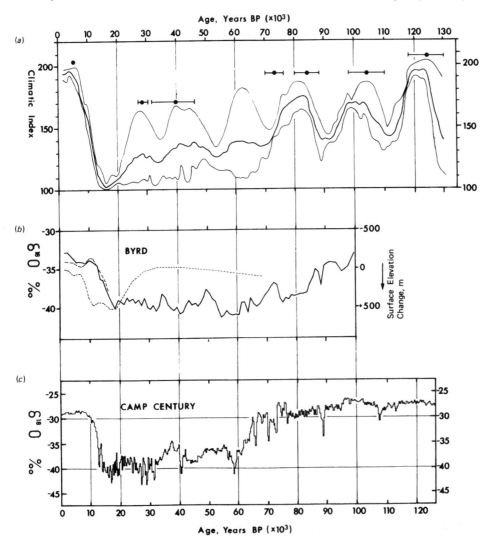

record by radio-echo layering looks satisfactory to about 11 000 a BP. Prior to this, considerable shrinking of the ice sheet from the glacial maximum suggests that dating errors from about 15 000–60 000 a BP will be more variable, and dates should be viewed with caution. Below that level we have already suggested (see section 3.4) that the sudden changes of δ value indicate shearing in the ice, irregular flow and no continuity of time scale.

Dating of the record of the climatic index in Figure 6.10(a) is satisfactory to about 30 000 a BP, but thereafter is less reliable. It is only since 20 000 a BP that all three climatic indicators appear to provide similar information about climatic change, so we discuss this section first.

Isotopic records from both Camp Century and Byrd show minimum values at approximately the same time as the climatic index of Figure 6.10(a). However, the warming trend at Byrd appears to start around 3000 a earlier than shown by the climatic index and Camp Century record. This continues so that the warmest period in the Antarctic has started by 11 000 a BP, in advance of the climatic index that continues rising to about 5000 a BP. The fall of the ice surface elevation curve at Byrd lags behind the warming trend by some 5000 a. This feature is also clearly present at Camp Century, D10 in Terre Adélie and on Law Dome, that is at all locations where both total gas content and isotopic profiles are available. As stated in section 5.3, this indicates that as the global climate becomes warmer and northern hemisphere ice sheets retreat northwards, atmosphere transport carries more water vapour to high latitudes causing ice sheets there to thicken. Subsequently it appears that rising sea level or other factors increases the flow of ice sheets sufficiently to decrease their size to near present-day levels.

Between 30 000 a BP and 60 000 a BP both ice cores indicate that ice age conditions were maintained over both polar regions although the climatic index, presumably due mainly to sea-level changes, suggests somewhat warmer conditions at lower latitudes. The trends of the different curves do not match closely over this period, except that both ice cores and the minimum index suggest that at about 50 000 a BP some warming (about 2 °C) took place after the ice age had attained a severity at about 60 000 a BP that was similar to that at 18 000 a BP.

Dating prior to 60 000 a BP is so unreliable that little should be written. Possibly the warmer ice dated at 100 000 a BP at Byrd should be dated at the interglacial around 120 000 a BP. In this case the Byrd ice core would indicate similar trends to the climatic index of Figure 6.10(a), with cooling from 120 000–70 000 a BP separated by two periods of limited warming.

6.4

Conclusion

This monograph has shown how profiles of temperature, isotopic ratio and total gas content, plus other glaciological data, make it possible to understand the flow and past history of ice sheets. It has also clarified the limits of reliability of information derived from these studies.

Other geochemical studies of ice cores have been mentioned briefly. Such studies will be especially important in the future as cores of almost pure ice provide an extremely effective way of recording past variations of atmospheric composition and contamination. In the future, deep ice cores from east Antarctica should provide a well-dated record of how atmospheric conditions varied back to at least 200 000 a BP and less well-dated information over several times this time span. This will make it possible to check many possible causes and effects of ice ages. Such studies are made possible by the knowledge of polar ice sheets which is described in this monograph.

Much of the progress in the glaciology of polar ice sheets has been achieved during the three decades of 1950–80. Many individual scientists, engineers and logistic personnel have played their part. We express our admiration and thanks for their very considerable achievements, often carried out under extremely unpleasant conditions in polar regions.

A major impetus to the study of polar ice sheets came from International Geophysical Year (IGY) and this has continued under the Scientific Committee on Antarctic Research (SCAR). Since 1969 the International Antarctic Glaciological Project has provided further stimulus and coordination to studies of the ice sheet of East Antarctica. The Greenland Ice Sheet Programme has played a similar role in Greenland over the past decade. Great appreciation is due to all, including governments who supported these programmes.

APPENDIX

Symbols used in monograph

A Accumulation rate; A_0 at origin of flowline, A_s at surface of ice column

A_δ Amplitude of δ value cycle; $A_{\delta s}$ at surface, $A_{\delta t}$ at time t

A   Flow law multiplier for ice deformation

B Thickness balance

B   Sliding law constant

C Positive constant

D, D_0, D_L Depth to bed below sea level

D_v Volume diffusion coefficient

E Surface elevation above sea level

F Thinning factor or freezing rate (section 3.10)

F_1 Filter; $\bar{F}_1$ Fourier transform

G Constant (section 2.3)

H Height of surface at centre of ice sheet

H   Horizontal temperature gradient within ice sheet

I Base temperature difference from pressure melting point

J Conversion factor for mechanical equivalent of heat

K Thermal conductivity

K_0, K_1 Profile constants (section 2.2)

L Half-width of ice cap

L_0, L_t, L_f, L_i Diffusion length; L_0 total mean in firn, L_t total after modification by strain at time t, L_f component due to diffusion in firn after modification by strain, L_i in ice

L   Latent heat

M Melt rate

M   Total depth of meltwater

N Ablation rate

P Atmospheric pressure

P_0 Standard atmospheric pressure (1013.25 mbar)

P_c Atmospheric pressure at pore close-off

$P(\ )$ Vapour pressure of substance in brackets, e.g. $P(H_2O)$, $P(HDO)$

Q Rate of internal heat generation

Q_x Surface mass balance velocity

R Constant advection term

R Specific gas constant for dry air

S Surface warming rate

T Inversion thickness

T   Time in years

U Average velocity of ice column

$\bar{U}$ Balance velocity of ice column

V Total gas content; V_c pore volume per gram of ice at close-off

W Water flux in layer beneath ice sheet; W_i input, W_o output

X Arbitrary function

Y Distance between pair of flowlines

Z Ice thickness

Z   Depth below surface

a, b, c, d Used as constants in equations

a   Half thickness of slab (section 3.4)

b   Bedrock elevation

c   Specific heat

$d\ (= 1/\rho_c - 1/\rho_i)$ Density function

f, f^* Conversion constants $(\theta = f\delta + f^*)$

g   Gravitational constant

h   Height of surface distant x from centre of ice sheet

i   left (l) or right (r) (section 2.2)

k   Mean thermal diffusivity

k   Wave number $(= 2\pi/\lambda = 2\pi/A)$

l   Left (subscript in section 2.4)

m   Sliding law constant

n   Glen's law constant, or integral number

p   Change of melting point with pressure

p   Gradient of surface pressure with elevation $(= dP_{surface}/dE)$

r   Right (subscript in section 2.4)

r   Correlation coefficient

s   Surface (used as a suffix)

t   Time

u, v, w Velocity components in orthogonal coordinate system x, y, z respectively

v_c Pore volume at close-off of individual bubble in ice

v   Velocity of temperature wave

w_s Vertical velocity near surface

w   Vertical velocity in transformed system

x, y, z Orthogonal coordinates; x is horizontal and positive in flow direction, z is vertically upwards (from base or sea level)

x   Position

z   Height above ice-rock interface

z   Depth below surface and coordinate vertically downwards

$\alpha, \beta, \gamma, \delta, \epsilon, \phi$ Constants used in sections 5.2 and 5.5

α Surface slope; also isotopic fractionation factor using subscript for isotope, e.g. α_D for deuterium, α_{18} for ^{18}O

β Bedrock slope

γ Thermal gradient; γ_G geothermal, γ_b calculated basal, γ_s near surface, γ_i lapse-rate within atmospheric inversion layer, γ_a lapse-rate of atmosphere above surface inversion near sea level (Numerical estimates of γ_G assume rock conductivity is the same as that of ice.)

δ Isotopic ratio

$\dot{\epsilon}$ Strain rate; $\dot{\epsilon}_{xy}$ etc.

$\dot{\epsilon}_{\mathrm{o}}$ Octahedral shear strain rate

$\dot{\epsilon}_{\mathrm{c}}$ Constant to fit experimental strain data

ζ Relative coordinate, depth

θ Temperature; θ_0 standard temperature, θ_{m} pressure melting point, θ_{c} temperature at pore close-off, θ_{a} mean atmosphere temperature above surface inversion, θ_{s} mean surface temperature

$\theta*$ Temperature profile relative to pressure melting point

λ Layer thickness; λ_0 at time of deposition, λ_t at time t

ν Constant to fit experimental data

ρ Density; ρ_{i} of pure ice, ρ_{c} of firn-ice at pore close-off

σ Cross-sectional area; standard deviation from mean: σ_θ of temperature, σ_δ of δ value, $\sigma_{\delta\theta}$ of $\delta-\theta$ noise, $\sigma_{\mathrm{N}\delta(\ )}$ deposition noise component of σ_δ

τ Shear stress magnitude, components τ_{xy}, τ_{yz}, τ_{zx}

τ_{b} Shear stress, basal

τ_{c} Shear stress, constant to fit experimental data; also critical value of yield stress in section 2.2

τ_{o} Shear stress, octahedral

ω Angular frequency, temperature wave

Γ Geothermal heat flux

Δ Denotes $\Delta\rho = (\rho_{\mathrm{w}} - \rho_{\mathrm{i}})$ in section 2.3

Λ Gradient of mean temperature with elevation

Φ Mass flux

REFERENCES

Aamot, H. W. C. (1968). Instrumented probes for deep glacial investigations. *Journal of Glaciology*, 7, 321–8.

Ackley, S. F. & Keliher, T. E. (1979). Ice sheet internal radio-echo reflections and associated physical property changes with depth. *Journal of Geophysical Research*, 84, B10, 5675–80.

Aegerter, S., Oeschger, H., Renaud, A. & Schumacher, E. (1969). Studies based on the tritium content of the samples. *Meddelelser om Grønland*, 177, 2, 76–88.

Aldaz, L. & Deutsch, S. (1967). On a relationship between air temperature and oxygen isotope ratio of snow and firn in the South Pole region. *Earth and Planetary Science Letters*, 3, 267–74.

Allison, I. (1979). The mass budget of the Lambert Glacier drainage basin, Antarctica. *Journal of Glaciology*, 22, 223–35.

Alt, J., Astapenko, P. & Ropar, N. J. (1959). Some aspects of the Antarctic atmospheric circulation in 1958. *International Geophysical Year General Report Series*, No. 4, 28 pp. (Available from World Data Center A, US Geological Survey, Tacoma, Washington.)

Anderson, D. L. & Benson, C. S. (1963). The densification and diagenesis of snow. In *Ice and Snow, Properties, Processes and Applications*, ed. W. D. Kingery, 391–411. Cambridge, Mass.: MIT Press.

Anderson, J. B. (1972). *The Marine Geology of the Weddell Sea. Contribution No. 35*, 222 pp. Florida State University (Sedimentological Research Laboratory of the Department of Geology).

Anderton, P. W. (1974). Ice fabrics and petrography, Meserve Glacier, Antarctica. *Journal of Glaciology*, 13, 285–306.

Andrews, J. T. & Falconer, G. (1969). Late glacial and post-glacial history and emergence of the Ottawa Islands, Hudson Bay, Northwest Territories: Evidence on the deglaciation of Hudson Bay. *Canadian Journal of Earth Sciences*, 6, 1263–76.

Aristarian, A. J. (1980). *Etude glaciologique de la calotte polaire de l'île James Ross (Peninsule Antarctique)*. Thèse de Troisième Cycle, Université Scientifique et Médicale de Grenoble. Publication no. 322 du Laboratoire de Glaciologie et Géophysique de l'Environnement du Centre National de la Recherche Scientifique, 130 pp. Grenoble: Laboratoire de Glaciologie.

Bader, H. (1965). Theory of densification of dry, bubbly glacier ice. *CRREL Research Report*, No. 141, 8 pp. Hanover, New Hampshire: US Army Cold Regions Research and Engineering Laboratory.

Bailey, J. T., Evans, S. & Robin, G. de Q. (1964). Radio echo soundings of polar ice sheets. *Nature, London*, 204, 420–1.

Bardin, V. I. (1982). Composition of East Antarctic moraines and some problems of Cenozoic history. In *Antarctic Geoscience*, ed. C. Craddock, 1069–76. Madison: University of Wisconsin Press.

Barkov, N. I., Gordiyenko, F. G., Korotkevich, Ye. S. & Kotlyakov, V. M. (1974). First results of the study of an ice core from the borehole at Vostok Station (Antarctica) by the oxygen-isotope method. *Doklady Akademii Nauk SSSR*, 214, 1383–6. (In Russian.)

Barkov, N. I., Gordiyenko, F. G., Korotkevich, Ye. S. & Kotlyakov, V. M. (1976). Isotopic studies of an ice core from Vostok Station (Antarctica) to a depth of 950 m. *Doklady Akademii Nauk SSSR*, 230, 656–9. (In Russian.)

Barkov, N. I., Korotkevich, E. S., Gordiyenko, F. G. & Kotlyakov, V. M. (1977). The isotope analysis of ice cores from Vostok Station (Antarctica), to the depth of 950 m. *International Association of Hydrological Sciences Publication*, No. 118, 382–7.

Barkov, N. I. & Uvarov, N. N. (1973). Geophysical studies in the Vostok Station borehole in 1970. *Informatsionnyy Byulleten' Sovetskoy Antarkticheskoy Ekspeditsii*, No. 85, 29–34 (in Russian), and *Soviet Antarctic Expedition Information Bulletin*, Vol. 8, No. 7, 380–5 (of translation by American Geophysical Union).

Barkov, N. I., Vostretsov, R. N. & Putikov, O. F. (1975). Temperature measurements in the Vostok Station borehole. *Antarctic Journal of the United States*, 10, 57–9.

Bauer, A. (1968). Mouvement et variation d'altitude de la zone d'ablation ouest. *Meddelelser om Grønland* 174, 1, 1–79.

Behrendt, J. C., Wold, R. J. & Dowling, F. L. (1962). Ice surface elevation of central Marie Byrd Lane. *Journal of Glaciology*, 4, 121–3.

Benoist, J. P., Jouzel, J., Lorius, C., Merlivat, L. & Pourchet, M. (1982). Isotope climatic record over the last 2,500 years from Dome C (Antarctica) ice cores. *Annals of Glaciology*, Vol. 3, 17–22.

Benson, C. S. (1962). Stratigraphic studies in the snow and firn of the Greenland ice sheet. *Snow, Ice and Permafrost Research Establishment Research Report*, No. 70, 93 pp. Copy at US Army Cold Regions Research and Engineering Laboratory, Hanover, New Hampshire.

Bentley, C. R. (1964). The structure of Antarctica and its ice cover. In *Research in Geophysics*, ed. H. Odishaw, Vol. 2, 335–89. Cambridge, Massachusetts: MIT Press.

Bentley, C. R. (1971a). Secular increase of gravity at South Pole station. *Antarctic Research Series* (American Geophysical Union), 16, 191–8.

Bentley, C. R. (1971b). Seismic evidence for moraine within the basal Antarctic ice sheet. *Antarctic Research Series* (American Geophysical Union) 16, 89–129.

Bentley, C. R. (1972). Seismic-wave velocities in anisotropic ice: a comparison of measured and calculated values in and around the deep drill hole at Byrd Station, Antarctica. *Journal of Geophysical Research*, 77, 4406–20.

Bentley, C. R. (1975). Advances in geophysical exploration of ice sheets and glaciers. *Journal of Glaciology*, 15, 113–34.

Bentley, C. R., Cameron, R. L., Bull, C., Kojima, K. & Gow, A. J. (1964). Physical characteristics of the Antarctic ice sheet. *Antarctic Map Folio Series*, No. 2. New York: American Geographical Society.

Bentley, C. R. & Chang, F.-K. (1971). Geophysical exploration in Marie Byrd Land, Antarctica. *Antarctic Research Series* (American Geophysical Union), 16, 1–38.

Bentley, C. R., Robertson, J. D. & Greischar, L. L. (1982). Isostatic gravity anomalies on the Ross Ice Shelf, 1077–81. In *Antarctic Geoscience*, ed. C. Craddock Madison: University of Wisconsin Press.

Berggren, W. A. (1972). Late Pliocene–Pleistocene glaciation. In *Initial Reports of the Deep Sea Drilling Project*, eds.

A. S. Laughton, W. A. Berggren *et al.*, Vol. XII, 953–63. Washington: US Government Printing Office.

Birchfield, G. E. & Weertman, J. (1978). A note on the spectral response of a model continental ice sheet. *Journal of Geophysical Research*, **83**, 4123–5.

Bird, I. G. (1976). Thermal ice drilling: Australian developments and experience. In *Ice-core drilling*, ed. J. F. Splettstoesser, 1–18. Lincoln & London: University of Nebraska Press.

Bishop, B. C. (1957). Shear moraines in the Thule area, northwest Greenland. *Snow, Ice and Permafrost Research Establishment, Research Report*, No. 17, 46 pp. Copy at US Army Cold Regions Research and Engineering Laboratory, Hanover, New Hampshire.

Blake Jr., W. (1972). Climatic implications of radiocarbon-dated driftwood in the Queen Elizabeth Islands, Arctic Canada. In *Climatic change in arctic areas during the last ten thousand years*, eds. Y. Vasari, H. Hyvärinen & S. Hicks, *Acta Universitatis Ouluensis Series A. Scientiae Rerum Naturalium*, **3**, Geologica 1, 77–104.

Blake Jr., W. (1977). Glacial sculpture along the east-central coast of Ellesmere Island, Arctic Archipelago. Project 750063. *Geological Survey of Canada Paper 77-1C*, 105–15. Ottawa: Geological Survey of Canada.

Blicks, H., Dengel, O. & Riehl, N. (1966). Diffusion von Protonen (Tritonen) in reinen und dotierten Eis-Einkristallen. *Physik der kondensierten Materie*, **4**, 375–81.

Bogoslovskiy, V. N. (1958). The temperature conditions (regime) and movement of the Antarctic glacial shield. *International Association of Scientific Hydrology Publication*, No. 47, 287–305.

Boulton, G. S. (1970). On the origin and transport of englacial debris in Svalbard glaciers. *Journal of Glaciology*, **9**, 213–29.

Boulton, G. S. (1972). The role of thermal regime in glacial sedimentation. In *Polar Geomorphology. Institute of British Geographers Special Publication*, No. 4, 1–19, compilers R. J. Price & D. E. Sugden. London: Institute of British Geographers.

Boulton, G. S. (1975). Processes and patterns of subglacial sedimentation: a theoretical approach. In *Ice Ages: Ancient and Modern*, eds. A. E. Wright & F. Moseley, 7–42. Geological Journal Special Issue No. 6. Liverpool: Seel House Press.

Boulton, G. S. (1979). Boulder shapes and grain-size distributions of debris as indicators of transport paths through a glacier and till genesis. *Sedimentology*, **25**, 773–99.

Boulton, G. S., Morris, E. M., Armstrong, A. A. & Thomas, A. (1979). Direct measurement of stress at the base of a glacier. *Journal of Glaciology*, **22**, 3–24.

Boutron, C. & Lorius, C. (1979). Trace metals in Antarctic snows since 1914. *Nature, London*, **277**, 551–4.

Bradley, R. S. & England, J. (1977). Past glacial activity in the high Arctic. Contribution No. 31 *Amherst, Massachusetts, University of Massachusetts*, Dept. of Geology and Geography.

Brady, H. & Martin, H. (1979). Ross Sea region in the Middle Miocene: a glimpse into the past. *Science*, **203**, 437–8.

Brooks, R. L. (1982). SEASAT altimeter results over East Antarctica. *Annals of Glaciology*, **3**, 32–35.

Brooks, R. L., Campbell, W. J., Ramseier, R. O., Stanley, H. R. & Zwally, H. J. (1978). Ice sheet topography by satellite altimetry. *Nature, London*, **274**, 539–43.

Bryson, R. A., Wendland, W. M., Ives, J. D. & Andrews, J. T. (1969). Radiocarbon isochrones on the disintegration of the Laurentide Ice Sheet. *Arctic and Alpine Research*, **1**, 1–13.

Budd, W. F. (1966a). Glaciological studies in the region of Wilkes, Eastern Antarctica, 1961. *Australian National Antarctic Research Expeditions Scientific Reports, Series A (IV) Glaciology, Publication*, No. 88, 152 pp. Melbourne, Australia: Antarctic Division, Department of Science.

Budd, W. F. (1966b). The dynamics of the Amery Ice Shelf. *Journal of Glaciology*, **6**, 335–58.

Budd, W. F. (1968). The longitudinal velocity profile of large ice masses. *International Association of Scientific Hydrology Publication*, No. 79, 58–77.

Budd, W. F. (1969). The dynamics of ice masses. *Australian National Antarctic Research Expeditions Scientific Reports, Series A (IV) Glaciology, Publication*, No. 108, 216 pp. Melbourne, Australia: Antarctic Division, Department of Science.

Budd, W. F. (1970a). The longitudinal stress and strain-rate gradients in ice masses. *Journal of Glaciology*, **9**, 19–27.

Budd, W. F. (1970b). The Wilkes Ice Cap Project. *International Association of Scientific Hydrology Publication*, No. 86, 414–29.

Budd, W. F. (1971). Stress variations with ice flow over undulations. *Journal of Glaciology*, **10**, 177–95.

Budd, W. F. (1972). The development of crystal orientation fabrics in moving ice. *Zeitschrift für Gletscherkunde und Glazialgeologie*, **8**, 65–105.

Budd, W. F. (1975). A first simple model for periodically self-surging glaciers. *Journal of Glaciology*, **14**, 3–21.

Budd, W. F. (1977). Recent work: Australia. *Ice* (News Bulletin of the International Glaciological Society), No. 55, 3–4.

Budd, W. F. & Carter, D. B. (1971). An analysis of the relation between the surface and bedrock profiles of ice caps. *Journal of Glaciology*, **10**, 197–209.

Budd, W. F., Jacka, D., Jenssen, D., Radok, U. and Young, N. W. (1982). *Derived physical characteristics of the Greenland Ice Sheet, Mark 1* Meteorology Department Publication No. 23, University of Melbourne, Australia.

Budd, W. F., Jenssen, D. & Radok, U. (1970). The extent of basal melting in Antarctica. *Polarforschung*, **6**, 293–306.

Budd, W. F., Jenssen, D. & Radok, U. (1971a). Derived physical characteristics of the Antarctic ice sheet. *Australian National Antarctic Research Expeditions Interim Reports, Series A (IV) Glaciology, Publication*, No. 120, 178 pp. Melbourne, Australia: Antarctic Division, Department of Science.

Budd, W. F., Jenssen, D. & Radok, U. (1971b). Reinterpretation of deep ice temperatures, *Nature, Physical Science*, **232**, 84–5.

Budd, W. F., Jenssen, D. & Young, N. W. (1973). Temperature and velocity interaction in the motion of ice sheets. In *First Australasian Conference on Heat and Mass Transfer*, Section I, 17–24. Melbourne, Australia: Monash University.

Budd, W. F., Jenssen, D. & Young, N. W. (1975). Calculation of the temperature profile on the basis of data from the Vostok Station borehole. *Informatsionnyy Byulleten' Sovetskoy Antarkticheskoy Ekspeditsii*, No. 90, 50–8 (in Russian) and *Soviet Antarctic Expedition Information Bulletin*, Vol. 8, No. 12, 668–73 (of translation by American Geophysical Union).

Budd, W. F., Landon-Smith, I. & Wishart, E. (1967). The Amery Ice Shelf. In *Physics of snow and ice*, ed. H. Oura, 447–67. Sapporo: Institute of Low Temperature Science, Hokkaido University.

Budd, W. F. & McInnes, B. J. (1974). Modeling periodically surging glaciers. *Science*, **186**, 925–7.

Budd, W. F. & McInnes, B. J. (1974). Modelling periodically and periodic surging of the Antarctic ice sheet. In *Climatic Change and Variability*, eds. A. B. Pittock, L. A. Frakes, D. Jenssen, J. A. Peterson & J. W. Zillman, 228–68. Cambridge, England: Cambridge University Press.

Budd, W. F. & Morgan, V. I. (1973). Isotope measurements as indications of ice flow and Palaeo-climates. *Palaeoecology of Africa and of the surrounding islands and Antarctica*, **8**, 5–22.

Budd, W. F. & Morgan, V. I. (1977). Isotopes, climate and ice sheet dynamics from core studies on Law Dome, Antarctica. *International Association of Hydrological Sciences Publication*, No. 118, 312–25.

Budd, W. F., Young, N. W. & Austin, C. R. (1976). Measured and computed temperature distributions in the Law Dome Ice Cap, Antarctica. *Journal of Glaciology*, 16, 99–109.

Bull, C. (1957). Observations in north Greenland relating to theories of the properties of ice. *Journal of Glaciology*, 3, 67–72.

Bull, C. (1971). Snow accumulation in Antarctica. In *Research in the Antarctic*, ed. L. Quam. *American Association for the Advancement of Science Publication*, No. 93, 367–421.

Bull, C. & Webb, P. N. (1973). Some recent developments in the investigation of the glacial history and glaciology of Antarctica. *Palaeoecology of Africa and the surrounding islands and Africa*, 8, 55–84.

Cameron, R. L. (1971). Glaciological studies at Byrd Station, Antarctica. *Antarctic Research Series* (American Geophysical Union), 16, 317–32.

Carslaw, H. S. & Jaeger, J. C. (1959). *Conduction of Heat in Solids*. Second edition. 510 pp. Oxford: Clarendon Press.

Chapman, W. H. & Jones, W. J. (1970). Analysis of ice movement at the Pole Station, Antarctica. *US Geological Survey Professional Paper* 700-C, 242–6.

Chappell, J. (1978). Theories of Upper Quaternary ice ages. In *Climatic Change and Variability*, eds. A. B. Pittock *et al.*, 211–28. Cambridge University Press.

Clark, J. A. & Lingle, C. S. (1977). Future sea level changes due to West Antarctic ice sheet fluctuations. *Nature*, 269, 206–9.

Classen, D. F. & Clarke, G. K. C. (1972). Thermal drilling and ice temperature measurements in the Rusty Glacier. In *Icefield Ranges Research Project, Scientific Results*, eds. V. C. Bushnell & R. H. Ragle, Vol. 3, 103–16. New York: American Geographical Society, and Montreal: Arctic Institute of North America.

Clausen, H. B. (1973). Dating of polar ice by ^{32}Si. *Journal of Glaciology*, 12, 411–16.

Clausen, H. B. & Dansgaard, W. (1977). Less surface accumulation on the Ross Ice Shelf than hitherto assumed. *International Association of Hydrological Sciences Publication*, No. 118, 172–6.

Clausen, H. B., Dansgaard, W., Nielsen, J. O. & Clough, J. W. (1979). Surface accumulation on Ross Ice Shelf. *Antarctic Journal of the United States*, 14, 68–74.

Cline, R. M. & Hays, J. D. (eds.) (1976). Investigation of late Quaternary paleooceanography and paleoclimatology. *Geological Society of America Memoir 145*, 464 pp.

Clough, J. W. (1977). Radio-echo sounding: reflections from internal layers in ice sheets. *Journal of Glaciology*, 18, 3–14.

Clough, J. W. & Hansen, B. Lyle. (1979). The Ross Ice Shelf Project. *Science*, 203, 433–4.

Coachman, L. K., Hemmingsen, E. & Scholander, P. F. (1956). Gas enclosures in a temperate glacier. *Tellus*, 8, 415–23.

Colbeck, S. C. & Gow, A. J. (1979). The margin of the Greenland ice sheet at Isua. *Journal of Glaciology*, 24, 155–65.

Collins, I. F. (1968). On the use of the equilibrium equations and flow law in relating the surface and bed topography of glaciers and ice sheets. *Journal of Glaciology*, 7, 199–204.

Crabtree, R. D. & Doake, C. S. M. (1980). Flow lines on Antarctic ice shelves. *Polar Record*, 31–7.

Cragin, J. H., Herron, M. M. & Langway, Jr., C. C. (1975). The chemistry of 700 years of precipitation at Dye 3, Greenland. *CRREL Research Report*, No. 341, 18 pp. Hanover, New Hampshire: US Army Cold Regions Research and Engineering Laboratory.

Craig, H. (1961). Isotopic variations in meteoric waters. *Science*, 133, 1702–3.

Craig, H. & Gordon, L. I. (1965). Deuterium and Oxygen-18 variations in the ocean and marine atmosphere. In *Stable Isotopes in Oceanographic Studies and Paleotemperatures*, ed. E. Tongiorgi, 9–130. Pisa: Consiglio Nazionale delle Ricerche, Laboratorio di Geologia Nucleare.

Crary, A. P. (1961). Glaciological studies at Little America Station, Antarctica, 1957 and 1958. *IGY Glaciological Report*, No. 5, 197 pp. American Geographical Society: IGY World Data Center A for Glaciology.

Crary, A. P. (1966). Mechanism for fiord formation indicated by studies of an ice-covered inlet. *Geological Society of America Bulletin*, 77, 911–30.

Crozaz, G. (1969). Fission products in Antarctic snow, an additional reference level in January 1965. *Earth and Planetary Science Letters*, 6, 6–8.

Crozaz, G. & Langway, C. C., Jr. (1966). Dating Greenland firn–ice cores with Pb-210. *Earth Planetary Science Letters*, 1, 194–6.

Crozaz, G., Langway, C. C., Jr. & Picciotto, E. (1966). Artificial radio-activity reference horizons in Greenland firn. *Earth and Planetary Science Letters*, 1, 42–8.

Dalrymple, P. C. (1966). A physical climatology of the Antarctic Plateau. *Antarctic Research Series* (American Geophysical Union), 9, 195–231.

Dansgaard, W. (1953). The abundance of ^{18}O in atmospheric water and water vapour, *Tellus*, 5, 461–9.

Dansgaard, W. (1964). Stable isotopes in precipitation. *Tellus*, 16, 436–68.

Dansgaard, W. (1969). Oxygen-18 analysis of water. *Meddelelser om Grønland*, 177, 2, 33–6.

Dansgaard, W., Barkov, N. I. & Splettstoesser, J. (1977). Stable isotope variations in snow and ice at Vostok, Antarctica. *International Association of Hydrological Sciences Publication*, No. 118, 204–9.

Dansgaard, W. & Johnsen, S. J. (1969a). A flow model and a time scale for the ice core from Camp Century, Greenland. *Journal of Glaciology*, 8, 215–23.

Dansgaard, W. & Johnsen, S. J. (1969b). Comment on paper by J. Weertman, 'Comparison between measured and theoretical temperature profiles on the Camp Century, Greenland, borehole.' *Journal of Geophysical Research*, 74, 1109–10. (Also see 'Errata', J. G. R., Vol. 74.)

Dansgaard, W., Johnsen, S. J., Clausen, H. B. & Gundestrup, N. (1973). Stable isotope glaciology. *Meddelelser om Grønland*, 197, 2, 1–53.

Dansgaard, W., Johnsen, S. J., Clausen, H. B., Hammer, C. U. & Langway, C. C., Jr. (1977). Stable isotope profile through the Ross Ice Shelf at Little America V, Antarctica. *International Association of Hydrological Sciences Publication*, No. 118, 322–5.

Dansgaard, W., Johnsen, S. J., Clausen, H. B. & Langway, C. C. (1972). Speculations about the next glaciation. *Quaternary Research*, 2, 396–8.

Dansgaard, W., Johnsen, S. J. & Møller, J. (1969). One thousand centuries of climatic record from Camp Century on the Greenland ice sheet. *Science*, 166, 377–81.

Dansgaard, W., Johnsen, S. J., Møller, J. & Langway, C. C., Jr. (1970). Oxygen isotope analysis of a core representing a complete vertical profile of a polar ice sheet. *International Association of Scientific Hydrology Publication*, No. 86, 93–4.

Dansgaard, W., Johnsen, S. J., Reeh, N., Gundestrup, N., Clausen, H. B. & Hammer, C. U. (1975). Climatic changes, Norsemen and modern man. *Nature, London*, 255, 24–8.

Dansgaard, W. & Tauber, H. (1969). Glacier oxygen-18 content and Pleistocene ocean temperatures. *Science*, 166, 499–502.

Davies, W. E., Krinsley, D. B. & Nicol, A. H. (1963). Geology of the North Star Bugt area, northwest Greenland. *Meddelelser om Grønland*, 162, 12, 1–66.

Delibaltas, P., Dengel, P., Helmreich, D., Riehl, N. & Simon, H. (1966). Diffusion von ^{18}O in Eis-Einkristallen. *Physik der kondensierten Materie*, 5, 166–70.

Delmas, R. J., Ascencio, J.-M. & Legrand, M. (1980). Polar ice evidence that atmospheric CO_2 20,000 yr BP was 50% of present. *Nature*, 284, 155–7.

Delmas, R. & Boutron, C. (1978). Sulfate in Antarctic snow: spatio-temporal distribution. *Atmospheric Environment*, 12, 723–8.

Denton, G. H., Armstrong, R. L. & Stuiver, M. (1971). The late Cenozoic glacial history of Antarctica. In *Late Cenozoic Glacial Ages*, ed. K. K. Turekian, 267–306. New Haven and London: Yale University Press.

Denton, G. H. & Borns, H. W., Jr. (1974). Former grounded ice sheets in the Ross Sea. *Antarctic Journal of the United States*, 9, 167–8.

Denton, G. H., Borns, H. W., Jr., Grosswald, M. G., Stuiver, M. & Nichols, R. L. (1975). Glacial history of the Ross Sea. *Antarctic Journal of the United States*, 10, 160–4.

Denton, G. H. & Hughes, T. J. (1981). *The Last Great Ice Sheets*, 484 pp. New York: John Wiley and Sons.

DiLabio, R. N. W. & Shilts, W. W. (1979). Composition and dispersal of debris by modern glaciers, Bylot Island, Canada. In *Moraines & Varves*, ed. Ch. Schlüchter, 145–55. Rotterdam: A. A. Balkema.

Doake, C. S. M. (1975). Glacier sliding measured by a radio-echo technique. *Journal of Glaciology*, 15, 89–91.

Dobrin, M. B. (1976). *Introduction to Geophysical Prospecting*. Third edition. 630 pp. New York, etc: McGraw-Hill Book Company.

Dorrer, E. (1970). Movement determination of the Ross Ice Shelf, Antarctica. *International Association of Scientific Hydrology Publication*, No. 86, 467–71.

Drewry, D. J. (1975a). Initiation and growth of the East Antarctic ice sheet. *Journal of the Geological Society (London)*, 131, 255–73.

Drewry, D. J. (1975b). Radio echo sounding map of Antarctica (≈90° E–180°). *Polar Record*, 17, 359–74.

Drewry, D. J. (1975c). Comparison of electromagnetic and seismic-gravity ice thickness measurements in East Antarctica. *Journal of Glaciology*, 15, 137–49.

Drewry, D. J. (1976). Deep-sea drilling from *Glomar Challenger* in the Southern Ocean. *Polar Record*, 18, 47–77.

Drewry, D. J. (1978). Aspects of the early evolution of West Antarctic ice. In *Antarctic Glacial History and World Palaeoenvironments*, ed. E. M. van Zinderen Bakker, 25–32. Rotterdam: A. A. Balkema.

Drewry, D. J. (1979). Late Wisconsin reconstruction for the Ross Sea region, Antarctica. *Journal of Glaciology*, 24, 231–43.

Drewry, D. J. (1980). Pleistocene bimodal response of Antarctic ice. *Nature, London*, 287, 214–16.

Drewry, D. J. & Cooper, A. P. R. (1981). Processes and models of glacio-marine sedimentation. *Annals of Glaciology*, 2, 117–22.

Emiliani, C. (1955). Pleistocene temperatures. *Journal of Geology*, 63, 149–58.

England, J. (1976). Postglacial isobases and uplift curves from the Canadian and Greenland High Arctic. *Arctic and Alpine Research*, 8, 61–78.

Epstein, S. & Mayeda, T. (1953). Variation of ${}^{18}O$ content of waters from natural sources. *Geochimica et Cosmochimica Acta*, 4, 213–24.

Epstein, S. & Sharp, R. P. (1967). Oxygen- and hydrogen-isotope variations in a firn core, Eights Station, Western Antarctica. *Journal of Geophysical Research*, 72, 5595–8.

Epstein, S., Sharp, R. P. & Goddard, I. (1963). Oxygen-isotope ratios in Antarctic snow, firn, and ice. *Journal of Geology*, 71, 698–720.

Epstein, S., Sharp, R. P. & Gow, A. J. (1970). Antarctic Ice Sheet: stable isotope analysis of Byrd Station cores and interhemispheric climatic implications. *Science*, 168, 1570–2.

Evans, S. & Smith, B. M. E. (1969). A radio echo equipment for depth sounding in polar ice sheets. *Journal of Scientific Instruments (Journal of Physics E), Series 2*, 2, 131–6.

Fairbridge, R. W. (1961). Eustatic changes in sea level. *Physics and Chemistry of the Earth*, 4, 99–189.

Farrell, W. E. & Clark, J. A. (1976). On postglacial sea level. *Geophysical Journal of the Royal Astronomical Society*, 46, 647–67.

Fillon, R. H. (1975). Late Cenozoic Paleo-oceanography of the Ross Sea, Antarctica. *Geological Society of America Bulletin*, 86, 839–45.

Fireman, E. L., Rancitelli, L. A. & Kirsten, T. (1979). Terrestrial ages of four Allan Hills meteorites: consequences for Antarctic ice. *Science*, 203, 453–4.

Fisher, D. A. (1979). Comparison of 10^5 years of oxygen isotope and insoluble impurity profiles from the Devon Island and Camp Century ice cores. *Quaternary Research*, 11, 299–305.

Flint, R. F. (1971). *Glacial and Quaternary Geology*. 892 pp. New York: Wiley.

Flohn, J. (1978). Abrupt events in climatic history. In *Climatic Change and Variability*, ed. A. B. Pittock *et al.*, 124–34. Cambridge University Press.

Friedman, I. (1953). Deuterium content of natural water and other substances. *Geochimica et Cosmochimica Acta*, 4, 89–103.

Garfield, D. E. & Ueda, H. T. (1976). Resurvey of the 'Byrd' station, Antarctica, drill hole. *Journal of Glaciology*, 17, 29–34.

Gates, W. L. & Imbrie, J. (1975). Climatic change. *Reviews of Geophysics and Space Physics*, 13, 3, 726–31.

Gillet, F. (1975). Steam, hot-water and electrical thermal drills for temperate glaciers. *Journal of Glaciology*, 14, 171–9.

Gillet, F., Donnou, D. & Ricou, G. (1976). A new electrothermal drill for coring in ice. In *Ice-core drilling*, ed. J. F. Splettstoesser, 19–27, Lincoln, Nebraska, & London: University of Nebraska Press.

Giovinetto, B. B. (1964). The drainage systems of Antarctica: accumulation. *Antarctic Research Series* (American Geophysical Union), 2, 127–55.

Giovinetto, M. B. & Schwerdtfeger, W. (1966). Analysis of a 200-year snow accumulation series from the South Pole. *Archiv für Meteorologie, Geophysik und Bioklimatologie*, Serie A, Vol. 15, 227–50.

Goldthwait, R. P. (1951). Development of end moraines in east-central Baffin Island. *Journal of Geology*, 59, 567–77.

Goldthwait, R. P. (1960). Study of ice cliff in Nunatarssuaq, Greenland. *CRREL Technical Report*, No. 39, 108 pp. Hanover, New Hampshire: US Army Cold Regions Research and Engineering Laboratory.

Goldthwait, R. P. (1971). Restudy of Red Rock ice cliff, Nunatarssuaq, Greenland. *CRREL Technical Report*, No. 224, 27 pp. Hanover, New Hampshire: US Army Cold Regions Research and Engineering Laboratory.

Gonfiantini, R. (1965). Some results on oxygen isotope stratigraphy in the deep drilling at King Baudouin station, Antarctica. *Journal of Geophysical Research*, 70, 1815–19.

Gonfiantini, R., Togliatti, V., Tongiorgi, E., de Breuck, W. & Picciotto, E. (1963). Geographical variations of oxygen-18/oxygen-16 ratio in surface snow and ice from Queen Maud Land, Antarctica. *Nature, London*, 197, 1096–8.

Gonfiantini. R. & Picciotto, E. (1959). Oxygen isotope variations in Antarctic snow samples. *Nature, London*, 184, 1557–8.

Gordiyenko, F. G. & Barkov, N. I. (1973). Variations of the ${}^{18}O$ content in the present precipitation of Antarctica. *Informatsionnyy Byulleten' Sovetskoy Antarkticheskoy Ekspeditsii*, No. 87, 40–3 (in Russian) and *Soviet Antarctic Expedition Information Bulletin*, Vol. 8, No. 9, 495–6 (of translation by American Geophysical Union).

Gordon, A. L. (1975). General ocean circulation. In *Numerical Models of Ocean Circulation*, 39–53. Washington DC: National Academy of Sciences.

Gould, L. M. (1940). Glaciers of Antarctica. *Proceedings of the American Philosophical Society*, 82, 835–77.

Gow, A. J. (1968). Deep core studies of the accumulation and densification of snow at Byrd Station and Little America V, Antarctica. *CRREL Research Report*, No. 197, 45 pp. Hanover, New Hampshire: US Army Cold Regions Research and Engineering Laboratory.

Gow, A. J. (1975). Time–temperature dependence of sintering in perennial isothermal snowpacks. *International Association of Hydrology Publication*, No. 114, 25–41.

Gow, A. J., de Blander, F., Crozaz, G. & Picciotto, E. (1972). Snow accumulation at 'Byrd' station, Antarctica. *Journal of Glaciology*, **11**, 59–64.

Gow, A. J., Epstein, S. & Sheehy, W. (1979). On the origin of stratified debris in ice cores from the bottom of the Antarctic ice sheet. *Journal of Glaciology*, **23**, 185–92.

Gow, A. J. & Williamson, T. (1975). Gas inclusions in the Antarctic ice sheet and their glaciological significance. *Journal of Geophysical Research*, **80**, 5101–8.

Gow, A. J. & Williamson, T. (1976). Rheological implications of the internal structure and crystal fabrics of the West Antarctic ice sheet as revealed by deep core drilling at Byrd Station. *CRREL Report*, 76-35, 25 pp. Hanover, New Hampshire: US Army Cold Regions Research and Engineering Laboratory.

Greischar, L. L. & Bentley, C. R. (1980). Isostatic equilibrium grounding line between the West Antarctic inland ice and the Ross Ice Shelf. *Nature*, **283**, 651–4.

Grindley, G. W. (1967). The geomorphology of the Miller Range, Transantarctic Mountains; with notes on the glacial history and neotectonics of East Antarctica, *New Zealand Journal of Geology and Geophysics*, **10**, 557–98.

Gripp, K. (1929). Glaziologische und geologische Ergebnisse der Hamburgischen Spitzbergen-Expedition 1927. *Abhandlungen des Naturwissenschaftlichen Vereins zu Hamburg*, **22**, 145–249.

Grosswald, M. G., Hughes, T. J. & Denton, G. H. (1978). Was there a late-Würm Arctic ice sheet? *Nature, London*, **266**, 596–602.

Gudmandsen, P. (1975). Layer echoes in polar ice sheets. *Journal of Glaciology*, **15**, 95–101.

Gudmandsen, P. & Overgaard, S. (1978). *Establishment of time horizons in polar ice sheets by means of radio echo sounding.* Electromagnetics Institute publication P. 312, 9 pp. Lyngby: Technical University of Denmark.

Haefeli, R. (1961). Contribution to the movement and the form of ice sheets in the Arctic and Antarctic. *Journal of Glaciology*, **3**, 1133–52.

Hammer, C. U. (1977). Past volcanism revealed by Greenland Ice Sheet impurities. *Nature, London*, **270**, 482–6.

Hammer, C. U. (1980). Acidity of polar ice cores in relation to absolute dating, past volcanism, and radio-echoes. *Journal of Glaciology*, **25**, 359–72.

Hammer, C. U., Clausen, H. B. & Dansgaard, W. (1980). Greenland ice sheet evidence of post-glacial volcanism and its climatic impact. *Nature, London*, **288**, 230–5.

Hammer, C. U., Clausen, H. B., Dansgaard, W., Gundestrup, N., Johnsen, S. J. & Reeh, N. (1978). Dating of Greenland ice cores by flow models, isotopes, volcanic debris and continental dust. *Journal of Glaciology*, **20**, 3–26.

Hansen, B. L. & Landauer, J. K. (1958). Some results of ice cap drill hole measurements. *International Association of Scientific Hydrology Publication*, No. 47, 313–7.

Hansen, B. L. & Langway, C. C. (1966). Deep core drilling in ice and core analysis at Camp Century, Greenland, 1961–1966. *Antarctic Journal of the United States*, **1**, 207–8.

Harrison, C. H. (1973). Radio echo sounding of horizontal layers in ice. *Journal of Glaciology*, **12**, 383–97.

Harrison, P. W. (1957). A clay-till fabric: its character and origin. *Journal of Geology*, **65**, 275–308.

Hayes, D. E. & Frakes, L. A. (1975). General synthesis, deep sea drilling project leg 28. *Initial Reports of the Deep Sea Drilling Project*, Vol. 28, 919–42.

Hays, J. D., Lozano, J., Shackleton, N. J. & Irving, G. (1976). Reconstruction of the Atlantic and western Indian Ocean sectors of the 18 000 B.P. Antarctic Ocean. In *Geological Society of America Memoir*, **145**, eds. R. M. Cline & J. D. Hays, 337–72.

Heezen, B. C., Tharp, M. & Bentley, C. R. (1972). Morphology of the Earth in the Antarctic and SubAntarctic. *Antarctic Map Folio Series*, No. 16. New York: American Geographical Society.

Hendy, C. H., Healy, T. R., Rayner, E. M., Shaw, J. & Wilson, A. T. (1979). Late Pleistocene glacial chronology of the Taylor Valley, Antarctica and the global climate. *Quaternary Research*, **11**, 172–84.

Herron, S. & Langway, C. C., Jr. (1979a). The debris-laden ice at the bottom of the Greenland ice sheet. *Journal of Glaciology*, **23**, 193–207.

Herron, M. M. & Langway, C. C., Jr. (1979b). Dating of Ross Ice Shelf cores by chemical analysis. *Journal of Glaciology*, **24**, 345–56.

Heuberger, J. C. (1954). Forages sur l'inlandsis. *Glaciologie, Groenland*, Vol. 1, 68 pp. Publication of Expéditions Polaires Francaises. Paris: Hermann & Cie.

Hobbs, P. V. (1974). *Ice Physics.* 837 pp. Oxford: Clarendon Press.

Hofmann, W. (1974). Die Internationale Glaziologische Grønland Expedition (EGIG). 2. Die Geodätische Lagemessung – Eisbewegung 1959–1967 in den EGIG Profilen. *Zeitschrift für Gletscherkunde und Glazialgeologie*, **10**, 217–24.

Holdsworth, G. (1973). Barnes Ice Cap and englacial debris in glaciers. *Journal of Glaciology*, **12**, 147–8. (Letter)

Holdsworth, G. (1974). Meserve Glacier, Wright Valley, Antarctica: Part I. Basal Processes. *Institute of Polar Studies Report*, No. 37, 104 pp. (Ohio State University)

Hollin, J. T. (1962). On the glacial history of Antarctica. *Journal of Glaciology*, **4**, 173–95.

Hollin, J. T. (1969). The Antarctic ice sheet and the Quaternary history of Antarctica. *Palaeoecology of Africa and of the Surrounding Islands and Antarctica*, **5**, 109–38.

Hollin, J. T. (1970). Antarctic glaciology, glacial history and ecology. In *Antarctic Ecology*, ed. M. W. Holdgate, Vol. 1. 15–19. London and New York: Academic Press.

Holtzscherer, J. J. (1954). Contribution à la connaissance de l'inlandsis du Groenland. Premier partie: measures seismiques. *International Association of Hydrology Publication*, No. 39, 244–70.

Hooke, R. L., Dahlin, B. B. & Kauper, M. T. (1972). Creep of ice containing dispersed fine sand. *Journal of Glaciology*, **11**, 327–36.

Hughes, T. (1973). Is the West Antarctic Ice Sheet disintegrating? *Journal of Geophysical Research*, **78**, 7884–910.

Hughes, T. (1975). The West Antarctic ice sheet: instability, disintegration, and initiation of ice ages. *Reviews of Geophysics and Space Physics*, **13**, 502–26.

Hughes, T. J., Denton, G. H., Anderson, B. G., Schilling, D. H., Fastook, J. L. & Lingle, C. S. (1981). The last great ice sheets: a global view. In *The Last Great Ice Sheets*, eds. G. H. Denton & T. J. Hughes, 263–317. New York, etc: John Wiley & Sons.

Itagaki, K. (1964). Self diffusion in single crystals of ice. *Journal of the Physical Society of Japan*, **19**, 1081.

Jarvis, G. T. & Clarke, G. K. C. (1974). Thermal effects of crevassing on Steele Glacier, Yukon Territory, Canada. *Journal of Glaciology*, **13**, 243–54.

Jenssen, D. (1977). A three-dimensional polar ice-sheet model. *Journal of Glaciology*, **18**, 373–89.

Jenssen, D. & Straede, J. (1969). The accuracy of finite difference analogues of simple differential operators. In *Proceedings of the WMO/IUGG Symposium on Numerical Weather Prediction of Nov.–Dec. 1968* Japan Meteorological Agency (Tokyo) Technical Report No. 67, Session VII, pp. 59–76.

Johnsen, S. J. (1977a). Stable isotope homogenization of polar firn and ice. *International Association of Hydrological Sciences Publication*, No. 118, 210–19.

Johnsen, S. J. (1977b). Stable isotope profiles compared with temperature profiles in firn with historical temperature records. *International Association of Hydrological Sciences Publication*, No. 118, 388–92.

Johnsen, S. J., Dansgaard, W., Clausen, H. B. & Langway, C. C., Jr. (1972). Oxygen isotope profiles through the Antarctic and Greenland ice sheets. *Nature, London,* **235**, 429–34.

Jones, A. S. (1978). The dependence of temperature profiles in ice sheets on longitudinal variations in velocity and surface temperature. *Journal of Glaciology,* **20**, 31–9.

Jouzel, J., Merlivat, L., Pourchet, M. & Lorius, C. (1979). A continuous record of artificial tritium fallout at the South Pole (1954–1978). *Earth and Planetary Science Letters,* **45**, 188–200.

Kamb, W. B. & La Chapelle, E. R. (1964). Direct observation of the mechanism of glacier sliding over bedrock. *Journal of Glaciology,* **5**, 159–72.

Kato, K. (1978). Factors controlling oxygen isotopic composition of fallen snow in Antarctica. *Nature,* **272**, 46–8.

Kato, K., Watanabe, O. & Satow, K. (1978). Oxygen isotopic composition of the surface snow in Mizuho Plateau. In *National Institute of Polar Research Special Issue,* No. 7, eds. T. Ishida *et al.*. 245–54. Tokyo: NIPR.

Kellogg, T. B. & Kellogg, D. E. (1981). Pleistocene sediments beneath the Ross Ice Shelf. *Nature,* **293**, 130–3.

Kellogg, T. B., Truesdale, R. S. & Osterman, L. E. (1979). Late Quaternary extent of the West Antarctic Ice Sheet: new evidence from Ross Sea Cores. *Geology,* **7 (5)**, 249–53.

Kennett, J. P. (1970). Pleistocene palaeoclimates and foraminiferal biostratigraphy in subantarctic deep-sea cores. *Deep-Sea Research,* **17**, 125–40.

Kirch, R. (1966). Temperaturverhältnisse in der Arktis während der letzten 50 Jahre. *Institut für Meteorologie und Geophysik der Freien Universität Berlin: Meteorologische Abhandlungen,* Vol. 69, No. 3, 102 pp.

Koch, J. P. & Wegener, A. (1917). Die glaciologischen Beobachtungen der Danmark-Expedition. *Meddelelser om Grønland,* **46, 1,** 1–77.

Koerner, R. M. (1971). A stratigraphic method of determining the snow accumulation rate at Plateau station, Antarctica, and application to South Pole-Queen Maud Land Traverse 2, 1965–1966. *Antarctic Research Series* (American Geophysical Union), **16**, 225–38.

Koerner, R. M. (1979). Accumulation, ablation, and oxygen isotope variations on the Queen Elizabeth Islands ice caps, Canada. *Journal of Glaciology,* **86**, 25–41.

Koerner, R. M. & Fisher, D. A. (1979). Discontinuous flow, ice texture, and dirt content in the basal layers of the Devon Island Ice Cap. *Journal of Glaciology,* **23**, 209–20.

Koerner, R. M., Paterson, W. S. B. & Krouse, H. R. (1973). $\delta^{18}O$ profile in ice formed between the equilibrium and firn lines. *Nature, Physical Science,* **245**, 137–40.

Koerner, R. & Russell, R. D. (1979). $\delta^{18}O$ variations in snow on the Devon Island ice cap, Northwest Territories, Canada. *Canadian Journal of Earth Sciences,* **16**, 1419–27.

Kohnen, H. (1971). The relation between seismic firn structure, temperature and accumulation. *Zeitschrift für Gletscherkunde und Glazialgeologie,* Vol. VII, Nos. 1–2, 141–51.

Korotkevich, Ye. S. & Kudryashov, B. B. (1976). Ice sheet drilling by Soviet Antarctic expeditions. In *Ice Core Drilling,* ed. J. F. Splettstoesser, 63–70. Lincoln & London: University of Nebraska Press.

Kotlyakov, V. M. (1961). The snow cover of the Antarctic and its role in the present-day glaciation of the continent. *Rezul'taty Issledovaniy po Programme Mazhdunarodnogo Geofizicheskogo Goda. Glyatsiologiya. IX razdel programmy MGG (Results of Studies in the International Geophysical Year Programme. Glaciology. Section IX of the IGY Programme),* No. 7. Translated for National Science Foundation, Washington, DC by Israel Program for Scientific Translations. 256 pp. (of translation).

Krige, L. J. (1939). Borehole temperatures in the Transvaal and Orange Free State. *Proceedings of the Royal Society of London, Series A.* **173**, 450–74.

Kukla, G., Berger, A., Lotti, R. & Brown, J. (1981). Orbital signature of interglacials. *Nature, London,* **290**, 295–300.

Lamb, H. H. (1972). Atmospheric circulation and climate in the Arctic since the last ice age. In *Climatic Changes in Arctic Areas during the last Ten-Thousand Years,* ed. Y. Vasari, H. Hyvärinen, and S. Hicks. *Acta Universitatis Ouluensis, A3, Geologica 1,* 455–98. Oulu, Finland: University of Oulu.

Lambert, G., Ardouin, B., Sanak, J., Lorius, C. & Pourchet, M. (1977). Accumulation of snow and radioactive debris in Antarctica: a possible refined radiochronology beyond reference levels. *International Association of Hydrological Sciences Publication,* No. 118, 146–58.

Langway, C. C., Jr. (1958). Bubble pressures in Greenland glacier ice. *International Association of Scientific Hydrology Publication,* No. 47, 336–49.

Langway, C. C., Jr. (1970). Stratigraphic analysis of a deep ice core from Greenland. *Geological Society of America Special Paper,* No. 125, 186 pp.

Langway, C. C., Jr., Klouda, G. A., Herron, M. M. & Cragin, J. H. (1977). Seasonal variations of chemical constituents in annual layers of Greenland deep ice deposits. *International Association of Hydrological Sciences Publication,* No. 118, 302–6.

Lawson, D. E. (1979). Sedimentological analysis of the western terminus region of the Matanuska Glacier, Alaska. *CRREL Report 79-9,* 112, pp. Hanover, New Hampshire: US Army Cold Regions Research and Engineering Laboratory.

Lebel, B. (1978). *Porosité et teneur en gaz de la glace polaire récente; applications a l'étude des carottes prélevées en profondeur.* Thèse de troisieme cycle, Université Scientifique et Médicale de Grenoble. Publication no. 255 du Laboratoire de Glaciologie du Centre de la Recherche Scientifique, 80 pp. Grenoble: Laboratoire de Glaciologie.

LeMasurier, W. E. (1976). Intraglacial volcanoes in Marie Byrd Land. *Antarctic Journal of the United States,* **11**, 269–70.

Levanon, N. & Bentley, C. R. (1979). Ice elevation map of Queen Maud Land, Antarctica, from balloon altimetry. *Nature, London,* **278**, 842–4.

Liebert, J. (1973). Astronomische Ortsbestimmungen in den Antarktisstationen Wostok und Mirny. *Vermessungstechnik* (Berlin), Jahrg. 21, Hft. 10, 381–2.

Liebert, J. & Leonhardt, G. (1974). Astronomic observations for determining ice movement in the Vostok area. *Informatsionnyy Byulleten' Sovetskoy Antarkticheskoy Ekspeditsii* (in Russian) and *Soviet Antarctic Expedition Information Bulletin,* Vol. 8, No. 10, 569–70 (of translation by American Geophysical Union).

Lile, R. C. (1978). The effect of anisotropy on the creep of polycrystalline ice. *Journal of Glaciology,* **21**, No. 85, 474–83.

Lile, R. C. (unpublished). *Rheology of polycrystalline ice.* Ph.D. Thesis, Meteorology Department, University of Melbourne, 1979.

Limbert, D. W. S. (1974). Variations in the mean annual temperature for the Antarctic Peninsula, 1904–72. *Polar Record,* **17**, 303–30.

Lingle, C. S. & Clark, J. A. (1979). Antarctic ice-sheet volume at 18 000 years B.P. and Holocene sea-level changes at the West Antarctic margin. *Journal of Glaciology,* **24**, 213–30.

List, R. J. (1949). *Smithsonian Meteorological Tables* (sixth edition), 527 pp. Washington: Smithsonian Institution Press.

Lorius, C. (1963). Le Deuterium, possibilités d'application aux problèmes de recherche concernant la neige, le névé et la glace dans l'Antarctique. *Comité National Francais des Recherches Antarctiques* [publication] 8, 102 pp. Paris: CNFRA.

Lorius, C. (1968). A physical and chemical study of the coastal ice sampled from a core drilling in Antarctica. *International Association of Scientific Hydrology Publication,* No. 79, 141–8.

Lorius, C. & Briat, M. (1976). Teneur en elements traces dans la glace; variations temporelles liées à l'activité humaine et au climat. *Société Hydrotechnique de France. XIVes*

Journées de l'Hydraulique (Paris 1976). Question 1. Rapport 2, 8 pp.

Lorius, C., Lambert, G., Hagemann, R., Merlivat, L. & Ravoire, J. (1970). Dating of firn layers in Antarctica: application to the determination of the rate of snow accumulation. *International Association of Scientific Hydrology Publication*, No. 86, 3–16.

Lorius, C. & Merlivat, L. (1977). Distribution of mean surface stable isotope values in East Antarctica: observed changes with depth in the coastal area. *International Association of Hydrological Sciences Publication*, No. 118, 127–37.

Lorius, C., Merlivat, L. & Hagemann, R. (1969). Variation in the mean deuterium content of precipitations in Antarctica. *Journal of Geophysical Research*, 74, 7027–31.

Lorius, C., Merlivat, L., Jouzel, J. & Pourchet, M. (1979). A 30,000-year isotope climatic record from Antarctic ice. *Nature, London*, 280, 644–8.

Lorius, C., Raynaud, D. & Dollé, L. (1968). Densité de la glace et étude des gaz en profondeur dans un glacier antarctique. *Tellus*, 20, 449–59.

McCall, J. G. (1960). The flow characteristics of a cirque glacier and their effect on glacial structure and cirque formation. In *Investigations on Norwegian Cirque Glaciers. Royal Geographical Society Research Series*, No. 4, 39–62. London: John Murray.

McIntyre, A., Kipp, N. G., Bé, A. W. H., Crowley, T., Kellog, T., Gardner, J., Prell, W. & Ruddiman, W. F. (1976). The glacial North Atlantic 18 000 years ago: a CLIMAP reconstruction. In *Geological Society of America Memoir* 145, eds. R. M. Cline & J. D. Hays, 449–64.

MacKinnon, P. K. (1980). Ice cores. *Glaciological Data. Report GD-8*, 139 pp. Boulder, Colorado: Institute of Arctic and Alpine Research (World Data Center A for Glaciology).

McLaren, W. A. (1968). A study of the local ice cap near Wilkes, Antarctica. *Australian National Antarctic Research Expedition Scientific Reports. Series A (IV). Glaciology*, Publication No. 103, 82 pp. Melbourne: Antarctic Division, Department of External Affairs [now Department of Science].

Mae, S. & Naruse, R. (1978). Possible causes of ice sheet thinning in the Mizuho Plateau. *Nature, London*, 273, 291–2.

Majoube, M. (1971a). Fractionation of oxygen-18 between ice and water vapour. *Journal de Chimie Physique et de Physico-Chimie Biologique*, 68, 625–36.

Majoube, M. (1971b). Fractionation of oxygen-18 and of deuterium between water and its vapour. *Journal de Chimie Physique et de Physico-Chimie Biologique*, 68, 1423–36.

Manabe, S. & Stouffer, R. J. (1980). Sensitivity of a global climate model to an increase of CO_2 concentration in the atmosphere. *Journal of Geophysical Research*, 85, 5529–54.

Mathews, W. H. (1974). Surface profiles of the Laurentide Ice Sheet in its marginal areas. *Journal of Glaciology*, 13, 37–43.

Matsuo, S. & Miyake, Y. (1966). Gas composition in ice samples from Antarctica. *Journal of Geophysical Research*, 71, 5235–41.

Mayewski, P. A. (1975). Glacial geology and late Cenozoic history of the Transantarctic Mountains, Antarctica. *Institute of Polar Studies Report*, No. 56, 168 pp. (Ohio State University)

Mayewski, P. A., Attig, J. W., Jr. & Drewry, D. J. (1979). Pattern of ice surface lowering for Rennick Glacier, Northern Victoria Land, Antarctica. *Journal of Glaciology*, 22, 53–65.

Mercer, J. H. (1968). Glacial geology of the Reedy Glacier area, Antarctica. *Bulletin of the Geological Society of America*, 79, 471–86.

Mercer, J. H. (1973). Cainozoic temperature trends in the southern hemisphere: Antarctic and Andean glacial evidence. *Palaeoecology of Africa and of the surrounding islands and Antarctica*, 8, 85–114.

Mercer, J. H. (1978). Glacial development and temperature trends in the Antarctic and South America. In *Antarctic*

Glacial History and World Palaeoenvironments, ed. E. M. van Zinderen Bakker, 25–32. Rotterdam: A. A. Balkema.

Merlivat, L. & Nief, G. (1967). Fractionnement isotopique lors des changements d'état solide-vapeur et liquide-vapeur de l'eau a des températures inférieures à 0 °C. *Tellus*, 19, 122–7.

Merlivat, L., Ravoire, J., Vergnaud, J. P. & Lorius, C. (1973). Tritium and deuterium content of the snow in Greenland. *Earth and Planetary Science Letters*, 19, 235–40.

Milankovitch, M. (1938). Die Chronologie des Pleistozans. *Bulletin, Academy of Natural Sciences and Mathematics, Belgrade*, 4, 49.

Millar, D. H. M. (1981a). Radio-echo layering in polar ice sheets and past volcanic activity. *Nature, London*, 292, 441–3.

Millar, D. H. M. (1981b). *Radio-echo layering in polar ice sheets*. Unpublished Ph.D. thesis, University of Cambridge.

Millar, D. H. M. (1982). Acidity levels in ice sheets from radio echo sounding. *Annals of Glaciology*, 3, 199–203.

Miller, K. J. (1978). *The physical properties and fabrics of two 40 m deep ice cores from interior Greenland*. M.A. thesis, 73 pp. Buffalo, NY: State University of New York.

Mock, S. J. (1963). Tellurometer traverse for a surface movement survey in N. Greenland. *International Association of Scientific Hydrology Publication*, No. 61, 147–53.

Mock, S. J. (1965). Glaciological studies in the vicinity of Camp Century, Greenland. *CRREL Research Report*, No. 157, 20 pp. Hanover, New Hampshire: US Army Cold Regions Research & Engineering Laboratory.

Mock, S. J. (1967). Accumulation patterns on the Greenland ice sheet. *CRREL Research Report*, No. 233, 11pp. Hanover, New Hampshire: US Army Cold Regions Research & Engineering Laboratory.

Mock, S. J. (1968). Snow accumulation studies on the Thule Peninsula, Greenland. *Journal of Glaciology*, 7, 59–76.

Mock, S. J. & Weeks, W. F. (1966). The distribution of 10 metre snow temperatures on the Greenland ice sheet. *Journal of Glaciology*, 6, 23–41.

Moore, T. C., Jr. (1973). Late Pleistocene–Holocene oceanographic changes in the Northeastern Pacific. *Quaternary Research*, 3, 99–109.

Morgan, P. J. (1970). Ice surface movement in Marie Byrd Land. (Ohio State University). *Institute of Polar Studies Research Report*, No. 38, 43 pp.

Morgan, V. I. (1972). Oxygen isotope evidence for bottom freezing on the Amery Ice Shelf. *Nature*, 238, 393–4.

Morgan, V. I. (1980). *Oxygen isotope analysis of Antarctic snow and ice*. Unpublished M.Sc. thesis. Australia: University of Melbourne.

Morgan, V. I. & Budd, W. F. (1975). Radio echo sounding of the Lambert Glacier basin. *Journal of Glaciology*, 15, 103–11.

Murozumi, M., Chow, T. J. & Patterson, C. (1969). Chemical concentrations of pollutant lead aerosols, terrestrial dusts and sea salts in Greenland and Antarctic snow strata. *Geochimica et Cosmochimica Acta*, 33, 1247–94 (p. 1285).

Naruse, R. (1978). Surface flow and strain of the ice sheet measured by a triangulation chain in Mizuho Plateau. In *National Institute of Polar Research Special Issue*, No. 7, eds. T. Ishida *et al.*, 198–226. Tokyo: NIPR.

National Academy of Sciences (1975). *Understanding climatic change: a program for action*. 239 pp. Washington DC: Panel on Climatic Variation, US Committee for the Global Atmospheric Research Program, National Academy of Sciences.

Neal, C. S. (1979). The dynamics of the Ross Ice Shelf revealed by radio echo-sounding. *Journal of Glaciology*, 24, 295–307.

Neftel, A. *et al.* (1982). Ice core sample measurements give atmospheric CO_2 content during the past 40,000 yr, by A. Neftel, H. Oeschger, J. Schwander, B. Stauffer & R. Zumbrunn. *Nature*, 295, 220–3.

Nief, G. (1969). Determination de la concentration en deuterium des échantillons. *Meddelelser om Grønland*, 177, 2, 36–42.

Ninkovich, D. & Shackleton, N. J. (1975). Distribution, stratigraphic position and age of ash layer 'L' in the Panama Basin region. *Earth and Planetary Science Letters, 27,* 20–34.

Nishiizumi, K. (1979). Measurements of ^{36}Cl in Antarctic meteorites and Antarctic ice using a Van de Graaff accelerator, by K. Nishiizumi and 8 others. *Earth and Planetary Science Letters, 45,* No. 2, 285–92.

Nye, J. F. (1952). The mechanics of glacier flow. *Journal of Glaciology, 2,* 82–93.

Nye, J. F. (1960). The response of glaciers and ice-sheets to seasonal and climatic changes. *Proceedings of the Royal Society of London, A,* 559–84.

Nye, J. F. (1967). Plasticity solution for a glacier snout. *Journal of Glaciology, 6,* 695–715.

Nye, J. F. (1969). The effect of longitudinal stress on the shear stress at the base of an ice sheet. *Journal of Glaciology, 8,* 207–13.

Oeschger, H., Alder, B., Loosli, H., Langway, C. C., Jr. & Renaud, A. (1966). Radio carbon dating of ice. *Earth and Planetary Science Letters, 1,* No. 2, 49–54.

Olausson, E. (1965). Evidence of climatic changes in North Atlantic deep-sea cores, with remarks on isotopic palaeotemperature analysis. *Progress in Oceanography, 3,* 221–52.

O'Neill, J. R. (1968). Hydrogen and oxygen isotopic fractionation between ice and water. *Journal of Physical Chemistry, 72,* 3683–4.

Orvig, S. (1953). On the variation of the shear stress on the bed of an ice cap. *Journal of Glaciology, 2,* 242–7.

Oswald, G. K. A. & Robin, G. de Q. (1973). Lakes beneath the Antarctic ice sheet. *Nature, London, 245,* 251–4.

Overgaard, S. & Gudmandsen, P. (1978). *Radioglaciology. Surface and bedrock contour maps at Dye 3.* 7 pp. Lyngby: Technical University of Denmark, Electromagnetics Institute, Report R 199.

Paren, J. G. & Robin, G. de Q. (1975). Internal reflections in polar ice sheets. *Journal of Glaciology, 14,* 251–9.

Paterson, W. S. B. (1968). A temperature profile through the Meighen Ice Cap, Arctic Canada. *International Association of Scientific Hydrology Publication,* No. 89, 440–9.

Paterson, W. S. B. (1969). *The Physics of Glaciers.* (First edition) 250 pp. Oxford and New York: Pergamon Press.

Paterson, W. S. B. (1972). Laurentide Ice Sheet: estimated volumes during late Wisconsin. *Reviews of Geophysics and Space Physics, 10,* 885–917.

Paterson, W. S. B. (1976). Vertical strain-rate measurements in an Arctic ice cap and deductions from them. *Journal of Glaciology, 17,* 3–12.

Paterson, W. S. B. (1977). Extent of the Late-Wisconsin glaciation in northwest Greenland and northern Ellesmere Island. *Quaternary Research, 8,* 180–90.

Paterson, W. S. B. (1981). *The Physics of Glaciers.* Second edition. Oxford: Pergamon Press.

Paterson, W. S. B. & Clarke, G. K. C. (1978). Comparison of theoretical and observed temperature profiles in Devon Island ice cap, Canada. *Geophysical Journal of the Royal Astronomical Society, 55,* 615–32.

Paterson, W. S. B., Koerner, R. M., Fisher, D., Johnsen, S. J., Clausen, H. B., Dansgaard, W., Bucher, P. & Oeschger, H. (1977). An oxygen-isotope climatic record from the Devon Island ice cap, arctic Canada. *Nature, London, 266,* 508–11.

Peel, D. A. & Clausen, H. B. (1982). Oxygen-isotope and total beta-radioactivity measurements on 10 m ice cores from the Antarctic Peninsula. *Journal of Glaciology, 28,* 43–55.

Peterson, D. N. (1970). Glaciological investigations on the Casement Glacier, southeast Alaska. *Institute of Polar Studies Report,* No. 36, 161 pp. (Ohio State University.)

Philberth, K. (1962). Une méthode pour mesurer les températures à l'interieur d'un inlandsis. *Comptes Rendus Hebdomadaires des Séances de l'Académie des Sciences, 254,* 3881–3.

Philberth, K. (1970). Thermische Tiefbohrung in Zentralgrönland. *Umschau in Wissenschaft und Technik, 16,* 515–6.

Philberth, K. & Federer, B. (1971). On the temperature profile and the age profile in the central part of cold ice sheets. *Journal of Glaciology, 10,* 3–14.

Phillpot, H. R. & Zillman, J. W. (1970). The surface temperature inversion over the Antarctic continent. *Journal of Geophysical Research, 75,* 4161–9.

Picciotto, E. (1967). Geochemical investigations of snow and firn samples from East Antarctica. *Antarctic Journal of the United States, 2,* 236–40.

Picciotto, E., Cameron, R., Crozaz, G., Deutsch, S. & Wilgain, S. (1968). Determination of the rate of snow accumulation at the Pole of Relative Inaccessibility, Eastern Antarctica: a comparison of glaciological and isotopic methods. *Journal of Glaciology, 7,* 273–87.

Picciotto, E., Crozaz, G. & De Breuck, W. (1971). Accumulation on the South Pole–Queen Maud Land traverse, 1964–1968. *Antarctic Research Series* (American Geophysical Union), 16, 257–315.

Picciotto, E., de Maere, X. & Friedman, I. (1960). Isotopic composition and temperature of formation of Antarctic snows. *Nature, London, 187,* 857–9.

Picciotto, E., Deutsch, S. & Aldaz, L. (1966). The summer 1957–1958 at the South Pole; an example of an unusual meterological event recorded by the oxygen isotope ratios in the firn. *Earth and Planetary Science Letters, 1,* 202–4.

Prest, V. K. (1969). Retreat of Wisconsin and recent ice in North America. *Geological Survey of Canada, Map 1257A.* Ottawa: Geological Survey of Canada.

Punning, J. M. K., Vaikmyae, R. A., Kotlyakov, V. M. & Gordiyenko, F. G. (1980). Oxygen-isotope investigations of ice core from the ice divide of Grönfjord and Fritjof glaciers, West Spitsbergen. *Materialy Glyatsiologicheskikh Issledovaniy. Khronika. Obsuzhdeniya.* Publication 37, 173–7 (in Russian with English summary). Moscow: Academy of Sciences of the USSR (Glaciology Section of the Soviet Geophysical Committee).

Putnins, P. (1970). The climate of Greenland. In *Climates of the Polar Regions* (World Survey of Climatology, Vol. 14), ed. S. Orvig, 3–128. Amsterdam: Elsevier Publishing Company.

Radok, U., Jenssen, D. & Budd, W. (1970). Steady-state temperature profiles in ice sheets. *International Association of Scientific Hydrology Publication,* No. 86, 151–65.

Radok, U. & Lile, R. C. (1977). A year of snow accumulation at Plateau Station. *Antarctic Research Series* (American Geophysical Union), 25, 17–26.

Ragone, S. E. & Finelli, R. (1972). Use of atomic absorption spectroscopy in the determination of μg/liter concentrations of Na^+, K^+, Ca^{2+}, and Mg^{2+}. *CRREL Special Report 174,* 5 pp. Hanover, New Hampshire: US Army Cold Regions Research and Engineering Laboratory.

Raynaud, D. (1976). *Les inclusions gazeuses dans la glace de glacier; leur utilisation comme indicateur du site de formation de la glace polaire; applications climatiques et rheologiques.* Doctorat d'Etat thesis, Université Scientifique et Médicale de Grenoble. Publication No. 214 du Laboratoire de Glaciologie du Centre National de la Recherche Scientifique, 104 pp. Grenoble: Laboratoire de Glaciologie.

Raynaud, D. & Delmas, R. (1977). Composition des gaz contenus dans la glace polaire. *International Association of Hydrological Sciences Publication,* No. 118, 377–81.

Raynaud, D., Delmas, R., Ascencio, J. M. & Legrand, M. (1982). Gas extraction from polar ice cores: a critical issue for studying the evolution of atmospheric CO_2 and ice sheet surface elevation. *Annals of Glaciology, 3,* 265–8.

Raynaud, D., Duval, P., Lebel, B. & Lorius, C. (1979). Crystal size and total gas content of ice: two indicators of the climatic evolution of polar ice sheets. In *Colloque international. Evolution des atmosphères planétaires et climatologie de la terre*, Nice, 1978, 83–94. Toulouse: Centre National d'Etudes Spatiales.

Raynaud, D. & Lebel, B. (1979). Total gas content and surface elevation of polar ice sheets: new evidence. *Nature, London*, 281, 289–91.

Raynaud, D. & Lorius, C. (1973). Climatic implications of total gas content in ice at Camp Century. *Nature, London*, 243, 283–4.

Raynaud, D. & Lorius, C. (1977). Total gas content in polar ice: rheological and climatic implications. *International Association of Hydrological Sciences Publication*, No. 118, 326–34.

Raynaud, D., Lorius, C., Budd, W. F. & Young, N. W. (1979). Ice flow along an I.A.G.P. flow line and interpretation of data from an ice core in Terre Adélie, Antarctica. *Journal of Glaciology*, 24, 103–15.

Raynaud, D. & Whillans, I. M. (1982). Air content of the Byrd core and past changes in the West Antarctic ice sheet. *Annals of Glaciology*, 3, 269–73.

Reeh, N., Clausen, H. B., Dansgaard, W., Gundestrup, N., Hammer, C. U. & Johnsen, S. J. (1978). Secular trends of accumulation rates at three Greenland stations. *Journal of Glaciology*, 20, 27–30.

Reeh, N., Clausen, H. B., Gundestrup, N., Johnsen, S. J. & Stauffer, B. (1977). $\delta(^{18}O)$ and accumulation rate distribution in the Dye 3 area, south Greenland. *International Association of Hydrological Sciences Publication*, No. 118, 177–81.

Reeh, N. & Fisher, D. A. (in prep.). Noise in accumulation rate and $\delta(^{18}O)$ time series as determined from comparison of adjacent Greenland and Devon Island ice cap cores.

Ritz, C., Lliboutry, L. & Rado, C. (1982). Analysis of a 870 m deep temperature profile at Dome C. *Annals of Glaciology*, 3, 284–9.

Robin, G. de Q. (1955). Ice movement and temperature distribution in glaciers and ice sheets. *Journal of Glaciology*, 2, 523–32.

Robin, G. de Q. (1958). Glaciology III. Seismic shooting and related investigations. *Norwegian–British–Swedish Antarctic Expedition, 1949–52. Scientific Results*, Vol. 5, 134 pp. Oslo: Norsk Polarinstitutt.

Robin, G. de Q. (1964). Glaciology. *Endeavour*, 23, 102–7.

Robin, G. de Q. (1967). Surface topography of ice sheets. *Nature, London*, 215, 1029–32.

Robin, G. de Q. (1970). Stability of ice sheets as deduced from deep temperature gradients. *International Association of Scientific Hydrology Publication*, No. 86, 141–51.

Robin, G. de Q. (1972). Radio-echo sounding applied to the investigation of the ice thickness and sub-ice relief of Antarctica. In *Antarctic Geology and Geophysics*, ed. R. J. Adie, 675–82. Oslo: Universitetsforlaget.

Robin, G. de Q. (1975a). Ice shelves and ice flow. *Nature, London*, 253, 168–72.

Robin, G. de Q. (1975b). Radio-echo sounding: glaciological interpretations and applications. *Journal of Glaciology*, 15, 49–63.

Robin, G. de Q. (1976). Reconciliation of temperature–depth profiles in polar ice sheets with past surface temperatures deduced from oxygen-isotope profiles. *Journal of Glaciology*, 16, 9–22.

Robin, G. de Q. (1977). Ice cores and climatic change. *Philosophical Transactions of the Royal Society of London B*, 280, 143–68.

Robin, G. de Q. (1979). Formation, flow and disintegration of ice shelves. *Journal of Glaciology*, 24, 259–71.

Robin, G. de Q. (1981). Climate into ice: the isotopic record in polar ice sheets. *International Association of Hydrological Sciences Publication*, No. 131, 207–16.

Robin, G. de Q., Drewry, D. J. & Meldrum, D. T. (1977). International studies of ice sheet and bedrock. *Philosophical*

Transactions of the Royal Society of London, B, 279, 185–96.

Robin, G. de Q., Evans, S. & Bailey, J. T. (1969). Interpretation of radio echo sounding in polar ice sheets. *Philosophical Transactions of the Royal Society of London, B*, 265, 437–505.

Robin, G. de Q. & Millar, D. H. M. (1982). Flow of ice sheets in the vicinity of sub-glacial peaks. *Annals of Glaciology*, 3, 290–4.

Robin, G. de Q., Swithinbank, C. W. M. & Smith, B. M. E. (1970). Radio echo exploration of the Antarctic ice sheet. *International Association of Scientific Hydrology Publication*, No. 86, 97–115.

Robinson, P. H. (1980). *An investigation into the processes of debris entrainment, transportation and deposition of debris in polar ice, with special reference to the Taylor Glacier, Antarctica*. Unpublished Ph.D. thesis. Wellington, New Zealand: Victoria University.

Rose, K. E. (1978). *Radio echo sounding studies of Marie Byrd Land, Antarctica*. Unpublished Ph.D. thesis, 204 pp. Cambridge, UK: University of Cambridge.

Rose, K. E. (1979). Characteristics of ice flow in Marie Byrd Land, Antarctica. *Journal of Glaciology*, 24, 63–74.

Rose, K. E. (1982). Radio echo studies of bedrock in southern Marie Byrd Land, West Antarctica. In *Antarctic Geoscience*, ed. C. Craddock, 985–92. Madison: University of Wisconsin Press.

Rundle, A. S. (1973). Glaciology of the Marr Ice Piedmont, Anvers Island, Antarctica. *Institute of Polar Studies Report*, No. 47, 237 pp. Ohio State University.

Russell-Head, D. S. & Budd, W. F. (1979). Ice-sheet flow properties derived from bore-hole shear measurements combined with ice-core studies. *Journal of Glaciology*, 24, 117–30.

Sanak, J. & Lambert, G. (1977). Lead 210 or climatic change at South Pole? *Geophysical Research Letters*, 4, 357–9.

Schaefer, T. G. (1972). Radio echo sounding in western Dronning Maud Land, 1971 – a preview. *South African Journal of Antarctic Research*, No. 2, 53–6.

Schofield, J. C. & Thompson. H. R. (1964). Postglacial sea levels and isostatic uplift. *New Zealand Journal of Geology & Geophysics*, 7, 359–70.

Scholander, P. F., Dansgaard, W., Nutt, D. C., de Vries, H., Coachman, L. K. & Hemmingsen, E. A. (1962). Radiocarbon age and oxygen-18 content of Greenland icebergs. *Meddelelser om Grønland*, 165, 1, 1–26.

Scholander, P. F., Hemmingsen, E. A., Coachman, L. K. & Nutt, D. C. (1961). Composition of gas bubbles in Greenland icebergs. *Journal of Glaciology*, 3, 813–22.

Schwerdtfeger, W. (1969). Ice crystal precipitation on the Antarctic Plateau. *Antarctic Journal of the United States*, 4, 221–2.

Schwerdtfeger, W. (1970). The climate of the Antarctic. In *World Survey of Climatology*, ed. H. E. Landsberg, Vol. 14, 253–355. Amsterdam, etc.: Elsevier Publishing Company.

Schwerdtfeger, W. (1975). The effect of the Antarctic Peninsula on the temperature regime of the Weddell Sea. *Monthly Weather Review*, 103, 45–51.

Schytt, V. (1960). Glaciology II. Snow and ice temperatures in Dronning Maud Land. *Norwegian–British–Swedish Antarctic Expedition, 1949–52. Scientific Results*, Vol. 4 D, 157–79. Oslo: Norsk Polarinstitutt.

Schytt, V. (1974). Inland ice sheets – recent and Pleistocene. *Geologiska Föreningens i Stockholm Förhandlingar*, 96, 299–309.

Scott Polar Research Institute (in press). *Antarctic Glaciological and Geophysical Map Folio*. To be published by SPRI, Cambridge, 1982.

Seckel, H. (1977). Höhenänderungen im grönländischen Inlandeis zwischen 1959 und 1968. *Meddelelser om Grønland*, 187, 4, 1–58.

Shackleton, N. J. (1967). Oxygen isotope analyses and Pleistocene temperatures re-assessed. *Nature*, 215, 15–17.

Shackleton, N. J. (1974). Attainment of isotopic equilibrium
between ocean water and the benthonic foraminifera
genus *Uvigerina:* isotopic changes in the oceans during the
last glacial. *Colloques Internationaux du CNRS No. 219.*
Paris: Centre National de la Recherche Scientifique.

Shackleton, N. J. & Kennett, J. P. (1975). Palaeotemperature
history of the Cenozoic and the initiation of Antarctic
glaciation: oxygen and carbon isotope analyses in DSDP
Sites 277, 279 and 281. In *Initial Reports of the Ôeep
Sea Drilling Project,* eds. J. P. Kennett, R. E. Houtz, *et al.,*
Vol. 29, 743–55. Washington: US Government Printing
Office.

Shackleton, N. J. & Kennett, J. P. (1975). Palaeotemperature
palaeomagnetic stratigraphy of equatorial Pacific core
V28–238: oxygen isotope temperatures and ice volumes
on a 10^5 and 10^6 year scale. *Quaternary Research, 3,*
39–55.

Shackleton, N. J. & Opdyke, N. D. (1976). Oxygen-isotope and
palaeomagnetic stratigraphy of Pacific core V28–239: late
Pliocene to latest Pleistocene. In *Geological Society of
America Memoir 145,* eds. R. M. Cline & J. D. Hays, 449–64.

Shackleton, M. J. & Opdyke, N. D. (1977). Oxygen isotope and
palaeomagnetic evidence for early northern hemisphere
glaciation. *Nature, London,* 270, 216–19.

Shumsk[i]y, P. A. (1970). The Antarctic ice sheet. *International
Association of Hydrology Publication,* No. 86, 327–47.

Souchez, R. A. (1966). The origin of morainic deposits and the
characteristics of glacial erosion in the western Sør-
Rondane, Antarctica. *Journal of Glaciology,* 6, 249–54.

Souchez, R. A. (1971). Ice-cored moraines in south-western
Ellesmere Island, N.W.T., Canada. *Journal of Glaciology,*
10, 245–54.

Splettstoesser, J. F., ed. (1976). *Ice-core drilling.* 189 pp. Lincoln,
Nebraska, & London: University of Nebraska Press.

Stauffer, B. & Berner, W. (1978). CO_2 in natural ice. *Journal of
Glaciology,* 21, 291–9.

Stewart, M. K. (1975). Stable isotope fractionation due to
evaporation and isotopic exchange of falling water drops:
applications to atmospheric processes and evaporation of
lakes. *Journal of Geophysical Research,* 80, 1133–46.

Suyetova, I. A. (1966). The dimensions of Antarctica. *Polar
Record,* 13, 344–7.

Sverdrup, H. U., Johnson, M. W. & Fleming, R. H. (1942).
The Oceans, 1087 pp. New York: Prentice Hall.

Swinzow, G. K. (1962). Investigation of shear zones in the
ice sheet margin, Thule area, Greenland. *Journal of
Glaciology,* 4, 215–29.

Swithinbank, C. W. M. (1963). Ice movement of valley glaciers
flowing into the Ross Ice Shelf, Antarctica. *Science,* 141,
523–4.

Tedrow, J. C. F. (1970). Soil investigations in Inglefield Land,
Greenland. *Meddelelser om Grønland,* 188, 3, 1–93.

Thomas, R. H. (1973). The dynamics of the Brunt Ice Shelf,
Coates Land, Antarctica. *British Antarctic Survey,
Scientific Reports,* No. 79, 45 pp.

Thomas, R. H. (1976). Thickening of the Ross Ice Shelf and
equilibrium state of the West Antarctic ice sheet. *Nature,*
259, 180–3.

Thomas, R. H. (1979). West Antarctic ice sheet: present-day
thinning and Holocene retreat of the margins. *Science,*
205, 1257–8.

Thomas, R. H. & Bentley, C. R. (1978*a*). The equilibrium state
of the eastern half of the Ross Ice Shelf. *Journal of
Glaciology,* 20, 509–18.

Thomas, R. H. & Bentley, C. R. (1978*b*). A model for Holocene
retreat of the West Antarctic ice sheet. *Quaternary
Research,* 10, 150–70.

Thomas, R. H. & MacAyeal, D. R. (1977). Glaciological measurements on the Ross Ice Shelf. *Antarctic Journal of the
United States,* 12, 114–5.

Thomas, R. H., MacAyeal, D. R., Eilers, D. H. & Gaylord, D. R.
(in press). Glaciological studies on the Ross Ice Shelf,

Antarctica, 1973–1978. *Antarctic Research Series*
(American Geophysical Union).

Thompson, L. G. (1973). Analysis of the concentration of
microparticles in an ice core from Byrd Station,
Antarctica. *Institute of Polar Studies Report,* No. 46,
44 pp. Ohio State University.

Thompson, L. G., Hamilton, W. L. & Bull, C. (1975). Climatological implications of microparticle concentrations in
the ice core from 'Byrd' station, western Antarctica.
Journal of Glaciology, 14, 433–44.

Tolstikov, Ye. I., ed. (1966). *Atlas of Antarctica,* Vol. 1, 225 pp.
Moscow and Leningrad: Main Administration of Geodesy
and Cartography of the Ministry of Geology, USSR.
Translation of legend matter and explanatory text in
Soviet Geography (1967) 8, 261–507 (American
Geographical Society).

Trail, D. S. (1964). The glacial geology of the Prince Charles
Mountains. In *Antarctic Geology,* ed. R. J. Adie, 143–51.
Amsterdam: North Holland Publishing Company.

Ueda, H. T. & Garfield, D. E. (1969). Core drilling through
the Antarctic ice sheet. *CRREL Technical Report,* No.
231, 19 pp. Hanover, New Hampshire: US Army Cold
Regions Research and Engineering Laboratory.

Ueda, H. T. & Garfield, D. E. (1970). Deep core drilling at
Byrd Station, Antarctica. *International Association
of Scientific Hydrology Publication,* No. 86, 53–68.

USGS (1976). South Pole ice moves 30 ft/yr. *E.O.S. Transactions American Geophysical Union,* Vol. 57, No. 1.
21.

Vella, P., Ellwood, B. B. & Watkins, N. D. (1975). Surface-water
temperature changes in the Southern Ocean southwest of
Australia during the last one million years. In *Quaternary
Studies,* eds. R. P. Suggate & M. M. Creswell, 297–309.
Wellington: The Royal Society of New Zealand.

Vernekar, A. D. (1972). Long-period global variations of
incoming solar radiation. *Meteorological Monographs,*
Vol. 12, No. 34, 20 pp.

Vilenskiy, V. D., Teys, R. V. & Kochetkova, S. N. (1970).
Determination of the snow accumulation rate in the
Mirny station area from variations in oxygen isotope
composition. *Informatsionnyy Byulleten' Sovetskoy
Antarkticheskoy Ekspeditsii,* No. 79, 30–4, (in Russian)
and *Soviet Antarctic Expedition Information Bulletin,*
Vol. 8, No. 1, 15–17, (of translation by American
Geophysical Union).

Vilenskiy, V. D., Teys, R. V. & Kochetkova, S. N. (1974).
Isotopic composition of oxygen in the snow cover of
some regions of eastern Antarctica. *Geokhimiya,* No.
1, 39–44 (in Russian).

Vivian, R. & Bocquet, G. (1973). Subglacial cavitation phenomena under the Glacier d'Argentière, Mont Blanc, France.
Journal of Glaciology, 12, 439–51.

Voronov, P. S. (1960). Attempt to reconstruct the ice sheet of
Antarctica at the time of maximum glaciation on Earth.
*Informatsionnyy Byulleten' Sovetskoy Antarkticheskoy
Ekspeditsii,* No. 23, 15–19 (in Russian) and *Soviet
Antarctic Expedition Information Bulletin,* Vol. 3 (1965),
88–93 (of translation by Elsevier Publishing Company:
Amsterdam, London, New York).

Vyalov, S. S. (1958). Regularities of glacial shields movement
and the theory of plastic viscous flow. *International
Association of Scientific Hydrology Publication 47,*
266–75.

Walcott, R. I. (1972). Late Quaternary vertical movements
in eastern North America: quantitative evidence of
glacio-isostatic rebound. *Reviews of Geophysics and
Space Physics,* 10, 849–84.

Walford, M. E. R. (1972). Glacier movement measured with
a radio echo technique. *Nature,* 239, 93–5.

Walford, M. E. R., Holdorf, P. C. & Oakberg, R. G. (1977).
Phase-sensitive radio-echo sounding at the Devon Island
Ice Cap, Canada. *Journal of Glaciology,* 18, 217–29.

References   208

Ward, W. H. (1952). The physics of deglaciation of Central
    Baffin Island. *Journal of Glaciology,* **2**, 9–22.

Weertman, J. (1957). On the sliding of glaciers. *Journal of
    Glaciology,* **3**, 33–8.

Weertman, J. (1961a). Equilibrium profile of ice caps. *Journal
    of Glaciology,* **3**, 953–64.

Weertman, J. (1961b). Mechanism for the formation of inner
    moraines found near the edge of cold ice caps and ice
    sheets. *Journal of Glaciology,* **3**, 965–78.

Weertman, J. (1966). Effect of a basal water layer on the
    dimensions of ice sheets. *Journal of Glaciology,* **6**, 191–
    207.

Weertman, J. (1968). Comparison between measured and
    theoretical temperature profiles of the Camp Century,
    Greenland, borehole. *Journal of Geophysical Research,*
    73, 2691–700.

Weertman, J. (1973). Position of ice divides and ice centers on
    ice sheets. *Journal of Glaciology,* **12**, 353–60.

Weertman, J. (1974). Stability of the junction of an ice sheet
    and an ice shelf. *Journal of Glaciology,* **13**, 3–11.

Weertman, J. (1976). Sliding – no sliding zone effect and age
    determination of ice cores. *Quaternary Research,* **6**, 203–7.

Weidick, A. (no date). *Quarternary map of Greenland.* Copenhagen:
    Geological Survey of Greenland.

Weidick, A. (1976). Glaciation and the Quaternary of Greenland.
    In *Geology of Greenland,* eds. A. Escher & W. Stuart Watt,
    431–58. Copenhagen: Geological Survey of Greenland.

Weller, G. & Schwerdtfeger, P. (1970). Thermal properties and
    heat transfer processes of the snow of the central Antarctic
    plateau. *International Association of Hydrology Publica-
    tion,* No. 86, 284–98.

Wellman, P. & Tingey, R. J. (1981). Glaciation, erosion and
    uplift over parts of East Antarctica. *Nature,* **291**, 142–4.

Weyant, W. S. (1967). The Antarctic atmosphere: climatology
    of the surface environment. *Antarctic Map Folio Series,*
    No. 8. New York: American Geographical Society.

Whillans, I. M. (1973). State of equilibrium of the West
    Antarctic inland ice sheet. *Science,* **182**, 476–9.

Whillans, I. M. (1975). Effect of inversion winds on topographic
    detail and mass balance on inland ice sheets. *Journal of
    Glaciology,* **14**, 85–90.

Whillans, I. M. (1976). Radio-echo layers and the recent stability
    of the West Antarctic ice sheet. *Nature,* **264**, 152–5.

Whillans, I. M. (1977). The equation of continuity and its
    application to the ice sheet near 'Byrd' Station,
    Antarctica. *Journal of Glaciology,* **18**, 359–71.

Whillans, I. M. (1978). Inland ice sheet thinning due to
    Holocene warmth. *Science,* **201**, 1014–16.

Whillans, I. M. (1979). Ice flow along the Byrd Station strain
    network, Antarctica, *Journal of Glaciology,* **24**, 15–26.

Whillans, I. M. (1981). Reaction of the accumulation zone
    portions of glaciers to climatic change. *Journal of
    Geophysical Research,* **86**, C5, 4274–82.

Wilson, A. T. (1964). Origin of ice ages: an ice shelf theory
    for Pleistocene glaciation. *Nature,* **201**, 147–9.

Yamada, T., Okukira, F., Yokoyama, K. & Watanabe, O. (1978).
    Distribution of accumulation measured by the snow
    stake method in Mizuho Plateau. *National Institute of
    Polar Research Memoirs, Special Issue,* No. 7, 125–39.
    Tokyo.

Yevteyev, S. A. (1959). Determination of the amount of morainic
    material carried by glaciers of the east Antarctic coast.
    *Informatsionnyy Byulleten' Sovetskoy Antarkticheskoy
    Ekspeditsii,* No. 11, 14–16 (in Russian) and *Information
    Bulletin of the Soviet Antarctic Expedition,* Vol. 2, 7–9
    (of translation by Elsevier Publishing Company:
    Amsterdam, London, New York).

Yiou, F. & Raisbeck, G. M. (1981). The age of sediments beneath
    the Ross Ice Shelf as implied by cosmogenic ^{10}Be concen-
    trations. *EOS, Transactions of the American Geophysical
    Union, Washington DC,* **62**, (17), 297.

Zhavoronkov, N. M., Uvarov, O. V. & Sevryugova, N. N. (1955).
    Some physico-chemical constants of heavy oxygen water.
    In *Primeneniye mechenykh atomov v analiticheskoy
    khimii,* ed. A. P. Vinogradov, 223–33. Moscow. Academy
    of Sciences of the USSR (Institute of Geochemistry and
    Analytical Chemistry).

Zotikov, I. A. (1961). The thermal regime of the ice cap in
    central Antarctica. *Informatsionnyy Byulleten' Sovetskoy
    Antarkticheskoy Ekspeditsii,* No. 28, 16–21 (in Russian)
    and *Information Bulletin of the Soviet Antarctic Expedi-
    tion,* Vol. 3, 289–93. (Translation by Elsevier Publishing
    Company, Amsterdam, London, New York.)

Zwally, H. J. & Gloersen, P. (1977). Passive microwave images
    of the polar regions and research applications. *Polar
    Record,* **18**, 431–50.

INDEX

References to figures and major tabulations of data are in italic